NATURAL POLYMERS

*Perspectives and Applications
for a Green Approach*

NATURAL POLYMERS

*Perspectives and Applications
for a Green Approach*

Edited by
Jissy Jacob
Fernando Gomes, PhD
Józef T. Haponiuk, PhD
Nandakumar Kalarikkal, PhD
Sabu Thomas, PhD, DSc, FRSC

First edition published 2022

Apple Academic Press Inc.
1265 Goldenrod Circle, NE,
Palm Bay, FL 32905 USA

4164 Lakeshore Road, Burlington,
ON, L7L 1A4 Canada

CRC Press
6000 Broken Sound Parkway NW,
Suite 300, Boca Raton, FL 33487-2742 USA

2 Park Square, Milton Park,
Abingdon, Oxon, OX14 4RN UK

© 2022 by Apple Academic Press, Inc.

Apple Academic Press exclusively co-publishes with CRC Press, an imprint of Taylor & Francis Group, LLC

Reasonable efforts have been made to publish reliable data and information, but the authors, editors, and publisher cannot assume responsibility for the validity of all materials or the consequences of their use. The authors, editors, and publishers have attempted to trace the copyright holders of all material reproduced in this publication and apologize to copyright holders if permission to publish in this form has not been obtained. If any copyright material has not been acknowledged, please write and let us know so we may rectify in any future reprint.

Except as permitted under U.S. Copyright Law, no part of this book may be reprinted, reproduced, transmitted, or utilized in any form by any electronic, mechanical, or other means, now known or hereafter invented, including photocopying, microfilming, and recording, or in any information storage or retrieval system, without written permission from the publishers.

For permission to photocopy or use material electronically from this work, access www.copyright.com or contact the Copyright Clearance Center, Inc. (CCC), 222 Rosewood Drive, Danvers, MA 01923, 978-750-8400. For works that are not available on CCC please contact mpkbookspermissions@tandf.co.uk

Trademark notice: Product or corporate names may be trademarks or registered trademarks and are used only for identification and explanation without intent to infringe.

Library and Archives Canada Cataloguing in Publication

Title: Natural polymers : perspectives and applications for a green approach / edited by Jissy Jacob, Fernando Gomes, PhD, Józef T. Haponiuk, PhD, Nandakumar Kalarikkal, PhD, Sabu Thomas, PhD, DSc, FRSC.
Names: Jacob, Jissy, editor. | Gomes, Fernando (Lecturer), editor. | Haponiuk, Józef T., editor. | Kalarikkal, Nandakumar, editor. | Thomas, Sabu, editor.
Description: First edition. | Includes bibliographical references and index.
Identifiers: Canadiana (print) 20210286105 | Canadiana (ebook) 20210286156 | ISBN 9781771889605 (hardcover) | ISBN 9781774639375 (softcover) | ISBN 9781003130765 (ebook)
Subjects: LCSH: Biopolymers.
Classification: LCC TP248.65.P62 N38 2021 | DDC 572/.33—dc23

Library of Congress Cataloging-in-Publication Data

Names: Jacob, Jissy, editor. | Gomes, Fernando (Lecturer), editor. | Haponiuk, Józef T., editor. | Kalarikkal, Nandakumar, editor. | Thomas, Sabu, editor.
Title: Natural polymers : perspectives and applications for a green approach / edited by Jissy Jacob, Fernando Gomes, Józef T. Haponiuk, Nandakumar Kalarikkal, Sabu Thomas.
Description: First edition. | Palm Bay, FL : Apple Academic Press, 2022. | Includes bibliographical references and index. | Summary: "This new volume, Natural Polymers: Perspectives and Applications for a Green Approach, covers the synthesis, characterizations, and properties of natural polymeric systems, including their morphology, structure, and dynamics. It also introduces the most recent innovations and applications of natural polymers and their composites in the food, construction, electronics, biomedical, pharmaceutical, and engineering industries. Natural polymers provide a striking substitute for various applications as compared to synthetic polymers obtained from petrochemicals because they are biocompatible, biodegradable, and easily available and fall within the budget of many industries. The applications of natural polymers in pharmaceutical industries are large in comparison to synthetic polymers and are also wide in scope in the food and cosmetic industries. This new volume provides the information needed to design new applications for natural polymers. This book is a valuable reference for researchers, academicians, chemists, pharmacists, researchers, scientists, industrialists dealing with applications of natural polymers and people working in field of natural polymers"-- Provided by publisher.
Identifiers: LCCN 2021037677 (print) | LCCN 2021037678 (ebook) | ISBN 9781771889605 (hardback) | ISBN 9781774639375 (paperback) | ISBN 9781003130765 (ebook)
Subjects: LCSH: Biopolymers.
Classification: LCC TP248.65.P62 N376 2022 (print) | LCC TP248.65.P62 (ebook) | DDC 572--dc23
LC record available at https://lccn.loc.gov/2021037677
LC ebook record available at https://lccn.loc.gov/2021037678

ISBN: 978-1-77188-960-5 (hbk)
ISBN: 978-1-77463-937-5 (pbk)
ISBN: 978-1-00313-076-5 (ebk)

About the Editors

Jissy Jacob

Jissy Jacob is currently working as a research scholar under the guidance of Prof. Sunny Kuriakose and Prof. Sabu Thomas at Mahatma Gandhi University, Kottayam, Kerala, India. She has expertise in polymer chemistry. Her research interests are in the field of biopolymer-based nanocomposites for food packaging application. Miss Jacob works toward development of mesoporous materials from sustainable resources via green synthesis process and the fabrication of several biopolymer composites with improved mechanical, thermal, and gas-barrier properties. Her research articles are published in reputed journals.

Fernando Gomes, PhD

Fernando Gomes, PhD, is Associate Professor at Biopolymers and Sensors Lab, Macromolecules Institute, Universidade Federal do Rio de Janeiro, Brazil. His expertise mainly focuses on the use of renewable resources and nanocomposites in sensors, drug delivery, and environmental recovery. He has coordinated ten research projects with financial support from government-sponsoring agencies, published 136 scientific articles, three chemistry books in Portuguese, thre books on biopolymers in English, and 197 papers at conferences and scientific meetings. He was the co-convener of the Sixth International Conference on Natural Polymers, Biopolymers and Biomaterials: Applications from Macro to Nanoscale (http://www.biopolymers.macromol. in/) and Chairman of the Fifth International Conference on Natural Polymers, Biopolymers and Biomaterials: Applications from Macro to Nanoscale (https://www.icnp2017rio.com/). He was the supervisor of over 100 undergraduate students, 28 MSc students, eight PhD students, and five postdoctoral students. Currently, he is the supervisor of four undergraduate students, two MSc students, 14 PhD students, and two postdoctoral students. Dr. Gomes is a member of the editorial board of *Current Applied Polymer Science* (since 2016), Associate Editor of the *MedCrave Online Journal (MOJ) Polymer Science* (since 2017), and Editor of the *Academic Journal of Polymer Science* (since 2018). He received the Young Scientist Award of Rio de Janeiro State (FAPERJ 2011 and 2014). He is a member of Postgraduate Program in Science and Technology of Polymers of the Federal University of Rio de Janeiro since 2008.

Jozef T. Haponiuk, PhD

Professor Józef T. Haponiuk has been the Head of the Polymers Technology Department at the Faculty of Chemistry of the Gdańsk University of Technology, Poland, since 2006. His special scientific interests include polyurethane chemistry and processing, polymer, and rubber recycling, biopolymers, the use of biomass derived raw materials in polymer technology, polymer composites, and nanocomposites. J.T. Haponiuk is the author or co-author of over 300 original scientific papers, including 90 published in periodicals from the JCR list, editor, and co-author of monographs in international publications (Elsevier, Apple Academic Press, and Royal Society of Chemistry), creator of 11 granted patents and 25 patent applications. Professor Józef T. Haponiuk is a valued academic teacher in the field of polymers chemistry and engineering. He was a supervisor in 12 PhD theses.

Prof. Nandakumar Kalarikkal

Professor Nandakumar Kalarikkal is the Director of School of Pure and Applied Physics and International and Inter University Centre for Nanoscience and Nanotechnology at Mahatma Gandhi University, Kottayam, Kerala, India. His research activities involve nanotechnology and nanomaterials, sol–gel synthesis of nanosystems, semiconducting glasses, ferroelectric ceramics, and nonlinear and electrooptic materials. He is the recipient of research fellowships and associateships from prestigious organizations such as the Department of Science and Technology and Council of Scientific and Industrial Research of the Government of India. He has collaborated with national and international scientific institutions in India, South Africa, Slovenia, Canada, and Australia, and is co-author of several books chapters, peer-reviewed publications, and invited presentations in international forums.

Prof. Sabu Thomas

Professor Sabu Thomas is currently Vice Chancellor of Mahatma Gandhi University and Director of School of Energy Materials. He is also the Founder Director and Professor of International and Interuniversity Centre for Nanoscience and Nanotechnology. He is also a full professor of Polymer Science and Engineering at School of Chemical Sciences, Mahatma Gandhi University, Kottayam, Kerala, India. Prof. Thomas is an outstanding leader with sustained international acclaims for his work in Nanoscience, Polymer Science and Engineering, Polymer Nanocomposites,

About the Editors

Elastomers, Polymer Blends, Interpenetrating Polymer Networks, Polymer Membranes, Green Composites and Nanocomposites, Nanomedicine, and Green Nanotechnology. Prof. Thomas has been conferred Honoris Causa DSc from the University of South Brittany, France and the University of Lorraine, France. He received many national and international awards and he published over 1000 peer reviewed research papers, reviews, and book chapters. He has co-edited 150 books and he is the inventor of 15 patents.

Contents

Contributors .... *xi*

Abbreviations .... *xiii*

Preface .... *xvii*

Introduction .... *xix*

1. **Polymeric Composites and Microplastics: Regulatory and Toxicological Perspectives** 1
   V. P. Sharma

2. **Natural Polymers: Applications, Biocompatibility, and Toxicity** 17
   N. S. Remya

3. **Development of Biocomposites from Industrial Discarded Fruit Fibers** 41
   J. S. Binoj, R. Edwin Raj, and N. Manikandan

4. **Characteristics and Prospects of Controlled Release of Fertilizers Using Natural Polymers and Hydrogels** 69
   Neetha John

5. **Chitosan Biopolymer for 3D Printing: A Comprehensive Review** 91
   Dhileep Kumar Jayashankar, Sachin Sean Gupta, Rajkumar Velu, and Arunkumar Jayakumar

6. **Potential Applications of Chitosan-Based Sorbents in Nuclear Industry: A Review** 117
   Anupkumar Bhaskarapillai

7. **Spray-Dried Chitosan and Alginate Microparticles for Application in Hemorrhage Control** 153
   Srikant Suman, Sandarbh Kumar, Ritvesh Gupta, Akhil Bhimireddy, and Devendra Verma

8. **High-Intensity Ultrasonication (HIU) Effect on Sunn Hemp Fiber-Reinforced Epoxy Composite: A Physicochemical Treatment Toward Fiber** 193
   Chinmayee Dash, Asim Das, and Dillip Kumar Bisoyi

Contents

9. **Tribological Behavior of Coconut Shell–Fly Ash–Epoxy Hybrid Composites: An Investigation** 225

Bibhu Prasad Ganthia, Maitri Mallick, Sushree Sasmita, Kaushik Utkal, and Ipsita Mohanty

10. **Cellulose Acetate-TiO$_2$-Based Nanocomposite Flexible Films for Photochromic Applications** 251

T. Radhika

11. **Effect of Thermal Treatment on Structure and Properties of Plasticized Starch–Polyvinyl Alcohol (PVA) Blend Films** 283

Subramaniam Radhakrishnan, Swapnil Thorat, Anagha Khare, and Malhar B. Kulkarni

Index *301*

Contributors

Anupkumar Bhaskarapillai
Water and Steam Chemistry Division, Bhabha Atomic Research Centre Facilities, Kalpakkam, Tamil Nadu 603102, India and Homi Bhabha National Institute, Anushakthi Nagar, Mumbai 400394, India. E-mail: anup@igcar.gov.in

Akhil Bhimireddy
Biotechnology and Medical Engineering Department, National Institute of Technology Rourkela, Rourkela, Odisha 769008, India. E-mail: akhilbhimireddy3559@gmail.com

J. S. Binoj
Micromachining Research Centre, Department of Mechanical Engineering, Sree Vidyanikethan Engineering College (Autonomous), Tirupati 517102, Andhra Pradesh, India. E-mail: binojlaxman@gmail.com

Dillip Kumar Bisoyi
Composite Laboratory, National Institute of Technology Rourkela, Rourkela 769008, Odisha, India. E-mail: dkbisoyi@nitrkl.ac.in

Asim Das
Composite Laboratory, National Institute of Technology Rourkela, Rourkela 769008, Odisha, India.

Chinmayee Dash
Composite Laboratory, National Institute of Technology Rourkela, Rourkela 769008, Odisha, India. E-mail: dash2016.chinmayee@gmail.com

Bibhu Prasad Ganthia
Electrical Engineering, IGIT, Sarang, Dhenkanal, Odisha, India. E-mail: jb.bibhu@gmail.com

Ritvesh Gupta
Biotechnology and Medical Engineering Department, National Institute of Technology Rourkela, Rourkela, Odisha 769008, India. E-mail: ritvesh95@gmail.com

Sachin Sean Gupta
Digital Manufacturing and Design Centre, Singapore University of Technology and Design, Singapore 487372.

Arunkumar Jayakumar
Department of Automobile Engineering, SRM Institute of Science and Technology, SRM Nagar, Kattankulathur 603203, Kanchipuram, Tamil Nadu, Chennai, India. E-mail: arunkumar.h2@gmail.com

Dhileep Kumar Jayashankar
Digital Manufacturing and Design Centre, Singapore University of Technology and Design, Singapore 487372. E-mail: jayashankar@sutd.edu.sg

Neetha John
CIPET: IPT-Kochi, Udyogamandal P.O, Eloor, Kochi, Kerala, India. E-mail: neethajob@gmail.com

Anagha Khare
Research Development and Innovation, Maharashtra Institute of Technology (MIT-WPU), S124, Paud Road Kothrud, Pune 411038, India. E-mail: anagha.khare@mitpune.edu.in

Malhar B. Kulkarni
Research Development and Innovation, Maharashtra Institute of Technology (MIT-WPU), S124, Paud Road Kothrud, Pune 411038, India. E-mail: malhari.kulkarni@mitpune.edu.in

Sandarbh Kumar
Biotechnology and Medical Engineering Department, National Institute of Technology Rourkela, Rourkela, Odisha 769008, India. E-mail: sandykmr97@gmail.com

Maitri Mallick
Civil Engineering, GIET, Bhubaneswar, Odisha, India.

N. Manikandan
Micromachining Research Centre, Department of Mechanical Engineering, Sree Vidyanikethan Engineering College (Autonomous), Tirupati 517102, Andhra Pradesh, India.

Ipsita Mohanty
School of Civil Engineering, KIITU, Bhubaneswar, Odisha, India.

Subramaniam Radhakrishnan
Research Development and Innovation, Maharashtra Institute of Technology (MIT-WPU), S124, Paud Road Kothrud, Pune 411038, India. E-mail: radhakrishnan.s@mitpune.edu.in

T. Radhika
Centre for Materials for Electronics Technology (C-MET), Ministry of Electronics and Information Technology (MEITY), Athani P.O, Shoranur Road, Thrissur, Kerala 680581, India. E-mail: rads12@gmail.com

R. Edwin Raj
Department of Mechanical Engineering, St. Xavier's Catholic College of Engineering, Nagercoil 629003, Tamil Nadu, India.

N. S. Remya
Division of Toxicology, Department of Applied Biology, Bio Medical technology Wing, Sree Chithra Institute for Medical sciences and technology, Thiruvananthapuram, Kerala 695012, India. E-mail: remya.bijoy@sctimst.ac.in

Sushree Sasmita
School of Civil Engineering, KIITU, Bhubaneswar, Odisha, India.

V. P. Sharma
Chief Scientist and Prof AcSIR, Regulatory Toxicology Group, CSIR-Indian Institute of Toxicology Research, 31, Vishvigyan Bhawan, M.G.Marg, Lucknow, 226001, India E mail: vpsitrc1@rediffmail.com; vpsharma@iitr.res.in

Srikant Suman
Biotechnology and Medical Engineering Department, National Institute of Technology Rourkela, Rourkela, Odisha 769008, India. E-mail: srikantsuman31@gmail.com

Swapnil Thorat
Research Development and Innovation, Maharashtra Institute of Technology (MIT-WPU), S124, Paud Road Kothrud, Pune 411038, India. E-mail: spthorat4@gmail.com

Kaushik Utkal
Electrical Engineering, GCE, Keonjhar, Odisha, India.

Rajkumar Velu
Digital Manufacturing and Design Centre, Singapore University of Technology and Design, Singapore 487372. E-mail: rajkumar7.v@gmail.com

Devendra Verma
Biotechnology and Medical Engineering Department, National Institute of Technology Rourkela, Rourkela, Odisha 769008, India. E-mail: dev.rivan@gmail.com

Abbreviations

AFH	areca fruit husk
AFHF	areca fruit husk fiber
AFHFC	areca fruit husk fiber composite
AM	additive manufacturing
APTT	activated partial thromboplastin time
ASTM	American Society for Testing and Materials
CA	cellulose acetate
CA	citric acid
CA-MBMMOC	cellulose acetate-modified metal oxide ceramics
CA-MMOC	cellulose acetate modified metal oxide ceramics
CA-MOC	cellulose acetate-metal oxide ceramics
ChCl	choline chloride
CI	crystallinity index
CMC	carboxymethyl cellulose
CMSs	ceramic matrix composites
CNF	cellulose nanofiber
CR	controlled release
CRED	criteria for reporting and evaluating ecotoxicity data
CRF	controlled release fertilizers
CS	chitosan
DA	degree of N-acetylation
DFU	diabetic foot ulcer
DIW	direct ink writing
DSC	differential scanning calorimeter
DTPA	diethylenetriaminepentaacetic acid
ESEM	environmental scanning electron microscope
FDM	fused deposition modeling
FF	flexible fabrics
FS	flexural strength
FTIR	Fourier Transmission Infrared
GFC	glass fiber composites
HIU	high-intensity ultrasonication
HyDCA	hybrid clay composite adsorbent
INR	international normalized ratio

IPN	interpenetrating polymer networks
LC	laminar composite
LD	linear dimension
MALS	multiangle light scattering
MB	methylene blue
mgCNF	magnetic cellulose nanofiber
ML	mass loss
MMCs	metal matrix composites
MMOC	modified metal oxide ceramics
MMs	matrix materials
MTT	3-(4,5-dimethylthiazol-2-yl)-2,5-diphenyltetrazolium bromide
NF	natural fiber
NFC	natural fiber composites
NFPC	natural fiber polymer composite
NFRC	natural fiber-reinforced composite
NPK	nitrogen, phosphorous, and potassium
NTA	nitrilotriacetic acid
OD	optical density
PBS	phosphate buffer saline
PBS	poly(butylene succinate)
PECs	polyelectrolyte complexes
PEG	polyethylene glycol
PLA	polylactic acid
PLC	plasticizer content
PMN	polymorphonuclear cells
POPs	persistent organic pollutants
PPP	platelet poor plasma
PT	prothrombin time
PVA	poly(vinyl alcohol)
PVP	polyvinyl pyrrolidone
RSH	Raw Sunn Hemp
SAP	super absorbent polymers
SC	structural composites
SEM	scanning electron microscopy
SLS	selective laser sintering
TEPA	tetraethylenepentamine
TF	tamarind fruit
TFF	tamarind fruit fiber
TFFC	tamarind fruit fiber composite

Abbreviations

TGA	thermogravimetric analysis
TPS	thermoplastic starch
TS	tensile strength
TXA	tranexamic acid
WVP	water vapor permeability
XRD	X-ray diffraction

Preface

Natural Polymers: Perspectives and Applications for a Green Approach is a book mainly devoted to the polymers from nature, which can be used in several applications as composites, fibers, and films, producing several effects on the environment, Earth, and even our lives.

This book is the materialization of the effort of a myriad of authors, editors, and reviewers devoted to the environmental cause.

We wish you all enjoy this book as much we are thrilled producing it.

Wishing you a pleasant reading,

—Dr. Fernando Gomes

Introduction

The word "polymer" means "many parts" (from the Greek, poly means "many," and meros means "parts"). Polymers are giant molecules that are made from same or different monomers. Both synthetic and natural polymers are available; around 80% of organic industries are devoted to the production of synthetic polymers such as plastics, fibers, and synthetic rubbers, but they creates a serious aesthetic problem in large urbanized areas of the world. Natural polymers are found in nature and commonly derived from corn, cellulose, potato, sugarcane, or they can be derivatives of animal sources, such as chitin, chitosan, and proteins. Several of these polymers are part of our diet and have been used in a variety of human applications. Recently, natural polymers present an attractive alternative for many applications compared to synthetic polymers derived from petrochemicals because they are biodegradable, biocompatible, economical, readily available, and nontoxic in nature. The applications of natural polymers in pharmaceutical industries are large in comparison to the synthetic polymers and have wide scope in food and cosmetic industry.

This book covers the synthesis, characterizations, and properties of natural polymeric systems including their morphology, structure, and dynamics. Moreover, it also introduces the most recent innovations and applications of natural polymers and their composites in the food, construction, electronics, biomedical, pharmaceutical, and engineering industries.

CHAPTER 1

Polymeric Composites and Microplastics: Regulatory and Toxicological Perspectives

V. P. SHARMA

Chief Scientist and Prof AcSIR, Regulatory Toxicology Group, CSIR-Indian Institute of Toxicology Research, 31, Vishvigyan Bhawan, M.G.Marg, Lucknow, 226001,India E mail: vpsitrc1@rediffmail.com; vpsharma@iitr.res.in

ABSTRACT

Generally, plastics are inert and persistent with extended lifetime. Polymers are made of monomers and chemical additives to attain desired properties or preselected specifications. Additives used are plasticizers, stabilizers, colorants, fillers, flame retardants, etc. Polymeric materials on exposure to environmental conditions may breakup into microplastics or new chemical entities with time and facilitate uptake by marine biota throughout the food cycle. There is a vital need to carry out focused scientific research to fill the knowledge gaps about the impacts of composites and microplastics. Composites are being engineered for target applications to address sustainability issues, quality control compliances as per EU/ISO/ IEC 17025:2017, OECD, American Society for Testing and Materials, Pharmacopoeias, ISO requirements and demonstrate global dependence on nonrenewable petroleum-based resources. Holistic studies are needed for standardization, validation in predesigned set conditions of plastics during simulated or landfill situations. Internationally exposure studies are in progress for devising environmental safe approaches and to explore better understanding of interactions.

1.1 INTRODUCTION

Plastics and polymeric products are integral part of our life and everyone is using the same in many forms. Plastics may be defined at molecular level as a kind of organic polymer which has molecules containing long carbon chains as the backbone with repeating units created via a process of polymerization.

Phthalates, which are well-known plasticizers, are known as endocrine disruptors. They are used at varying concentrations depending on the requirements of a particular product. These may be harmful at extremely low concentrations for organisms and pose potential risks to biodiversity. Globally, due to inherent characteristics and market demand, there is a continuous growth of plastics. The long term exposure, that is, chronic exposures are anticipated to have accumulative effect in dose-dependent manner.

The configuration is dependent on the structure of the repeating units and types of atoms or additives used to provide the desired plasticity to the finished polymeric product [1]. There is exponential growth in plastic consumption, production, and usage pattern since 1950s. The world production is projected to reach around 1800 million tons by 2050. Nowadays, plastics are vital asset for the humanity and they cannot be easily removed due to functionality and economical reasons. Several economic activities of the modern generation are attributed to benefits to the society through specifications, lightweight, variability in desired shapes and color, quality of life, jobs and robustness, etc. They have replaced the components of computers, automobiles, missiles, agricultural appliances, health care products. No doubt this has simultaneously given rise to electronic waste generation and other degradation moieties. These ultimately reach to the oceans, beaches, or other water bodies and become an integral part of soil sediments. Plastic beverages bottles during transportation and distribution are shown in Figure 1.1.

Bioplastics, degradable polymers, composites, and degradation moieties have spurred an unprecedented thrust in academicians so as to investigate their fate and impacts with limitation of standardized laboratory bioassays. Biopolymeric products have been serving the mankind in several ways [2, 4–9, 11, 13, 15, 16, 19, 20, 24]. As an effort towards sustainability, the lignocelluloses biomass has ever remained a choice due to its abundance on the planet. Lignin is a phenolic macromolecule and the prime biological source of aromatic structures or phenolic skeletons. Its complex structure may differ depending on plant species and the process of its isolation [10]. It is used generally to produce energy while some are used for low or medium value applications. Lignin is used extensively to provide structural support and as a barrier to microbial infestation with facilitation to have sufficient movement of water within the plant systems. For improved applications, we need to understand the structure of lignin monomers and their interactions within the biomass besides carbohydrate polymers deconstruction processes and valorization of all biopolymers of lignocelluloses residues.

FIGURE 1.1 Plastic beverages bottles during transportation and distribution—how safe?

1.2 MICROPLASTICS: CAUSE, IMPLICATIONS, AND R&D EFFORTS

The microplastic debris has a massive impact on marine life and is responsible for the extinction of few species [25]. Small finer particles of plastic waste or fragments are having adverse implications on flora and fauna [18]. The reach to the food chain and create ecological imbalances. They are particulates made of plastic material less than 5 mm in size and constitute microsized fraction of plastic marine litter. Thus, microplastics are now as niche areas of global R&D for policy decisions and pollution research [17]. Institutions such as Tianjin University, Chinese Academy of Sciences care currently identifying the potential sources of microplastic pollution in coastal environments or adjoining areas so as to guide regulatory, decision-making steps at international platforms [22]. Microplastic pollution in global aquatic environments has aroused increasing concern in recent years for selection of raw material for final products designing and marketing. They may enter natural ecosystems from several routes of exposure ranging from cosmetics as facial cleansers, clothing, occupational, or industrial processes. In some cases, their use in medicine as vectors for drugs. Scrubbers that are used in exfoliating hand cleansers and facial scrubs have replaced traditionally used natural ingredients, including ground almonds, oatmeal, and pumice.

Concentrations of plastic debris, size, and distribution pattern of the plastic debris due to terrestrial or land-based activities with continued fragmentation into nanosized particles is an issue of global concern. Modern analytical techniques introduce a great bias during detection and quantification of nanoparticles. Toxicological effects of plastic particles may be due to the plastic particles themselves, release of persistent organic pollutants adsorbed to the plastics leaching of additives of the plastics. Chemical and biological toxicity may occur due to the localized leaching of component monomers, endogenous additives, and adsorbed environmental pollutants. The microplastics have been determined by researchers from sediment samples in varied ranges, for example, 252.80 ± 25.76 particles m^{-2}. Moreover, bio-based novel materials, smart polymers, and/or bio-inspired biomaterials are gaining due attention with high-value biomedical, tissue repair, drug delivery, therapeutic usage, and pharmaceutical potentials.

1.2.1 MICROPLASTICS—AN INCREASING MENACE FOR ENVIRONMENT AND HEALTH

Nowadays, air blasting technology uses new techniques to remove rust and paint through specialized scrubbers. Awareness efforts are being taken for reducing the production of microbeads through super fragmentation and thus there are countless sources of both primary and secondary microplastics. Microplastic fibers may enter the environment from the washing of synthetic clothing. They may be classified into primary or secondary depending on size, mode of entry, and characteristics defined by the regulators. There is increased probability of microplastics being ingested, incorporated into, and accumulated in the bodies or tissues of many organisms. The entire cycle and movement of microplastics in the environment are not well understood and R&D efforts are in process at varied levels to investigate the same with holistic viewpoints. The exposure to microplastics (<10 μm) associated with plastic bottles mineral water consumption is the well documented

Polymeric Composites and Microplastics

quantitative study [25]. Every year approximately 18 billion pounds of plastic waste enters the world's ocean from coastal regions. There are biodiversity hotspots in the ocean region the Atlantic Ocean, the Caribbean Sea, and even other recognized highly biodiverse marine environments. Wagner [21] has assessed the endocrine disruptors in bottled mineral water.

1.2.2 MICROPLASTICS-BOTTLED WATER AND SALT

Microplastics have been detected and characterized in seafood products and in other foods and beverages such as beer, honey, and table salt including bottled mineral water [23, 25]. Most of the bioplastics are made from sugar and starch materials; to expand their use significantly raises the prospect of competition between growing crops to supply food or plastics, similarly to the diversion of food crops for the manufacture of primary biofuels. The applications of oxo-plastics are also under monitor as the European Union has moved to restrict their use due to environmental issues. The extensive segregation for reuse, collection, and recycling of plastic items at the end of their life is a critical aspect. This may have a correlation for the ideology to underpin the circular economy.

1.3 IMPORTANCE OF QUALITY ASSURANCE FOR IMPACT STUDIES OF MICROPLASTICS IN ECOSYSTEM

The studies related to ingestion of microplastics by marine organisms are essential so as to monitor potential risk associated through microplastics in the ecosystem. Microplastics have been detected in seawater of specific regions for bacteria and diatoms. Microbiological studies indicate microbial attachment to polyethylene terephthalate and polyethylene plastic bags. The bacterial association with plastic surfaces has been revealed in engineered ecosystems or drinking water systems with the need of detailed studies. The rapid bacterial colonization of low-density polyethylene microplastics in coastal sediment microcosms is interesting and informative to take a lead for future studies.

Although few studies are available for base development but lack in terms of comparability of results due to deficiency of standardization of methods. The quality assurance needs to check adopting checklist constituting of sampling method and strategy, sample size, sample processing and storage, laboratory preparation, clean air conditions, negative controls, etc. The impacts on marine species, human health through microplastic or particles <5 mm, combined with associated effects have raised concerns. They have been detected in oceans ranging from shallow coasts to the deep sea as it is ingested as meso or macroplastic by marine biota across multiple trophic levels. Ingested microplastic particles may induce a biological response through food web via biochemical or physical mechanisms and thus detailed studies are needed. The measurement of size or quantification of microplastics to appropriately assess the risk of such hazards is vital. The presence may lead to blockages or abrasions in the body system. They are also anticipated to be a carrier for other persistent organic pollutants (POPs) due to chemical characteristics. Food safety is also linked to microplastics as significant amount of aquatic organisms are taken as food throughout the globe and is more in coastal regions. In few studies, the Klimisch score or the criteria for reporting and evaluating ecotoxicity data (CRED) are being used for toxicological studies beside other statistical tools or scores. CRED expands the reliability assessment by requiring the scoring of aquatic ecotoxicity studies against a more extensive through specific array of reliability criteria and crucially also includes a discrete set of relevance criteria via enabling studies to be scored separately for reliability and/or relevance. These improve the reproducibility, transparency, and consistency of evaluations of aquatic ecotoxicity studies among regulatory frameworks, countries, institutes, and individual assessors. Nowadays, predicted-no-effect concentrations and environmental quality standards are derived in a large number of legal frameworks worldwide. Various colored bottles for packaging of foodstuffs and drinking water are shown in Figure 1.2. As an effort toward green chemistry concepts, plastic manufacturers are also required to fulfill the following aspects:

1. High resource effectiveness and high atom economy, maximizing the content of raw materials in the product.
2. Clean production processes, preventing wastes, and reducing greenhouse gas emissions with high safety standards.
3. Minimum usage of auxiliary/indirect additive(s) substances, namely, organic solvents, blocking groups, etc.
4. High-energy efficiency of materials manufacturing and applications.
5. No health and environmental hazards by minimizing toxicity.
6. Use of renewable resources and renewable energy.
7. Low carbon footprint and controlled product lifecycles with effective waste recycling.

Polymeric Composites and Microplastics

FIGURE 1.2 Various colored bottles for the packaging of foodstuffs and drinking water.

Biodegradation process may also generate water-soluble and even few toxic metabolites that are washed away by rain and reach to groundwater. The biodegradation and bioerosion generally render polymers brittle so that they readily disintegrate when exposed to mechanical stresses and form micron and nanometer-sized particles, which are carried away by wind or rain. Sometimes the biodegrading plastics are not visible to the eye and the resulting fine and invisible particles may accumulate in the air and cause inhalation hazards. Most biodegradable polymers are a coveted source of food for bacteria and other microorganisms such as fungal spores, insects, bugs, mice, etc.

Biodegradable plastics may take approximately 60–180 days to decompose completely and it depends on factors viz temperature, oxygen availability, and the presence of appropriate moisture content. The readily used raw material for developing the compostable plastics are corn starch, soybean, beet, potato, vegetable fats, oils, microbiota, etc., which may be converted into a polymer with similar properties as normal plastic products. These may provide biodegradability and consequently solution of waste disposal to some extent. Durable plant-based bioplastics may also be recycled as well as their conventional equivalents and thus helping the growth of a more sustainable world economy. Unlike traditional plastics and biodegradable plastics, the bioplastics commonly does not lead to a net increase in carbon dioxide gas on break down. Few bioplastics may degrade in few weeks and unable to meet the challenges of the stakeholders or demanding customers. The properties, namely, flexibility, durability, printability, transparency, barrier, heat resistance, etc., are to be considered as value-added features during the preparation of bio-based items or finished materials.

It is considered that degradable plastics are not clogging our land and oceans, threatening the health of humans and animals. It is suggested that policies that cut our plastic footprint need to be urgently implemented for sustainability measures. We must minimize microplastics in cosmetics, personal care products, detergents, and cleaning products and take concrete measures to tackle other sources of microplastics. They are associated with chemicals involved during manufacturing processes and that absorb from the surrounding environment as evidence regarding microplastic toxicity and epidemiological are emerging [3, 12].

1.4 MULTIFUNCTIONAL CAPABILITIES OF BIO-BASED COMPOSITES

The composites of the future are developed with multifunctional capabilities such as thermal, electrical, and fire resistance. Composites may be derived from corn stover and feather meals as double coating materials for controlled release and water retention urea fertilizers. Electrically conductive coated fiber may be embedded in a polymer and may have self-monitoring. The carbon-fiber-based composites with epoxy and selected polymer matrix are structural materials with lightweight and strong mechanical properties to be used in automobiles, aircraft, and other transportation vehicles. Sometimes the damage in carbon fiber structures may remain hidden below the surface and thus undetected during visual inspection and posing possibility of catastrophic failure. The strength of nanoparticles embedded fibers adhered to polymer matrix need to be systematically evaluated in accredited facility of repute.

1.5 STRATEGIES TOWARD RENEWABLE PLASTICS GENERATION

There are different strategies for generation of renewable plastics which may be useful for green economy and holistic development. Jute and cloth bags as alternatives to single-use plastics are shown in Figure 1.3. Natural fibers, namely, jute, hemp, sisal, etc., provide excellent fiber reinforcement for thermosets and thermoplastics. The efforts include:

1. Biorefining of biomass and chemical conversion of carbon dioxide are employed to produce synthetic crude oil and green monomers for highly resource- and energy-effective polymer manufacturing processes without impairing established recycling technologies.
2. Living cells may be converted into solar-powered chemical reactors with exploitation of genetic engineering and biotechnology routes to produce biopolymers as well as bio-based polymers.
3. Carbon dioxide is activated and polymerized.

FIGURE 1.3 Jute and cloth bags as alternatives to single-use plastics.

Novel families of thermosetting and thermoplastic wood/plastics composites, namely, injection moldable wood/plastic compounds are being developed to minimize processing costs and improve performance. During the last few decades, micro fibrillated cellulose from wood and cellulose nanowhiskers have been developed as nanofillers for polymer nanocomposites, as well as bio-based medical implants. Wood is composed of cellulose and is chemically bonded to lignin. As a phenolic resin, lignin protects cellulose against thermo-oxidative degradation and microbial attack. Most chemical processes for separating cellulose from lignin are economically viable and energy-intensive. Hydrolysis of lignocelluloses is considered as a preferred source of sugars. These are in demand as a food source for bacteria in various biotechnology and biorefinery processes, including biofuel production.

Due to the high water solubility of starch, several manufacturers have developed biohybrids using starch as a blend component with hydrophobic nondegradable polyolefins and compostable polyesters such as terephthalic acid/adipic acid/1,4-butanediol copolyesters. Chitosan contains glucosamine repeat units and is derived by partial or full deacetylation of chitin. This is extracted from the shells of shrimps, crabs, lobsters, etc. The usage of chitosan includes drug release, encapsulation, wound dressings, and water purification. Terpeneslikepoly (*cis*-1,4-isoprene) constitutes of more than 40% of the rubber market and fats and oils serve as valuable intermediates for the chemical industry for developing a variety of surfactants. Proteins such as collagen are abundant in mammals. Protein fibers such as silk and wool have been used in the textile industry for centuries. Gelatin, which is denatured collagen, is one of the preferred materials for drug encapsulation. Casein, produced from cow's milk, is used as a binder and as an adhesive. Bio polyesters such as poly(L-lactic acid), poly(hydroxybutyrate), and other poly-(hydroxyalkanoates), produced by bacteria as chemical energy storage in cells, are fully biodegradable and decompose to produce water, carbon dioxide, and humus when air and water are present. It is possible to tailor stiff, soft plastics or polyesters. Polylactic acid produced by bacteria or by lactide polymerization has been the well-accepted material of choice for resorbable sutures in surgery.

The bio-based polymers combine the advantages of a low carbon footprint, typical for renewable agricultural feedstocks with the recycling capability and high resource and energy-effectiveness of solvent-free melt and gas-phase polymerization processes. Polyolefin materials may serve as prominent for successful sustainable development. They are generated by highly effective solvent-free olefin n polymerization processes and simultaneously conserve the energy content and may be tailored to meet the demands of growing markets scenario. The nontoxic and biodegradable cyclic carbonates with their high boiling and flashpoints may be readily manufactured from oxiranes, namely, ethylene oxide and propylene oxide in the presence of tetrabutylammonium bromide.

Active catalyst generations have been developed for alternating copolymerization of propylene oxide with carbon dioxide to produce high-molecular-weight polypropylene carbonates. Amorphous polymers, for example, polypropylene carbonates with glass transition temperatures slightly above room temperature

Polymeric Composites and Microplastics

are soft materials and are of interest as blend components with a low carbon footprint [14].

1.6 RISK ASSESSMENTS OF NOVEL POLYMERIC FORMULATIONS

The guidance document for assessing the risks of exposure to nanomaterials, nanocomposites, biopolymers, and other complex materials from their usage in the food chain have been developed. There are uncertainties which are linked to detection, identification, characterization of various microplastics and nanoparticles in complex matrices, namely, food and water. The information on migration rates of nanoparticles into food or food simulants are sparse. The uptake and interactions may be related to the chemical composition and physicochemical properties of the nanoparticles or fibers. The migration may be higher into acidic matrices.

1.7 FUTURE OF NANOCOMPOSITES, QUANTIFICATION, AND INNOVATIONS

Plastics have greatly improved human living conditions and are indispensable and irreplaceable in modern life. Due to high energy, cost, eco, and resource-efficient production processes, facile processing, and high versatility in terms of tunable properties and a broad range of applications, they render high technology affordable life. The production of state of art plastics meets the demands of green chemistry for clean production by using solvent-free processes with efficient use of resources, waste management, and even exploitation of renewable resources. Sustainable developments are being made so as to meet the ever-increasing demand of food safety and quality. Biopolymer composites are emerging as new sustainable materials with improved mechanical and barrier properties with food-packaging solutions. Biopolymers obtained via renewable resources are expected to be competitive with fossil fuel-derived plastics as food-packaging materials and serve as alternatives to single-use plastics.

1.7.1 NEW AREAS FOR COMPOSITES, BIOCOMPATIBILITY, AND CHARACTERIZATION

Now niche areas are being explored for robust and green biodegradable nanocomposites as new opportunities for the entrepreneurs. The cellulosic nanomaterials are also being developed for food and nutraceuticals industry using renewable sources with biocompatibility and increased mechanical strength. In view of immobilization of various bioactive agents, enzymes, stabilization these have potential applications in intelligent or smart packaging. The pH indicators, pathogen detectors are upcoming areas in futuristic food sciences for safety to environment and health. Microscopic particles have been detected in sea sediments and new methods are being validated for quick and reliable detection using modern analytical techniques. The residual organic matter may cause interference in downstream Raman Spectroscopic analyses and it may be eliminated through use of hydrogen peroxide (15%) digestion step. The process may be applied to detect microplastic in marine samples and improvise the understanding. The accurate diffusion coefficient data of additives in a polymer are of paramount importance for estimating the migration of the additives in due course of time. According to study conducted at IIT, Guwahati and GIFU University, Japan, the formulation of cellulose nanofiber (CNF) or magnetic cellulose nanofiber (mgCNF) dispersed chitosan-based edible nanocoating with superior mechanical, thermal, optical, and texture properties. The fabrication of mgCNF is successfully achieved through a single-step co-precipitation route, where iron particles get adsorbed onto CNF. The thermal stability of mgCNF may be improved considerably, where ~17% reduction in weight whereas CNF degrades completely under identical conditions. Thermogravimetric analysis (TGA) analysis reveals that there is an improvement in thermal stability for both CNF- and mgCNF-reinforced chitin and chitosan (CS) nanocoatings, where mgCNF provides more heat dimensional stability than CNF-dispersed CS nanocoatings.

The synthetic polymers may be analyzed for molar mass distribution by size exclusion chromatography coupled to multiangle light scattering (MALS) detection. It has a limitation that for ultrahigh molar mass or branched polymers this method is not sufficient, because shear degradation and abnormal elution effects falsify the calculated molar mass distribution and information on branching. The high-temperature asymmetrical flow field-flow fractionation coupled to infrared, MALS, and viscometry detection has been accepted as a method for the characterization of high molecular weight polyethylene. The rheological behavior and processability of polyethylene materials may be governed by molecular weight fractions. The high-temperature size-exclusion chromatography may be useful in characterization, round-robin testing, repeatability, and intra laboratory reproducibility data generation for quality assurance.

1.8 REGULATIONS RELATED TO BIO-BASED COMPOSITES OR DEGRADABLE POLYMERS

Standardization is an effort by industrial stakeholders to define generally accepted criteria and guidelines for the description of products, services, and processes. The aim is to ease competition and commercial growth by overcoming trade barriers that result from unclear or inconsistent specifications and communication, to introduce benchmarks for desirable quality requirements, and to prevent fraudulent market behavior.

1. Degradable plastics may be either photodegradable or oxo-degradable plastics which disintegrate into small pieces when exposed to sunlight/oxygen;
2. Semibiodegradable blends of starch and polyethylene or
3. 100% biodegradable; refers only to those materials which are consumed by microorganisms such as bacteria, fungi, and algae.

Plastics which are fundamentally made from petrochemicals are not a product of nature and thus may not be broken easily via natural processes. The semidegraded, hydrophobic, high surface area plastic residues migrate into the water table and other compartments of the ecosystem causing irreparable harm to the environment. According to Central Pollution Control Board (CPCB) report in India, the life-cycle of petro-based plastics is incomplete and remains on the landscape for several years. Besides, plastic products can be recycled three to four times only and after each recycling the product quality deteriorates and ultimately, it is dumped on land-fill site, leading to burden on the earth and damaging environment due to nonbiodegradability. The petrochemical products are found in the entire spectrum of consumer usage items and thus traceable in every vista of life, namely, packaging, textiles, building construction, upholstery, defense, automobiles, fast-moving consumables market, agriculture, toys, packaging, medical appliances, electronics, electrical, etc. According to the study conducted by CPCB in 60 major cities of India, it has been concluded that around 4059 T/day of plastic waste is generated from these cities. The fraction of plastic waste in total municipal solid waste varies from 3.10% (Chandigarh) to 12.47% (Surat). As per the estimates over 150 million tons of plastics is in the ocean and expected to grow by extensively by 2050 if effective measures are not taken.

The R&D studies reveal that plastic debris around the globe may degrade away and end up as microscopic granular or fiber-like fragments, and that these fragments have been steadily accumulating in the oceans. Experiments demonstrate that marine animals consume microscopic bits of plastic, as seen in the digestive tract of an amphipod. Thus usage of biodegradable polymers derived from renewable resources has received considerable attraction in recent years. We need to implement applications for biodegradable plastics in different day-to-day life.

1.9 SALIENT GUIDELINES

IS/ISO 17088:2012 and ASTMD 6400 series of specifications are intended to establish the requirements for the labeling of plastic products and materials, including packaging made from plastics. The adherence to standards is voluntary and it depends on individual organizations to seek compliance with standard or not. Compliance with test methods alone cannot substantiate claims to conformity with hard-and-fast industry standards in the absence of pass/fail criteria. While there is no comprehensive European Union (EU) legislation specifically harmonizing standards for environmental and product marketing claims, the European Commission as well as national governments, ministries, and independent standardization institutes have issued a multitude of standards that may serve as a basis for evaluating claims related to bio-based plastic items. The salient standardization bodies are International Organization for Standardization (ISO), European Committee for Standardization and American Society for Testing and Materials (ASTM). Besides them, there are several individual nations' standardization organizations. The European norm EN 16640:2017, bio-based products–determination of the bio-based carbon content of products using the radiocarbon method describes how to measure the carbon isotope C14, that is, radiocarbon method, based on the provisions of the EU Directive 2009/28/EC (Renewable Energy Directive).

1.10 INFERENCE

The persistence of plastic debris in environment is well established in estuarine and marine ecosystems. They represent a threat to the ecological integrity due to their high potential to be ingested by wildlife and the capacity to transport POPs and plastic additives into marine food webs. The small-sized organisms may function as pioneering surface colonizers and effect ecosystem production processes, cycling, and biodegradation of anthropogenic pollutants. Composites are the need of the hour and several organizations which are working in the area of sustainability are looking forward to new bio-based composites and raw materials. The regulatory compliances are to be adhered as

Polymeric Composites and Microplastics

per the national/international norms. Moreover, they need to be cost-effective and readily available for increased acceptance by the community.

ACKNOWLEDGMENT

Author is thankful for the encouragement by CSIR—Indian Institute of Toxicology Research, Lucknow. The interactions with Prof Sabu Thomas at MG University, Kottayam, Kerala were a source of inspiration.

KEYWORDS

- **composites**
- **degradation**
- **fragmentation**
- **toxicity**
- **safety**
- **guidelines**

REFERENCES

1. Agency for Toxic Substances and Disease Registry (ATSDR)—DEHP 2002.
2. American Society for Testing and Materials (ASTM) D 7611 Standard Practice for Coding Plastics manufactured from Resins Identification; SPI Codeswww.astm.org; 2019.
3. Andrady, A. L. *Plastics and Environment*, Wiley–Intersciences: Germany, 2003.
4. Aresta, M. *Carbon Dioxide as Chemical Feedstock*, Wiley-VCH: Germany, 2010.
5. Belgacem, M. N.; Gandini, A. *Monomers, Polymers and Composites from Renewable Resources*. Elsevier: Oxford, 2008.
6. Central Pollution Control Board Report on Biodegradable Plastics—The Impact on Environment, 2010.
7. Central Pollution Control Board Report, Overview of Plastic Waste Management, 2013.
8. Composite Material https://en.wikipedia.org/wiki/Composite Material.
9. Costoya, A.; Concheiro, A.; Alvarez-Lorenzo, C. Electrospun Fibers of Cyclodextrins and Poly (cyclodextrins). *Molecules*. **2017,** *22*(2).
10. Guragain, Y. N.; Herrera, A. I.; Vadlani, P. V.; Prakash, O. Lignins of Bioenergy Crops: A Review? Nat. Prod. Commun. **2015,** *10*(1), 201–208.

11. Hermsen, E.; Mintenig, S.; Besseling, E.; Koelmans, A. A. Quality Criteria for the Analysis of Microplastic in Biota Samples: A Critical Review. Environ. Sci. Technol. **2018,** *52*(18), 10230–10240.
12. International Life Sciences Institute-India; ILSI Monograph, 2017.
13. Janssen, L.; Moscicki, L. *Thermoplastic Starch: A Green Material for Various Industries.* Wiley-VCH: Germany, 2009.
14. Kamphuis, A. J.; Picchioni, F.; Pescarmona, P. P. CO_2-Fixation into Cyclic and Polymeric Carbonates: Principles and Applications. *J Green Chem.* **2019,** 3, 406–448.
15. Koelmans, A. A.; Mohamed, N. N. H.; Hermsen, E.; Kooi, M.; Mintenig, S. M.; De, F. J. Microplastics in Freshwaters and Drinking Water: Critical Review and Assessment of Data Quality. Water Res. **2019,** *155*, 410–422.
16. Monteiro, R. C. P.; Ivar, J. A.; Costa, M. F. Plastic Pollution in Islands of the Atlantic Ocean. Environ. Pollut. **2018,** *238*, 103–110.
17. Obmann, B. E.; Sarau, G.; Holtmannspötter, H.; Pischetsrieder, M.; Christiansen, S.H.; Dicke, W. Small-Sized Microplastics and Pigmented Particles in Bottled Mineral Water. Water Res. **2018,** *141*, 307–316.
18. Silva, C. J. M.; Silva, A. L. P.; Gravato, C.; Pestana, J. L. T. Ingestion of Small-Sized and Irregularly Shaped Polyethylene Microplastics Affect Chironomusriparius Life-History Traits. *Sci. Total Environ.* **2019,** *672*, 862–868.
19. Smith, M.; Love, D. C.; Rochman, C. M.; Roni, N. A. Microplastics in Seafood and the Implications for Human Health. Curr. Environ. Health Rep. **2018,** *5*(3), 375–386.
20. Szente, L.; Fenyvesi, E. Cyclodextrin-Enabled Polymer Composites for Packaging. Molecules. • *23*(7), 1556; https://doi.org/10.3390/molecules23071556.
21. Wagner, M. Endocrine Disruptors in Bottled Mineral Water: Total Estrogenic Burden and Migration from Plastic Bottles. *Environ. Sci. Pollut. Res.* **2009,** *16*(3), 278–286.
22. Wang, M. H.; He, Y.; Sen, B. Research and Management of Plastic Pollution in Coastal Environments of China. Environ. Pollut. **2019,** *248*, 898–905.
23. Welle, F.; Franz, R. Microplastic in Bottled Natural Mineral Water—Literature Review and Considerations on Exposure and Risk Assessment. Food Addit. Contam. Part A Chem. Anal. Control Expo. Risk Assess. **2018,** *35*(12), 2482–2492.
24. Zhao, S.; Danley, M.; Ward, J. N.; Li, D; Mincer, T. J. An Approach for Extraction, Characterization and Quantitation of Microplastic in Natural Marine Snow Using Raman Microscopy. *Anal. Methods.* **2017,** 9, 1470–1478.
25. Zuccarello, P.; Ferrante, M.; Cristaldi, A.; Copat, C.; Grasso, A.; Sangregorio, D.; Fiore, M.; Oliveri, G.; Exposure to Microplastics (<10 μm) Associated to Plastic Bottles Mineral Water Consumption: The First Quantitative Study. *Water Res.* **2019,** *157*, 365–371.

CHAPTER 2

Natural Polymers: Applications, Biocompatibility, and Toxicity

N. S. REMYA

Division of Toxicology, Department of Applied Biology, Bio Medical Technology Wing, Sree Chithra Institute for Medical Sciences and Technology, Kerala, India. E-mail: remya.bijoy@sctimst.ac.in.

ABSTRACT

The term biocompatibility as defined by the ability of any material when interacted with living system should neither produce any adverse health effects nor induce local or systemic tissue reaction compromising the devices indented use and should not have tumourigenic potential. Natural polymers are prospective candidates to be used as biomaterials defined by their excellent biocompatibility. However the biosafety of biocompatible natural polymers further needs prediction, evaluation, and indication on the potential hazards associated with their formulations, processing techniques, storage, sterilization, and biodegradation. This chapter briefs about the commonly used natural polymers for biomedical applications, safety, and toxicity concerns associated with their use as potential biomaterials.

2.1 INTRODUCTION

Biomedical engineering, a multidisciplinary field that encompasses the principles of engineering, medicine, and biological sciences has remarkably advanced in the present era that has created a huge impact on improvising and managing human health [1]. Development of biocompatible prostheses, diagnostic, and therapeutic medical devices ranging from sophisticated equipment used clinically to microimplants, imaging equipment, scaffolds for tissue growth and regeneration, drugs, pharmaceutical, and other therapeutic products are the major applications of biomedical engineering [2]. With the advancement of novel technologies for biomedical applications such as tissue engineering, controlled drug delivery, etc. in the last few decades, there is a high demand for biomaterials with tuneable physico–chemical and degradation properties [3]. A biomaterial which is now defined as a substance that has been engineered to take a form which, alone or as part of a complex system, is used to direct, by control of interactions with components of living systems, the course of any therapeutic or diagnostic procedure, contribute a major portion of the resources underlying the innovation of any successful medical device [4].

The practice of biomaterials for health care applications dates back to ancient times [5]. However, over a period of time, their versatility and utility have considerably evolved to a extend that revolutionized the modern bioengineering concepts. Many biodegradable polymers have been proposed to be used potent novel powerful therapeutics as agents for sustained/direct delivery of drugs or as scaffolds for tissue engineering and regeneration purposes [6]. Controlled release systems designed by the biodegradable polymers are in direct and sustained contact with the tissues, which subsequently degrades in situ [7]. On the other hand, biodegradable polymers used in tissue engineering to provide temporary support for cell growth should degrade with time and when sufficient extracellular matrix is laid down, in a controlled way into products, that are eliminated by regular metabolic pathways in the body [8].

A great variety of biodegradable polymeric materials have been exploited for the above-said purposes. Biodegradable polymeric materials are classified into natural (biologically derived) and synthetic (chemically derived) based on their origin [9]. Compared to synthetic counterparts, natural polymers are preferred owing to their economic and environmental aspects [10]. Advantages of natural polymers are that many of them are natural bodily constituents that provide a natural adhesive surface for cells and carry the required information for their activity [11]. Furthermore, as their degradation products are metabolized by the enzymes present in the body and subsequently cleared from the system so are nontoxic [12]. Even though cell friendly, the risk of disease transmission, batch-to-batch variability, concerns regarding their animal origin, and low mechanical stability makes them inferior for use. Nevertheless, physical or chemical modification can be imparted to natural polymers in order to improve the properties making it ideal for desired purposes [13].

2.2 CLASSES OF NATURAL POLYMERS

A polymer, by definition is a macromolecule composed of multiple repeating units (monomers) with characteristic high relative molecular mass and distinct physicochemical properties [14]. Natural polymers are directly or indirectly formed during the growth cycle of living organisms by complex metabolic processes that are either enzyme-mediated or by activated monomers participating in chain growth reactions. Based on the chemical constituents, natural polymers are broadly classified into polysaccharide, polyamides, polynucleotides, and lipids. They could be obtained from plant or animal sources or could be synthesized by bacteria from small molecules.

2.2.1 *POLYSACCHARIDE POLYMERS*

Polysaccharide polymers are biopolymers composed of monosaccharide linked together by glycosidic linkages. As they are abundant in nature, hence could be derived from source materials in a very cost-effective manner. Furthermore, the physicochemical properties such as mechanical strength, solubility, viscosity, gelling potential, etc. are governed by the composition, degree of polymerization, and chain length of the participating monosaccharide units. They could be made into different forms such as films or membranes, foam or sponges, hydrogels, etc. Some of the abundantly used polysaccharide biopolymers are detailed below:

2.2.1.1 *CELLULOSE*

Cellulose, a major structural component of plant cell wall is the most abundant renewable resource of the earth. Besides plants, some species of bacteria and algae are capable of synthesizing cellulose [15]. Basically, cellulose is a linear homopolymer of beta-D glucose monomers linked together covalently by 1→4 glycosidic bonds. The hydroxyl groups at C1 of a glucose unit and C4 of the neighboring glucose unit is linked via condensation reaction [16] (Figure 2.1). Hence every ringed glucose monomers is rotated 180° with respect to its adjacent glucose unit along the fiber axis which makes it a dimer unit called cellobiose. Prevalence of intra and interchain hydrogen bonding is more in cellulose because of the high incidence of hydroxyl groups (three per unit). This hydrogen bonding along with Van der Waals forces that contributes to the stacking interaction, imparts crystalline nature to cellulose fibrils.

Crystallinity and organization of fibrils however source dependent. For example, highly crystalline (~80% crystallinity fibers can be obtained from cotton fiber where wood fibers yield cellulose of lower crystallinity (~50% crystallinity) [17]. In general, this biopolymer is fibrous and hydrophilic but insoluble in water. Bacteria derived cellulose often presents with ultrafine nanofibrous network structure and high hydrophilicity. Cellulose-based biomaterials are of particular use artificial skin and wound dressing, due to the high biocompatibility, tunable surface chemistry and structure, moisture retention capacity, and drug-releasing capabilities [18]. They are also been exploited for use as bioink for 3D bioprinting technologies. Cellulose derived from bacteria does not have intrinsic antibacterial properties, hence antimicrobial agents are incorporated for wound healing applications [19].

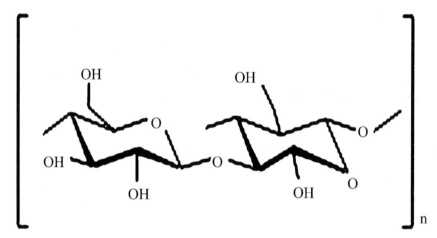

FIGURE 2.1 Structure of cellulose.

2.2.1.2 STARCH

Starch is a naturally occurring biopolymer generated by the photosynthetic activity of plants from carbon dioxide and water. Structurally starch is composed of 2 D glucose homopolymers. One is a linear amylase which is an αD(1,4')-glucan and a branched amylopectin which has mostly α-1, 6'-linkages [20] (Figure 2.2). Owing to the presence of a lot of free hydroxyl groups on starch chains, they are extremely hydrophilic and highly reactive. Depending upon the source, proportion of amylose and amylopectin varies ranging from 10% to 20% amylose and 80 to 90% amylopectin. Like other natural polymers, starch is also biodegradable. It is hydrolyzed into glucose moieties by microorganisms or other enzymes and then metabolized to its unit component carbon dioxide and water. Starch is modified by different physical or chemical methods to improve its structural and mechanical properties. Blending with other synthetic or natural polymers is often employed for the above purpose. Derivatives of starch can be synthesized by utilizing the free reactive hydroxyl groups present on starch surface retaining the biodegradability of starch and can be specially tuned for desired biomedical applications [21].

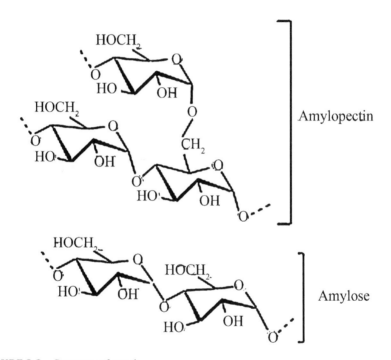

FIGURE 2.2 Structure of starch.

2.2.1.3 CHITOSAN

Chitosan is a linear polysaccharide consisting of β (1→4) linked D-glucosamine residues with a variable number of randomly located N-acetyl-glucosamine groups (Figure 2.3). It thus shares some characteristics with various glycosaminoglycans (GAGs) and hyaluronic acid present in articular cartilage. In terms of crystallinity, chitosan is considered as semicrystalline. The degree of crystallinity is determined by the degree of deacetylation [22]. Crystallinity

22 *Natural Polymers: Perspectives and Applications for a Green Approach*

is maximum for both chitin (i.e., 0% deacetylated) and fully deacetylated (i.e., 100%) chitosan. Intermediate degrees of deacetylation of chitosan give minimum crystallinity. Commercially available preparations have degrees of deacetylation ranging from 50% to 90%. Because of its crystalline structure, chitosan is normally insoluble in aqueous solutions above pH 7. However, in dilute acids, the free amino groups are protonated and the molecule becomes fully soluble below pH 5. Since the solubility of chitosan is pH dependent, processing at mild conditions is possible is cationic nature and high charge density makes chitosan an attractive candidate as a biomaterial. The high charge density in solution allows chitosan to form ionic complexes that are insoluble with water-soluble anionic polymers [23].

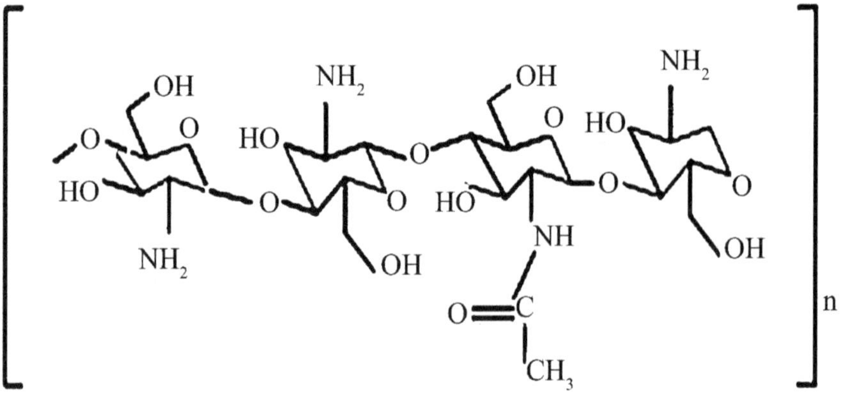

FIGURE 2.3   Structure of chitosan.

2.2.1.4   HYALURONIC ACID

Hyaluronic acid is a linear unsulfated glycosaminoglycan composed of repeating disaccharide units of D-glucuronic acid and *N*-acetyl glucosamine linked α-(1–4) and β-(1–3), respectively (Figure 2.4). This molecule is one of the major components in synovial fluid. Hyaluronic acid molecules are also present in the cartilage matrix as the backbone structure in proteoglycan aggregates [24]. In general, hyaluronic acid plays a major role as an organizer of the extracellular matrix. Purified hyaluronic acid has been employed as a structural biomaterial because of its high molecular weight and gel-forming ability. The properties of the molecule can be broadly altered by chemical modification to suit for different applications [25].

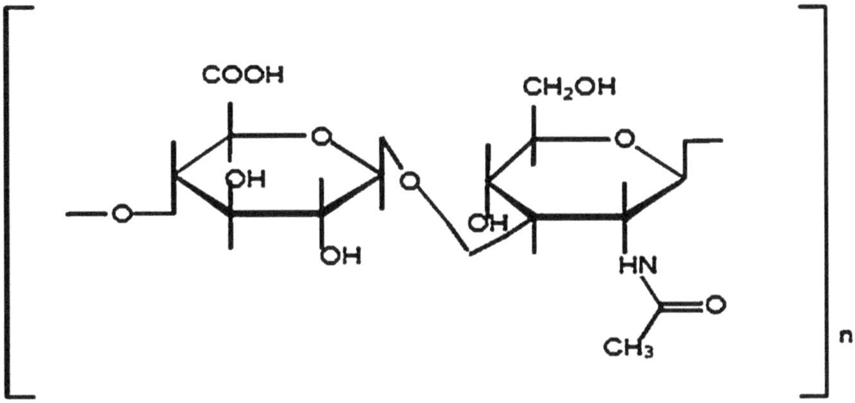

Natural Polymers: Applications, Biocompatibility

FIGURE 2.4 Structure of hyaluronic acid.

2.2.1.5 AGAROSE

Agarose is a linear polymer extracted from marine red algae. It is made up of (1→3)-β-D-galactopyranose-(1→4)-3,6-anhydro-α-L-galactopyranose units. Agarose has the ability to undergo gelling reaction when cooled to temperature below its transition temperature and is thermo irreversible [26]. When this occurs the coiled helical structure is converted to an infinite 3D network of agarose fibers. Agarose is a neutral polysaccharide that can be purified easily. The unique gelling kinetics of agarose allows a homogeneous distribution of cells/drugs hence widely used in cell/drug encapsulation technologies [27].

2.2.1.6 ALGINATE

Alginates are naturally occurring polysaccharides abundantly present as structural components of marine brown algae and certain bacterial species. They are linear unbranched polysaccharides constituted by varying amounts of (1-4)-linked β-D-mannuronic acid and α-L-guluronic acid (Figure 2.5). They are usually cross-linked by Ca ions to form stable hydrogels. Gelation reaction is mediated by the interaction between carboxylic acid groups of the polymer and cation that chelates it [28]. Chemical modification of carboxylic acid present in alginic acid can be done to obtain alginate for the desired application.

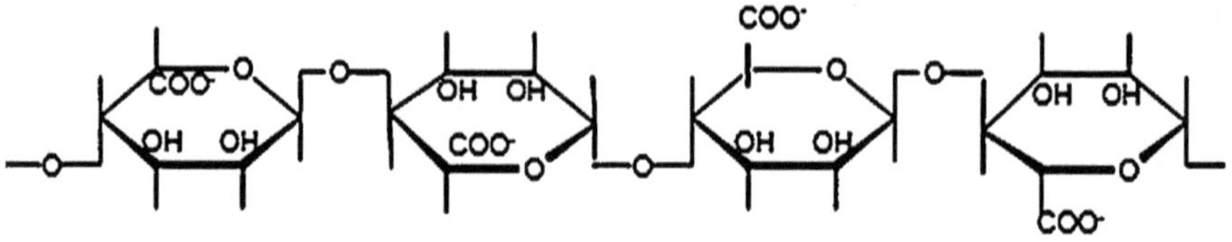

Natural Polymers: Perspectives and Applications for a Green Approach

FIGURE 2.5 Structure of alginate.

2.2.1.7 DEXTRAN

Dextran is a complex polysaccharide composed of d-glucose (d-glucopyranose) monomer units joined together through α-1,6-glycoside linkages with small α (1,3) branches (Figure 2.6). They are obtained from certain species of bacteria. They have high molecular weight and are water-soluble with high volume expansive capacity [29].

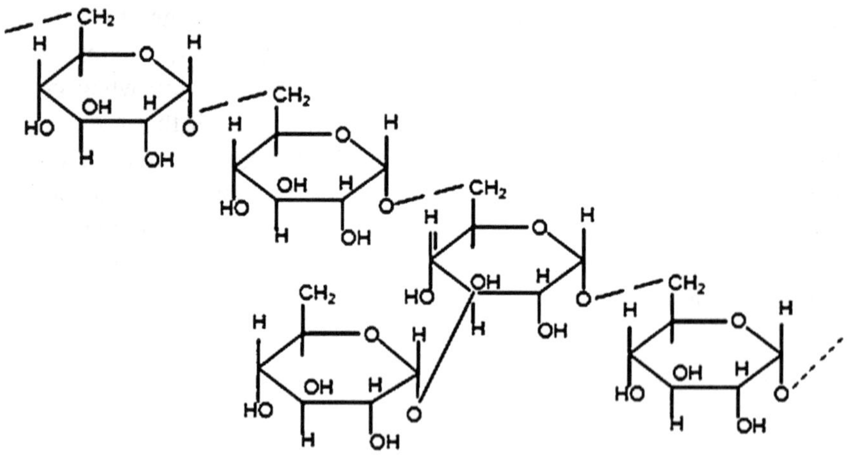

FIGURE 2.6 Structure of dextran.

2.2.1.8 COLLAGEN

Collagen is the most abundant proteinaceous natural polymer in the body. It is typically a constituent of the extracellular matrix of skin, blood vessels, and connective tissues like bone, cartilage, etc. Collagen fibers mechanically reinforce the extracellular matrix. Structurally each collagen

Natural Polymers: Applications, Biocompatibility 25

unit has a triple helical structure [30]. The variation in the length of the helix as well as the number and nature of carbohydrate attachment on the triple helix groups them into different types. The fibrillar type I collagen is present in skin, bone, and tendon. Type II collagen is the main collagen present in cartilage. Type III is the constituent of blood vessel wall. On the other hand, nonfibrillar type IV collagen is a constituent of basement membrane of epithelial tissue [31]. The primary structure of collagen consists of three polypeptide chains made up of predominantly glycine, proline, and hydroxyproline amino acids (Figure 2.7). Hydrogen bonding potential of peptide residues renders collagen its secondary structure. In the tertiary configuration polypeptide, secondary structures are packed together forming a physicochemically stable entity in solution namely the triple-helical collagen molecule. The quaternary structure comprises of several molecules packed in a quasi-hexagonal lattice that forms a collagen microfibril [32]. The isolation and purification of collagen molecule can be done by molecular technology wherein soluble collagen molecules from a collagenous tissue is obtained by using proteolytic enzymes that cleave the peptide linkage and then precipitating out with neutral salt. In the alternate fibrillar technology, other noncollagenous materials are removed from collagen-rich tissue using sequential extraction and enzymatic digestion procedures. Such purified collagen molecules can further be fabricated according to specific medical applications [33]. Collagen has intrinsic hemostatic potential which makes them an ideal candidate for hemostatic application. This hemostatic potential is dependent on the size and organization of collagen molecule [34].

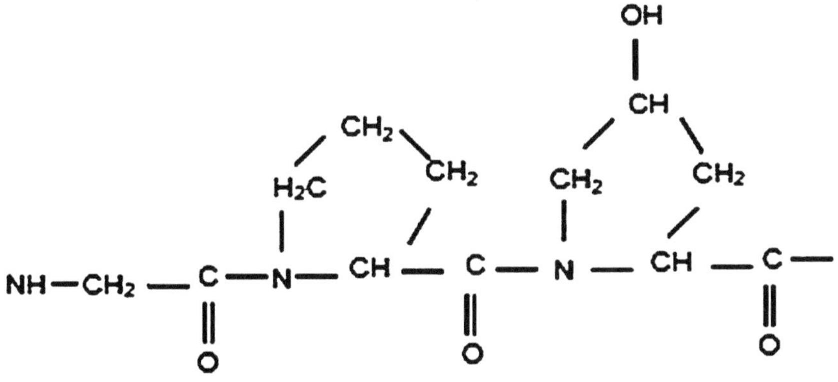

FIGURE 2.7 Structure of collagen.

2.2.1.9 GELATIN

Denaturation of collagen yields gelatin which is again another commonly used natural polymer for biomedical applications [35]. Amide groups present in collagen is hydrolyzed into carboxyl groups by either alkaline or acid process to obtain gelatin with different isoelectric points. Two types of gelatin namely Type A and Type B can thus be obtained depending on the different gelatin processing methods from native collagen. Type A gelatin has an isoelectric pH of 9.0 and is derived from acid hydrolysis of type 1 collagen while Type B gelatin is derived from alkaline hydrolysis with an isoelectric point of 5.0. In both cases, the amide groups of asparagine and glutamine amino acids of type 1 collagen are hydrolyzed into carboxyl groups resulting in aspartate and glutamate residues in the hydrolyzed product [36]. The triple helical structure of collagen molecule is broken down into single polypeptide chains during the hydrolysis (Figure 2.8). Noncovalent interactions such as van der Waal forces, hydrogen bonds, and electrostatic and hydrophobic interactions are also present between gelatin strands which influence its physicochemical properties. Gelatin is hydrophilic and water-soluble. Another unique property is that it can undergo sol–gel formation at a lower temperature to form hydrogels which are thermoreversible [37].

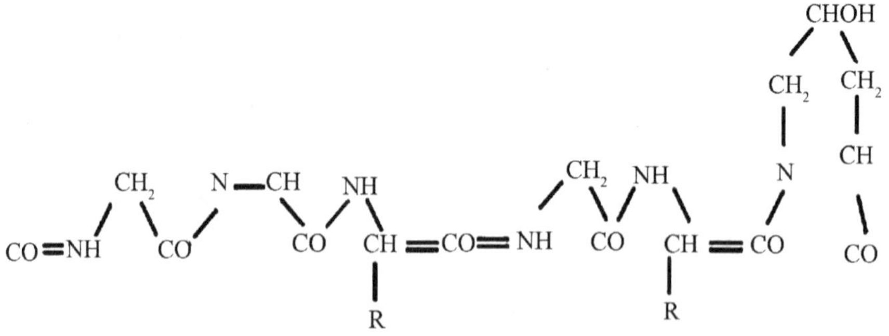

FIGURE 2.8 Structure of gelatin.

2.2.1.10 FIBRIN

Fibrin is a biopolymer that is formed in the body as an outcome of natural blood coagulation pathway from the monomer fibrinogen. Fibrin matrix can be fabricated in vitro by combining commercially available fibrinogen and thrombin. Fibrinogen molecule consists of two sets of three polypeptide chains joined together by disulfide bonds [38]. Thrombin cleaves the peptide bonds between the chains leading to conformational changes and initiates polymerization giving rise to fibrin matrix. The fibrin clot is then stabilized by factor 13 or fibrin stabilizing factor by covalently crosslinking the fibrin networks [39]. By controlling the concentration components, morphological, and other physicochemical characteristics of fibrin matrix can be tailored for desired purposes.

2.2.1.11 SILK

These are naturally occurring fibrous proteins produced by silkworms and certain spiders. These fibers have remarkable mechanical strength and have been successfully used as suture materials. Basically, the polymer consists of a core structural protein called fibroin which is coated with another glue-like proteinaceous substance called sericin which holds the fiber together [40]. Two major structural fibroin proteins are there, light and heavy chains. The fibers consist of polypeptide crystalline regions dominated by hydrophobic sequences arranged in layers of antiparallel beta-sheets.

2.3 APPLICATIONS OF NATURAL POLYMERS

Biomedical applications of natural polymers include burn dressings, skin substitutes, bioadhesives, orthopedic, dental applications, etc. [41]. All applications can be summarized into three main segments (a) wound care and management, (b) drug delivery, and (c) tissue engineering wound care and management products based on natural polymers is a rapidly growing field. Skin substitutes and skin-friendly materials from natural polymers form attractive option for advanced wound care treatment [42]. Hyaluronic acid, collagen, etc. are good examples of suture materials, absorbable sponges, hemostats, medicated lint-free wound dressings based on natural polymers are commercially available and is still a vast growing research area [43]. Drugs incorporated inside natural polymer carriers help in the controlled release of desired drugs at the region of interest [44]. Natural polymer as capsular material is able to incorporate the desired quantity of drugs and can be noninvasively introduced inside the body. Polysaccharides such as agarose, alginate, pectin, etc. have been exploited for this purpose. Another important area is in the field of tissue engineering. Tissue engineering is a multidisciplinary field that combines three-dimensional scaffold structures, appropriate cells, and soluble biochemical cues or growth factors to create a microenvironment favorable for the growth of desired cell types. Natural polymers provide advantages of being cell-friendly, biocompatible, and form nontoxic metabolic products, hence used for the engineering of different tissues and organs [45].

Collagen has been extensively used as wound dressing material. Collagen matrices in the form of sponges and membranes show good integration with the underlying tissues hence it has been widely exploited for skin tissue engineering applications [46]. Being cell-friendly and biomimetic, cell-seeded collagen constructs as skin substitutes are commercially available for use. An example is Graft Skin (APLI GRAF) which is a bilayered skin construct made from bovine collagen seeded with fibroblast and keratinocytes [47]. It has been observed that electrospun collagen fibers provide good adhesion, growth, and proliferation of keratinocytes. Besides, collagen sponges are also used for localized delivery of antibiotics [48]. The intrinsic clotting ability of collagen is made use in fabricating hemostatic sponges out of collagen material. Examples for collagen-based hemostats available in the market are Co stasis consisting of collagen, thrombin, and autologous plasma and Floseal made up of bovine thrombin and collagen [49]. The property of favoring cell attachment and growth makes them an ideal candidate as scaffold for tissue engineering applications. Several researches are available using collagen matrix with other composites for bone tissue engineering [50], as myocardial patch [51], small diameter vascular graft [52], bioink for 3D bioprinting [53], and cartilage tissue engineering [54].

Gelatin the denatured form of collagen is a versatile polymer used for different biomedical applications. Gelatin-based sponges incorporated with antibiotics, growth factors, etc. have been investigated for wound-healing applications [55]. Its use as an absorbable hemostatic sponge is also well documented [56]. The biomimetic property of gelatin is taken advantage by blending with other synthetic biomaterials for tissue engineering application [57]. Scaffolds made from such blends are used for cardiac tissue engineering [58]. Electrospun scaffolds of gelatin and poly(D,L-lactic-co-glycolic acid) (PLGA) blends were shown to give good results in vascular tissue engineering [59]. Gelatin-based hydrogels are used to deliver drug/growth factors/other molecules that help in bone formation and also for the treatment of bone infections by sustained release of antibiotics at the infection site [60]. Recreating the microenvironment of the liver by gelatin matrices is a promising approach for hepatic tissue engineering [61].

Several artificial skin substitutes are commercially available based on cellulose [62]. 3D bioprinting of skin particularly uses cellulose as a bioink

Natural Polymers: Applications, Biocompatibility

which allows mimicking the natural structure of skin [63]. Since cellulose resembles the collagen fibers of bone, they are used as scaffolding material in bone tissue engineering especially for the delivery of molecules promoting osteo differentiation [64]. Neural tissue engineering also makes use of cellulose as it is shown to enhance nerve cell proliferation and differentiation [65]. Blood compatibility of cellulose makes it suitable for vasculature engineering especially as vascular graft material [66]. Cellulose being chemically and biologically inert is considered as a promising material for capsule-based drug delivery [67].

Macro porous silk scaffolds are used for osteoid composites for bone tissue engineering applications [68]. Silk fibroin scaffolds as well as silk hydrogels are used as cartilage substitutes as they have been shown to reduce friction at the grafted site [69]. Silk fibroin scaffolds are utilized for fibrocartilaginous tissue development, hence it is of particular use in meniscal tissue engineering [70]. It is also been utilized for soft tissues requiring mechanical strength such as ligaments, tendons, etc. Literature reports show that silk-based scaffolds have the ability to promote angiogenesis and endothelialization; hence, it is exploited for vascular tissue engineering as vascular grafts [71]. Fabrication of nerve conduits from silk fibroin by incorporating other conductive polymers is also under research as silk fibroin is proved to have nerve regeneration properties [72].

Starch is extensively used in food and pharmaceutical industry. Recently, it has gained much interest in biomedical applications as well. The advantages of availability, biocompatibility, and nontoxicity make them attractive candidates for biomedical applications as well [73]. Starch has been used as carriers for drugs and other smaller molecules or hormones and their sustained release. Starch-based blends are used in bone tissue engineering with human osteoblasts [74].

Fibrin formed by the polymerization of fibrinogen found its application in biomedical field in many ways. Being an integral part of body natural clotting process, they have been widely used a hemosealant or bioadhesive [75]. The role of fibrin clot in normal wound healing process is also well documented, hence fibrin-based materials are designed and used for promoting wound healing. Fibrin matrices are reservoirs of growth factors hence these matrices are used for providing biochemical cues for the enhanced growth and proliferation of different type of cells for tissue engineering and regenerative medicine [76].

The applications of chitosan are well-reviewed. Having intrinsic antibacterial property and anti-inflammatory property, chitosan-based hydrogels films, and sponges are extensively tried for wound care applications [77].

30 *Natural Polymers: Perspectives and Applications for a Green Approach*

Hyaluronic acid-derived biomaterials are widely used in regenerative medicine for cell and molecular delivery [78]. This hyaluronan-based synthetic extracellular matrix can be customized for the growth of stem cells and other adult cell populations to be used for cellular therapy. Literature reports are available showing the use of hyaluronic acid for cartilage soft tissue, bone, and vascular tissue engineering [79].

2.4 BIODEGRADATION OF NATURAL POLYMERS

Biodegradation by definition is the breakdown of materials by living organisms or under physiological conditions leading to changes in its physical properties. When a polymer is introduced into a living system, proteins get adsorbed into surfaces that are in immediate contact with the body fluid, while soluble components get absorbed into the bulk of the material. Other cellular elements and enzymes also attack the polymer initiating the process of biodegradation [80].

For biomedical application, materials with controlled degradation rate are much preferred especially as resorbable sutures and other devices, tissue engineering scaffolds, drug delivery matrices, etc. The materials intended for the above use should degrade over time so that the natural functional tissues are set in place. And due to this, the performance of many biomaterials depends on their degradation rate and the products of degradation. All biopolymers are susceptible to biodegradation. The rate of degradation is governed by its physicochemical properties such as molecular weight, crystallinity, prevalence of hydrolyzable bonds, surface area, nature of biological environment, etc.

Under in vivo conditions, biodegradation is mainly by hydrolysis followed by oxidation. Hydrolysis is the breaking down of functional groups by reaction with water that are maybe catalyzed by acids, bases, or enzymes. In physiological conditions, the ions present in extracellular fluids can initiate ion-mediated hydrolysis of polymers. Localized pH changes at the site of implantation due to inflammatory reaction can further enhance the hydrolysis rates. Hydrolytic enzymes released by the phagocytic cells also contribute to enzyme-mediated hydrolysis. Oxidative degradation also follows owing to the release of many enzymes and reactive oxygen species having oxidative activity from the recruited inflammatory cells at the implanted site. Mechanical traction forces (loading, wear, and tear) experienced by the implant inside the body also contributes to the degradation of the material.

The degradation of natural polymers occurs in three different ways. Firstly, the main chain bonds are hydrolyzed yielding oligomers and then

Natural Polymers: Applications, Biocompatibility 31

monomers. Secondly, the side functional groups are hydrolyzed making the polymer water-soluble. And thirdly for cross-linked polymers, the cross-links are dissociated making the polymer soluble when the molecular weight of resulting products is very low enough (<50 kDa), they are cleared through the kidneys. Often the degradation products of natural polymers are physiological constituents hence they are recycled back in the body. Natural polymers may undergo enzymatic degradation inside the body. The enzyme hyaluronidase can cause degradation of hyaluronan-based biomaterials [81]. Similarly, glucuronidase, hexaminidase, and other glycosidase may specifically cleave polysaccharides to its corresponding monomer units. Fibrin is degraded by the enzyme plasminogen in the body fluid. Proteolytic enzymes like collagenase, trypsin, chymotrypsin, matrix metalloproteinase, and other proteases degrade protein-based biomaterials like collagen, silk fibroin, etc. However, the rate of degradation and its extent totally depend upon the scaffold morphology, enzyme-substrate ratio, and degradation time.

2.5   STERILIZATION OF NATURAL POLYMERS

Sterilization is a critical parameter that has to be considered when natural polymers are used for biomedical applications [82]. Since the biological materials are particularly regulated before clinical use by the regulatory agencies, these materials must be properly processed and sterilized before its usage. Hence it is imperative to review available sterilization modalities and their impact on structure and function of natural polymers. Sterilization is the process by which a device/material is freed of contamination from living microorganisms including viruses, yeast, and bacteria. Sterilization of natural-based products is highly challenging as it is difficult to maintain the structural and functional properties post sterilization. Also, sterilization resistance is yet another concern. Bacterial spores are found to be more resistant to currently practiced sterilization techniques. There are several effective methods of sterilization which include heat sterilization, ethylene oxide sterilization, ionizing radiations, and other disinfectants.

2.5.1   HEAT TREATMENT

Heat sterilization involves either dry heat or moist heat. Sterilization of powders, oil products, etc. are often done by dry heat methods, while moist heat is preferred for surgical instruments. Heated air is the sterilizing agent

32 *Natural Polymers: Perspectives and Applications for a Green Approach*

in case of dry heat sterilization that destroys microorganisms by denaturation and coagulation of proteins and nucleic acids. In moist heat sterilization or autoclaving, high temperature is reached quickly than dry heat which destroys the microorganisms. However, this method is unsuitable for natural polymers as most of the natural polymers have low glass transition temperatures. Also, the presence of water vapor in moist heat sterilization can initiate hydrolytic degradation of the material. High temperatures can still affect the structural property of protein-based biomaterials like collagen, fibrin, etc.

2.5.2 ETHYLENE OXIDE STERILIZATION

This method is commonly used for heat-sensitive materials, equipments, and biological materials. Ethylene oxide alkylates the functional groups present in nucleic acids and proteins thereby destroying the microorganisms. But the disadvantage is that devices whose surface is modified with proteins may lose their functionality. Often there is chance of formation of ethylene glycol a by-product which is toxic even in trace quantities. Proper aeration of scaffolds post sterilization is therefore advised for the complete removal of residual EtO.

2.5.3 GAMMA IRRADIATION STERILIZATION

The high energy particle that is emitted from the decay of cobalt 60 ([60] Co) or cesium isotopes (^{137}Cs) induces ionization and emission of the secondary electron which in turn creates free radicals that damages proteins and nucleic acids thereby destroying the microorganisms. Even though it is a highly effective technique, ionizing radiation can affect crosslinking, degradation kinetics, and material properties such as tensile strength, elastic modulus, color, etc. of the material.

2.5.4 ALTERNATIVE METHODS OF STERILIZATION

Other methods include the use of liquid chemicals such as ethanol, glutaraldehyde, peracetic acid, iodine, and formaldehyde. Vapor systems involving peracetic acid and hydrogen peroxide, glutaraldehyde vapor, etc. are also employed. High-intensity light, ultraviolet light, freeze-drying and supercritical carbon dioxide, membrane filtration, etc., are other alternate sterilization techniques that could be used for sterilizing natural polymers.

Natural Polymers: Applications, Biocompatibility 33

2.6 SOLUBILITY ISSUES—NEED FOR CROSSLINKING

Natural polymers have a disadvantage compared to synthetic counterparts; it is more soluble in water and has low thermal stability and mechanical strength. So in order to improve these qualities crosslinkers are employed. Crosslinking is the process by which a firm network is introduced in the matrix by connecting the functional groups of the polymer chain to one another through covalent or no covalent interactions, thereby improving the biomechanical properties of the material. Chemical crosslinkers are widely employed to improve the properties of natural polymers [83]. Glutaraldehyde is the most widely used crosslinker. It connects the amine or hydroxyl groups of polymers via Schiff's linkage forming a strong cross-linked network [84]. Formaldehyde also works on the same principle. But cellular toxicity of these crosslinkers is a major concern. Ethyl, 3-dimethyl aminopropyl-carbodiimide is another chemical that can react with a variety of functional groups, such as carboxyl, hydroxyl, and sulfhydryl groups and crosslink them [85]. Genipin [86], citric acid, and glyoxal are the other crosslinkers that are currently in use for the chemical crosslinking of natural polymers [87].

2.7 BIOCOMPATIBILITY OF NATURAL-BASED POLYMERS

A proper understanding on the in vivo response is mandatory for any biomaterial intended to be used for human application. Material–biological interactions mainly present as inflammatory reactions, foreign body reactions, immunological reactions systemic toxicity, blood surface interactions, implant-associated infections, tumorigenesis, etc. [88]. Generally a breach in host defense mechanism occurs when a biomaterial is introduced to the living system and the host response is the sequel of this intervention.

Normally following any implantation procedures, the trauma associated with the procedure may trigger an acute inflammatory response. Inflammation is generally defined as the reaction of any vascularized tissue to local injury. This response may typically take place in minutes to hours and may last up to days depending upon the severity of the injury. Upon creation of an injury, there is the activation of natural coagulation cascade. Irrespective of the pathways, complement activation occurs resulting in the secretion of chemotactic agents. Adsorption of proteins along the surface of the implant may further act as a strong chemoattractant for inflammatory cells predominantly polymorphonuclear (PMN) cells. After being recruited to the site of injury, PMN gets activated and undergo a "respiratory burst" which generates reactive oxygen species and undergoes degeneration [89]. Neutrophils predominate the early phase of acute inflammation later followed by monocytes which are recruited to injury site under the influence of chemotactic factors which continues the phagocytosis activity. They process the foreign material and present the antigens initiating specific immunological reactions by subsequent lymphocytic cells. The outcome of inflammatory reaction following an injury depends on many factors and especially with the presence of a biomaterial, normal course of healing process may be affected. The prolonged presence of implantable biomaterial and its rate and product degradation determine the progress of inflammatory response from acute to chronic. Chronic inflammation can last for weeks or months and can stimulate either fibrosis or granuloma formation [90].

In vivo responses to natural polymers can be assessed by different methodologies. The most commonly used methodology for biocompatibility evaluation is subcutaneous or intramuscular or intraperitoneal implantation of the material in animal models. Being an open cavity, intraperitoneal implantation allows the recruitment of circulating leukocytes as well as from nearby organs to the implantation site. The relatively high vascularization in skeletal muscles contributes to the faster recruitment of circulating cells to the site of implantation. As reviewed, modeling the kinetics of inflammatory cells recruitment, its activation, and production of cytokines are best done in intramuscular model of biocompatibility evaluation [91]. Subcutaneous implantation also provides useful information concerning the reaction of inflammatory cells as the implanted material is in contact with the deeper layers of skin as well as a portion of smooth muscle [92].

As most of natural polymers are mainly constituted by polysaccharides or proteins, they are normally recognized as natural invaders like microorganisms (nonself) or as body component (self). Some of the factors that influence the in vivo responses to natural polymers are:

1. Type of natural polymer (source)
2. Surface characteristics
3. Mechanical properties
4. Degradation rate
5. Degradation products
6. Animal models used
7. Implantation procedures.

Under physiological conditions, the natural polymers tend to undergo enzymatic degradation and the rate of degradation as well as degradation products can influence the type of host response [93]. The digestion products

Natural Polymers: Applications, Biocompatibility 35

following phagocytic activity is also linked with the response elicited by host tissues. Surface characteristics of the implanted material determine the type and amount of proteins adsorbed to the surface triggering inflammatory response when in contact with blood or other body fluids.

Systemic effects of biomaterials can also occur due to direct chemical toxicity, accumulation of degradation products, or by excess inflammatory responses. Presence of crosslinking agents in fabrication of natural biomaterials may often contribute to systemic toxicity of the material. Another type of host response is the delayed type of hypersensitivity reaction in contact with the natural-based biomaterials that are immunogenic. Here foreign body reactions are more intensified with the recurring contact due to the involvement of plasma cells [94]. Hypersensitivity immune response can also contribute to systemic toxic response which is observed in some natural polymers like chitosan, collagen, fibrin, etc. This also has to be evaluated thoroughly before projecting any natural-based material for biomedical applications.

2.8 CONCLUSION

Natural polymers have found many applications in the field of biomedical engineering. This chapter reviewed the application of the selected class of natural polymers and the concerns on biodegradation and biocompatibility of natural polymers. Overcoming the major disadvantages of natural polymers in terms of stability and batch-to-batch variation, sterility issues, etc. is quite challenging. However, deeper research and development in this area is essential to promote nature's choice of material that mimics human body constituents for biomedical applications.

KEYWORDS

- **biocompatibility**
- **natural polymers**
- **biodegradation**
- **biomaterials**
- **sterilization**
- **crosslinking**

REFERENCES

1. Chien, S.; Bashir, R.; Nerem, R. M.; Pettigrew, R. Engineering as a New Frontier for Translational Medicine. *Sci. Tranl. Med.* **2015**, *7* (281), 281fs13–281fs13.
2. Ibrahim, F.; Thio, T. H. G.; Faisal, T.; Neuman, M. The Application of Biomedical Engineering Techniques to the Diagnosis and Management of Tropical Diseases: A Review. *Sensors (Basel, Switzerland)* **2015**,*15* (3), 6947–6995.
3. Tibbitt, M. W.; Rodell, C. B.; Burdick, J. A.; Anseth, K. S. Progress in Material Design for Biomedical Applications. *PNAS* **2015**,*112* (47), 14444–14451.
4. Chen, F.-M.; Liu, X. Advancing Biomaterials of Human Origin for Tissue Engineering. *Prog. Polym. Sci.* **2016**, *53*, 86–168.
5. Bhat, S.; Kumar, A. Biomaterials and Bioengineering Tomorrow's Healthcare. *Biomatter.* **2013**, *3* (3), e24717.
6. Song, R.; Murphy, M.; Li, C.; Ting, K.; Soo, C.; Zheng, Z. Current Development of Biodegradable Polymeric Materials for Biomedical Applications. *Drug Des. Dev. Ther.* **2018**, *12*, 3117–3145.
7. Kamaly, N.; Yameen, B.; Wu, J.; Farokhzad, O. C. Degradable Controlled-Release Polymers and Polymeric Nanoparticles: Mechanisms of Controlling Drug Release. *Chem. Rev.* **2016**, *116* (4), 2602–2663.
8. Mano, J. F.; Silva, G. A.; Azevedo, H. S.; Malafaya, P. B.; Sousa, R. A.; Silva, S. S.; Boesel, L. F.; Oliveira, J. M.; Santos, T. C.; Marques, A. P.; Neves, N. M.; Reis, R. L. Natural Origin Biodegradable Systems in Tissue Engineering and Regenerative Medicine: Present Status and Some Moving Trends. *J. R. Soc. Interface* **2007**,*4* (17), 999–1030.
9. Vroman, I.; Tighzert, L. Biodegradable Polymers. *Materials.* **2009**, *2* (2), 307–344.
10. Bhatia, S. Natural Polymers vs Synthetic Polymer. *Natural Polymer Drug Delivery Systems: Nanoparticles, Plants, and Algae* ISBN 978-3-319-41129-3 2016; pp. 95–118.
11. Ige, O. O.; Umoru, L. E.; Aribo, S. Natural Products: A Minefield of Biomaterials. *ISRN Mat. Sci.* **2012**, *2012*, 20.
12. Gajre Kulkarni, V.; Butte, K.; Rathod, S. Natural Polymers- A Comprehensive Review. **2012**; *Int. J. Res. Pharm. Biomed. Sci.* 3(4), 1597–1613.
13. Vieira, M. G. A.; da Silva, M. A.; dos Santos, L. O.; Beppu, M. M. Natural-based Plasticizers and Biopolymer Films: A Review. *Eur. Polym. J.* **2011**, *47* (3), 254–263.
14. Stewart, A. S.; Domínguez-Robles, J.; Donnelly, F. R.; Larrañeta, E. Implantable Polymeric Drug Delivery Devices: Classification, Manufacture, Materials, and Clinical Applications. *Polymers.* **2018**, *10* (12), 1379–1403.
15. Li, S.; Bashline, L.; Lei, L.; Gu, Y. Cellulose Synthesis and its Regulation. *The Arabidopsis Book* 2014, *12*, e0169–e0169.
16. Credou, J.; Berthelot, T. Cellulose: From Biocompatible to Bioactive Material. *J. Mater. Chem. B.* **2014**, *2* (30), 4767–4788.
17. Moon, R. J.; Martini, A.; Nairn, J.; Simonsen, J.; Youngblood, J. Cellulose Nanomaterials Review: Structure, Properties and Nanocomposites. *Chem. Rev.* **2011**, *40* (7), 3941–3994.
18. Suarato, G.; Bertorelli, R.; Athanassiou, A. Borrowing From Nature: Biopolymers and Biocomposites as Smart Wound Care Materials. *Front. Bioeng. Biotechnol.* **2018**, *6*, 137.
19. Wang, J.; Vermerris, W. Antimicrobial Nanomaterials Derived from Natural Products—A Review. *Materials (Basel, Switzerland)* **2016**, *9* (4), 255.

Natural Polymers: Applications, Biocompatibility 37

20. Kaur, L.; Singh, J.; Liu, Q. Starch—A Potential Biomaterial for Biomedical Applications. In *Nanomaterials and Nanosystems for Biomedical Applications*; Mozafari, M. R., Ed. Springer: Netherlands, 2007; pp. 83–98.

21. Salam, A.; Lucia, L.; Jameel, H. Starch Derivatives that Contribute Significantly to the Bonding and Antibacterial Character of Recycled Fibers. *ACS Omega* **2018**, *3* (5), 5260–5265.

22. Elieh-Ali-Komi, D.; Hamblin, M. R. Chitin and Chitosan: Production and Application of Versatile Biomedical Nanomaterials. *Int. J. Adv. Res.* **2016**,*4* (3), 411–427.

23. Croisier, F.; Jérôme, C. Chitosan-Based Biomaterials for Tissue Engineering. *Eur. Polym. J.* **2013**,*49* (4), 780–792.

24. Yazdani, M.; Shahdadfar, A.; Jackson, J. C.; Utheim, P. T. Hyaluronan-Based Hydrogel Scaffolds for Limbal Stem Cell Transplantation: A Review. *Cells* **2019**, *8* (3), 245–257.

25. Borzacchiello, A.; Russo, L.; Malle, B. M.; Schwach-Abdellaoui, K.; Ambrosio, L. Hyaluronic Acid Based Hydrogels for Regenerative Medicine Applications. *BioMed Res. Int.* **2015**, *871218*, 12–17.

26. Zarrintaj, P.; Manouchehri, S.; Ahmadi, Z.; Saeb, M. R.; Urbanska, A. M.; Kaplan, D. L.; Mozafari, M. Agarose-Based Biomaterials for Tissue Engineering. *Carbohyd. Polym.* **2018**, *187*, 66–84.

27. Varoni, E.; Tschon, M.; Palazzo, B.; Nitti, P.; Martini, L.; Rimondini, L. Agarose gel as Biomaterial or Scaffold for Implantation Surgery: Characterization, Histological and Histomorphometric Study on Soft Tissue Response. *Connect. Tissue Res.* **2012**, *53* (6), 548–54.

28. Lee, K. Y.; Mooney, D. J. Alginate: properties and biomedical applications. *Prog. Polym. Sci.* **2012**, *37* (1), 106–126.

29. Wang, R. Dijkstra, P. J.; Karperien, M. Dextran. In *Biomaterials from Nature for Advanced Devices and Therapies* 2016 chapter 18.

30. Fernandes, H.; Moroni, L.; van Blitterswijk, C.; de Boer, J. Extracellular Matrix and Tissue Engineering Applications. *J. Mater. Chem.* **2009**, *19* (31), 5474–5484.

31. Deshmukh, S. N.; Dive, A. M.; Moharil, R.; Munde, P. Enigmatic Insight into Collagen. *JOMFP* **2016**, *20* (2), 276–283.

32. Shoulders, M. D.; Raines, R. T. Collagen Structure and Stability. *Annu. Rev. Biochem.* **2009**, *78*, 929–958.

33. Pacak, C. A.; MacKay, A. A.; Cowan, D. B. An Improved Method for the Preparation of Type I Collagen from Skin. *J. Vis. Exp.* **2014**, (83), e51011.

34. Samudrala, S. Topical Hemostatic Agents in Surgery: A Surgeon's Perspective. *AORN J.* **2008**, *88* (3), S2–11.

35. Meyer, M. Processing of Collagen Based Biomaterials and the Resulting Materials Properties. B*iomed. Eng. Online* **2019**, *18* (1), 24.

36. Mariod, A.; Fadul, H. Review: Gelatin, Source, Extraction and Industrial Applications. *Acta Sci. Pol. Technol. Aliment.* 12, 135–147.

37. Tan, H.; Marra, K. G. Injectable, Biodegradable Hydrogels for Tissue Engineering Applications. *Materials* **2010**, *3* (3), 1746–1767.

38. Mosesson, M. W.; Siebenlist, K.; Meh, D. The Structure and Biological Features of Fibrinogen and Fibrin. *Ann. N. Y. Acad. Sci.* **2001**, *936*, 11–30.

39. Smith, S. A.; Travers, R. J.; Morrissey, J. H. How It All Starts: Initiation of the Clotting Cascade. *Crit. Rev. Biochem. Mol.* **2015**, *50* (4), 326–336.

40. Mandal, B.; Kundu, S. Biospinning by Silkworms: Silk Fiber Matrices for Tissue Engineering Applications. *Acta Biomater.* **2010**, *6*(2), 360–371.

41. Kunduru, K. R.; Basu, A.; Domb, A. Biodegradable Polymers: Medical Applications. *Encyclopedia of Polymer Science and Technology*, 2016, pp. 1–22.

42. Andreu, V.; Mendoza, G.; Arruebo, M.; Irusta, S. Smart Dressings Based on Nanostructured Fibers Containing Natural Origin Antimicrobial, Anti-Inflammatory, and Regenerative Compounds. *Materials* **2015**,*8* (8), 5154–5193.

43. Pereira, R. F.; Bártolo, P. J. Traditional Therapies for Skin Wound Healing. *Adv. Wound Care* **2016**, *5* (5), 208–229.

44. Vashist, A.; Vashist, A.; Gupta, Y. K.; Ahmad, S. Recent Advances in Hydrogel Based Drug Delivery Systems for the Human Body. *J. Mater. Chem B* **2014**, *2* (2), 147–166.

45. Chan, B. P.; Leong, K. W. Scaffolding in Tissue Engineering: General Approaches and Tissue-Specific Considerations. *Eur. Spine J.* **2008**, *17(Suppl 4)*, 467–479.

46. Chattopadhyay, S.; Raines, R. T. Review Collagen-Based Biomaterials for Wound Healing. *Biopolymers* **2014**, *101* (8), 821–833.

47. Nicholas, M. N.; Jeschke, M. G.; Amini-Nik, S. Methodologies in Creating Skin Substitutes. *CMLS* **2016**, *73* (18), 3453–3472.

48. Gizaw, M.; Thompson, J.; Faglie, A.; Lee, S.-Y.; Neuenschwander, P.; Chou, S.-F. Electrospun Fibers as a Dressing Material for Drug and Biological Agent Delivery in Wound Healing Applications. *Bioengineering (Basel, Switzerland)* **2018**, *5* (1), 9.

49. Aziz, O.; Athanasiou, T.; Darzi, A. Haemostasis Using a Ready-to-Use Collagen Sponge Coated with Activated Thrombin and Fibrinogen. *Surg. Technol. Int.* **2005**, *14*, 35–40.

50. Ferreira, A. M.; Gentile, P.; Chiono, V.; Ciardelli, G. Collagen for Bone Tissue Regeneration. *Acta Biomater.* **2012**, *8* (9), 3191–3200.

51. Gao, J.; Liu, J.; Gao, Y.; Wang, C.; Zhao, Y.; Chen, B.; Xiao, Z.; Miao, Q.; Dai, J. A Myocardial Patch Made of Collagen Membranes Loaded with Collagen-Binding Human Vascular Endothelial Growth Factor Accelerates Healing of the Injured Rabbit Heart. *Tissue Eng. Part A* **2011**, *17* (21–22), 2739–2747.

52. Goissis, G.; Suzigan, S.; Parreira, D. R.; Maniglia, J. V.; Braile, D. M.; Raymundo, S. Preparation and Characterization of Collagen-Elastin Matrices From Blood Vessels Intended as Small Diameter Vascular Grafts. *Artif. Organs* **2000**, *24* (3), 217–223.

53. Drzewiecki, K. E.; Malavade, J. N.; Ahmed, I.; Lowe, C. J.; Shreiber, D. I. A Thermoreversible, Photocrosslinkable Collagen Bio-Ink for Free-Form Fabrication of Scaffolds for Regenerative Medicine. *Technology* **2017**, *5* (4), 185–195.

54. Deponti, D.; Di Giancamillo, A.; Gervaso, F.; Domenicucci, M.; Domeneghini, C.; Sannino, A.; Peretti, G. M. Collagen Scaffold for Cartilage Tissue Engineering: The Benefit of Fibrin Glue and the Proper Culture Time in an Infant Cartilage Model. *Tissue Eng. Part A* **2014**, *20* (5–6), 1113–1126.

55. Ulubayram, K.; Cakar, A.; Korkusuz, P.; Ertan, C.; Hasirci, N. EGF Containing Gelatin-Based Wound Dressing. *Biomaterials.* **2001**, *22* (11), 1345–1356.

56. Takagi, T.; Tsujimoto, H.; Torii, H.; Ozamoto, Y.; Hagiwara, A. Two-Layer Sheet of Gelatin: A New Topical Hemostatic Agent. *Asian J. Surg.* **2018**, *41* (2), 124–130.

57. Echave, M. C.; Saenz del Burgo, L.; Pedraz, J. L.; Orive, G. Gelatin as Biomaterial for Tissue Engineering. *Curr. Pharm. Des.* **2017**, *23* (24), 3567–3584.

58. Tijore, A.; Irvine, S. A.; Sarig, U.; Mhaisalkar, P.; Baisane, V.; Venkatraman, S. Contact Guidance for Cardiac Tissue Engineering Using 3D Bioprinted Gelatin Patterned Hydrogel. *Biofabrication* **2018**, *10* (2), 025003.

Natural Polymers: Applications, Biocompatibility

59. Han, J.; Lazarovici, P.; Pomerantz, C.; Chen, X.; Wei, Y.; Lelkes, P. Co-Electrospun Blends of PLGA, Gelatin, and Elastin as Potential Nonthrombogenic Scaffolds for Vascular Tissue Engineering. *Biomacromol.* **2011**, *12* (2), 399–440.
60. Dorati, R.; DeTrizio, A.; Modena, T.; Conti, B.; Benazzo, F.; Gastaldi, G.; Genta, I. Biodegradable Scaffolds for Bone Regeneration Combined with Drug-Delivery Systems in Osteomyelitis Therapy. *Pharmaceuticals (Basel, Switzerland)* **2017**, *10* (4), 96.
61. Mazza, G.; Al-Akkad, W.; Rombouts, K.; Pinzani, M. Liver Tissue Engineering: From Implantable Tissue to Whole Organ Engineering. *Hepatol. Commun.* **2018** *2* (2), 131–141.
62. Nicholas, M. N.; Yeung, J. Current Status and Future of Skin Substitutes for Chronic Wound Healing. *J. Cutan. Med. Surg.* **2017**, *21* (1), 23–30.
63. Derakhshanfar, S.; Mbeleck, R.; Xu, K.; Zhang, X.; Zhong, W.; Xing, M. 3D Bioprinting for Biomedical Devices and Tissue Engineering: A Review of Recent Trends and Advances. *Bioact. Mater.* **2018**, *3* (2):144–156.
64. Polo-Corrales, L.; Latorre-Esteves, M.; Ramirez-Vick, J. E. Scaffold Design for Bone Regeneration. *J. Nanosci. Nanotechnol.* **2014**, *14* (1), 15–56.
65. Subramanian, A.; Krishnan, U. M.; Sethuraman, S. Development of Biomaterial Scaffold for Nerve Tissue Engineering: Biomaterial Mediated Neural Regeneration. *J. Biomed. Sci.* **2009**, *16*, 108.
66. Fink, H.; Hong, J.; Drotz, K.; Risberg, B.; Sanchez, J.; Sellborn, A. An In Vitro Study of Blood Compatibility of Vascular Grafts Made of Bacterial Cellulose in Comparison with Conventionally-Used Graft Materials. *J. Biomed. Mater. Res. A.* **2011**, *97* (1), 52–58.
67. Jain, D.; Raturi, R.; Jain, V.; Bansal, P.; Singh, R. Recent Technologies in Pulsatile Drug Delivery Systems. *Biomatter.* **2011**, *1* (1), 57–65.
68. Joo Kim, H.; Kim, U.-J.; Kim, H.; Li, C.; Wada, M.; G Leisk, G.; Kaplan, D. Bone Tissue Engineering with Premineralized Silk Scaffolds. *Bone.* **2008** 42 (6):1226–1234.
69. Farokhi, M.; Mottaghitalab, F.; Fatahi, Y.; Reza Saeb, M.; Zarrintaj, P.; Kundu, S.; Khademhosseini, A. Silk Fibroin Scaffolds for Common Cartilage Injuries: Possibilities for Future Clinical Applications. *Eur. Polym. J.* **2019**, 251–267.
70. Gruchenberg, K.; Ignatius, A.; Friemert, B.; von Lübken, F.; Skaer, N.; Gellynck, K.; Kessler, O.; Dürselen, L. In Vivo Performance of a Novel Silk Fibroin Scaffold for Partial Meniscal Replacement in a Sheep Model. *Knee Surg. Sports Traumatol. Arthrosc.* **2015**, *23* (8), 2218–2229.
71. Kasoju, N.; Bora, U. Silk Fibroin in Tissue Engineering. *Adv. Healthc. Mater.* **2012**,*1* (4), 393–412.
72. Sobajo, C.; Behzad, F.; Yuan, X.-F.; Bayat, A. Silk: A Potential Medium for Tissue Engineering. *Eplasty* **2008**, *8*, e47.
73. Roslan, M. R.; Nasir, N. F. M.; Cheng, E. M.; Amin, N. A. M. In Tissue Engineering Scaffold Based on Starch: A Review, 2016 *International Conference on Electrical, Electronics, and Optimization Techniques (ICEEOT)*, 2016; pp. 1857–1860.
74. Carvalho, Pedro Pires; Rodrigues, Márcia T; Reis, Rui L. Starch-Based Blends in Tissue Engineering. In *Biomaterials from Nature for Advanced Devices and Therapies* 2016, pp. 244–257
75. Spotnitz, W. D. Fibrin Sealant: The Only Approved Hemostat, Sealant, and Adhesive; A Laboratory and Clinical Perspective. *ISRN Surg.* **2014**, *2014*, 28.
76. Ahmed, T. A.; Dare, E. V.; Hincke, M. Fibrin: A Versatile Scaffold for Tissue Engineering Applications. *Tissue Eng. Part B Rev.* **2008**, *14* (2), 199–215.

77. Ahmadi, F.; Oveisi, Z.; Samani, S. M.; Amoozgar, Z. Chitosan Based Hydrogels: Characteristics and Pharmaceutical Applications. *Res. Pharm. Sci.* **2015**, *10* (1), 1–16.
78. Prestwich, G. D. Hyaluronic Acid-Based Clinical Biomaterials Derived for Cell and Molecule Delivery in Regenerative Medicine. *J. Control Release* **2011**, *155* (2), 193–199.
79. Vindigni, V.; Cortivo, R.; Iacobellis, L.; Abatangelo, G.; Zavan, B. Hyaluronan Benzyl Ester as a Scaffold for Tissue Engineering. *Int. J. Mol. Sci.* **2009**, *10* (7), 2972–2985.
80. Ratajska, M.; Boryniec, S. Physical and Chemical Aspects of Biodegradation of Natural Polymers. *React. Funct. Polym.* **1998**, *38* (1), 35–49.
81. Park, S.; Park, K. Y.; Yeo, I. K.; Cho, S. Y.; Ah, Y. C.; Koh, H. J.; Park, W. S.; Kim, B. J. Investigation of the Degradation-Retarding Effect Caused by the Low Swelling Capacity of a Novel Hyaluronic Acid Filler Developed by Solid-Phase Crosslinking Technology. *Ann. Dermatol.* **2014**, *26* (3), 357–362.
82. Dai, Z.; Ronholm, J.; Tian, Y.; Sethi, B.; Cao, X. Sterilization Techniques for Biodegradable Scaffolds in Tissue Engineering Applications. *J. Tissue Eng.* **2016**, *7*, 2041731416648810.
83. Parhi, R. Cross-Linked Hydrogel for Pharmaceutical Applications: A Review. *Adv. Pharm. Bull.* **2017**, *7* (4), 515–530.
84. Barbosa, O.; Ortiz, C.; Berenguer-Murcia, Á.; Torres, R.; Rodrigues, R. C.; Fernandez-Lafuente, R. Glutaraldehyde in Bio-Catalysts Design: A Useful Crosslinker and a Versatile Tool in Enzyme Immobilization. *RSC Adv.* **2014**, *4* (4), 1583–1600.
85. Hafemann, B.; Ghofrani, K.; Gattner, H. G.; Stieve, H.; Pallua, N. Cross-Linking by 1-Ethyl-3- (3-Dimethylaminopropyl)-Carbodiimide (EDC) of a Collagen/Elastin Membrane Meant to be Used as a Dermal Substitute: Effects on Physical, Biochemical and Biological Features In Vitro. *J. Mater. Sci. Mater. Med.* **2001**, *12* (5), 437–446.
86. Manickam, B.; Sreedharan, R.; Elumalai, M. 'Genipin'—The Natural Water Soluble Cross-Linking Agent and Its Importance in the Modified Drug Delivery Systems: An Overview. *Curr. Drug Deliv.* **2014**, *11* (1), 139–145.
87. Yang, Q.; Dou, F.; Liang, B.; Shen, Q. Studies of Cross-Linking Reaction on Chitosan Fiber with Glyoxal. *Carbohyd. Polym.* **2005**, *59* (2), 205–210.
88. Anderson, J. M.; Rodriguez, A.; Chang, D. T. Foreign Body Reaction to Biomaterials. *Semin. Immunol.* **2008**, *20* (2), 86–100.
89. Mittal, M.; Siddiqui, M. R.; Tran, K.; Reddy, S. P.; Malik, A. B. Reactive Oxygen Species in Inflammation and Tissue Injury. *Antioxid. Redox Signal.* **2014**, *20* (7), 1126–1167.
90. Morais, J. M.; Papadimitrakopoulos, F.; Burgess, D. J. Biomaterials/Tissue Interactions: Possible Solutions to Overcome Foreign Body Response. *AAPS J.* **2010**, *12* (2), 188–196.
91. Darville, N.; van Heerden, M.; Erkens, T.; De Jonghe, S.; Vynckier, A.; De Meulder, M.; Vermeulen, A.; Sterkens, P.; Annaert, P.; Van den Mooter, G. Modeling the Time Course of the Tissue Responses to Intramuscular Long-acting Paliperidone Palmitate Nano-/Microcrystals and Polystyrene Microspheres in the Rat. *Toxicol. Pathol.* **2016**, *44* (2), 189–210.
92. Kastellorizios, M.; Tipnis, N.; Burgess, D. J. Foreign Body Reaction to Subcutaneous Implants. *Adv. Exp. Med. Biol.* **2015**, *865*, 93–108.
93. Azevedo, H.; Reis, R. L. Understanding the Enzymatic Degradation of Biodegradable Polymers and Strategies to Control Their Degradation Rate. In *Biodegradable Systems in Tissue Engineering and Regenerative Medicine*, CRC Press, **2007**, pp. 177–201
94. Mariani, E.; Lisignoli, G.; Borzì, R. M.; Pulsatelli, L. Biomaterials: Foreign Bodies or Tuners for the Immune Response? *Int. J. Mol. Sci.* **2019**, *20* (3), 636.

CHAPTER 3

Development of Bio-Composites From Industrial Discarded Fruit Fibers

J. S. BINOJ[1], R. EDWIN RAJ[2], and N. MANIKANDAN[1]

[1]*Associate Professor, Micromachining Research Center,*
Department of Mechanical Engineering, Sree Vidyanikethan Engineering
College (Autonomous), Tirupati–517102, Andhra Pradesh, India,
Mobile: +91-8754379212, E-mail: binojlaxman@gmail.com (J. S. Binoj)

[2]*Professor, National Rail and Transportation Institute*
(Deemed to be University), National Academy of Indian Railways (NAIR)
Campus, Vadodara–390004, Gujarat, India

ABSTRACT

Global pollution has caused serious health issues and environmental destruction. Researchers have helped to replace the popular hazardous artificial fiber reinforcement in polymer matrix composite with natural, sustainable, and eco-friendly fibers. Areca fruit husk (AFH) and tamarind fruit (TF) are industrially discarded fruit fibers studied for their microstructural, physical, chemical, morphological, mechanical, and thermal properties. Areca fruit husk fiber (AFHF) and tamarind fruit fiber (TFF) are removed by microbial degradation and combing. AFHFs and TFFs had improved density, surface characteristics, crystalline size and index, tensile strength, thermal stability, and cellulose content for reinforcing element strength.

The mechanical properties of AFHF and TFF reinforced polyester composites are optimized by varying the fiber content dispersed randomly, and the produced composite is morphologically and thermally characterized to support the study. Increased fiber content up to 40 wt.% resulted in revolutionary changes in mechanical properties; however, inadequate fiber bonding with the matrix resulted in a loss of strength. That causes fiber pull-out during loading due to weakening of the bonding. Comparing AFHF and

42 *Natural Polymers: Perspectives and Applications for a Green Approach*

TFF reinforcement in a polymer matrix to an artificial E-glass fiber polymer composite, 40% was found to be optimal. This is compared to the expected values using series and Hirsch's models, which are shown to be equivalent within the fiber bonding constraints. It was developed by reinforcing natural fruit fibers from Areca husk and TF in polymer matrices, which proved its promise as a substitute to damaging synthetic fiber composites in polymer composite industries.

3.1 INTRODUCTION

Polymers are now widely used in everyday life all around the world. To meet the growing demand for high specific strength materials, polymers must be made more versatile [1]. Encouraging the development of innovative eco-friendly materials for varied uses such as building, automotive, aerospace, and packaging. In this case, natural fiber-reinforced polymer matrix composites have been shown to be more promising [2, 3]. Moreover, natural fibers are preferred over synthetic fibers for reinforcing polymer composites because they are more sustainable and do not impair human health [4, 5]. A polymer composite's properties are influenced by the matrices, reinforcements, interfaces, and fiber dispersion [6]. Abaca, banana, bamboo, and other plants are used to make natural fibers. These fibers have similar characteristics to E-glass fiber-reinforced polymer composites [7, 8]. The mechanical properties of such composite materials are generally studied, such as flexural, impact, tensile, and hardness [9, 10]. Natural fibers are also less abrasive, eco-friendly, non-toxic, renewable, recyclable, and require less energy to process [11, 12].

Research concentration for extracting natural fibers from agro-waste leads to deforestation; therefore, the use of unwanted fiber wastes needs to be encouraged [13]. The use of lignocellulosic fiber waste from the tobacco and food processing industries as polymer composite reinforcement is a positive sign. India alone produces roughly 1,300,000 Mt of AFHF each year, which is generally squandered and pollutes the environment [14]. Similarly, tamarind is a dicotyledonous plant native to tropical Africa that is commonly cultivated in India for its edible sour fruit pulp. Its fruits and leaves are used to treat malaria, gastrointestinal ailments, bilious vomiting, datura poisoning, alcoholic intoxication, and eye diseases. With six to eight seeds, the fruit is reddish-brown in color and turns black when matured. The rigid pod shells are removed when the fruits are mature. After removing the seeds, the pulp is utilized as a primary acidulant in meals, seasonings, etc. They are linked together with the seeds by fruit fibers of 0.5–0.8 mm in

Development of Bio-Composites From Industrial 43

diameter. Straightened, the fibers range in length from 50 mm to 100 mm. These unusable fibers are thrown away, generating foul odors, and polluting the environment.

Using undesired natural fruit fibers from agro-food processing applications as a reinforcement for polymer composites requires thorough evaluation. The AFHF and TFF physicochemical, mechanical, and thermal characteristics are investigated. Then they are reinforced with unsaturated polyester (UPE) matrix at different compositions ranging from 10 wt.% to 50 wt.% in stages of 10 wt.% and tested for mechanical behavior to optimize fiber content. The failure pattern of areca fruit husk fiber composite (AFHFC) and tamarind fruit fiber composite (TFFC) is identified by SEM analysis of shattered specimen surfaces. TGA analysis was also used to assess temperature stability. The use of rejected fruit fibers in the polymer sector not only increases specific strength but also increases farmer income, product cost-effectiveness, worker health, and environmental sustainability.

3.2   AFHFC AND TFFC PREPARATION AND CHARACTERIZATION

3.2.1   CHARACTERIZATION OF AFHF AND TFF

The density of AFHF and TFF is determined by pouring a powder sample into a flask of known volume and measuring the mass volume. Before and after filling, the weight of the flask is recorded in order to calculate the weight of the fiber and the fiber mass density. Standard test protocols were used to determine the main biochemical elements of AFHF and TFF, such as cellulose, hemicelluloses, lignin, and wax content [15]. The ash content is tested using the ASTM E 1755-61 standard, whereas the moisture content is determined by drying the sample for 4 hours at 104°C. The mechanical characteristics of AFHF and TFF are determined using an INSTRON (5500R) universal testing machine and ASTM D3822-07 tensile test. Around 20 samples are evaluated at four different gauge lengths (10 mm, 20 mm, 30 mm, and 40 mm) while the crosshead speed is kept constant at 5 mm per minute under ambient atmospheric conditions to invalidate the results and identify the best length of the fiber.

3.2.2   AFHFC AND TFFC FABRICATION PROCEDURE

Farmers plant areca palm and tamarind trees to produce areca nuts and tamarind fruit (TF), which are used in the betel nut and food processing sectors, respectively. However, after removing the seeds, the husk casing the areca nuts and the fiber bordering the tamarind pulp are dumped as agrowaste by these agrocompanies. These discarded husks and fibers joined with tamarind pulp are gathered from the tobacco and food processing industries, washed, and soaked in water for five days, loosening the fibers and removing dust particles. They are then dried for three days to reduce moisture content before being brushed to extract fiber strands for further processing [16, 17]. The entire process of composite manufacture from its source is depicted in Figure 3.1(a) and (b).

The UPE resin matrix, methyl ethyl ketone peroxide (MEKP), the catalyst, and the accelerator, cobalt naphthenate, are combined in a 98:1:1 ratio and blended evenly with a mechanical mixer. The physical and mechanical properties of the resin are tested before and after curing, and the findings are listed in Table 3.1. The mild steel mold, which measures 300 mm × 150 mm × 3 mm, is used to create specimens with varying fiber concentrations by weight percentage ranging from 10 wt.% to 50 wt.% in 10 wt.% increments at random fiber orientation. The short AFHFs and TFFs are evenly dispersed around the mold and squeezed to form a sheet. To avoid air traps, the degassed matrix solution is poured over the compressed fiber sheet and rolled over using a grooved roller. The mold is then sealed with a lid and kept under 400 kN pressure for 24 hours in a hydraulic press.

3.2.3 AFHFC AND TFFC CHARACTERIZATION

3.2.3.1 MECHANICAL ANALYSIS OF AFHFC AND TFFC

Tensile, flexural, impact, and hardness test specimens are made from AFHFC and TFFC plates according to ASTM specifications [18]. The INSTRON S-Series H25K-S UTM performs tensile and flexural tests in accordance with ASTM standards ASTM D 3039 M-95 and ASTM D790-10, respectively [19]. For both tensile and flexural tests, the cross-head speed is maintained at 1 mm/min. The tensile test gauge length is 100 mm, whereas the three-point flexural test gauge length is 50 mm. The energy pre-occupation capability of AFHFC and TFFC is determined using an Izod impact test in accordance with ASTM D256-10, and hardness is determined using a digital Rockwell hardness testing machine in accordance with ASTM D785-98.

Development of Bio-Composites From Industrial 45

FIGURE 3.1a AFHFC fabrication movement diagram (a) palm tree (b) collected agro waste (c) extracted fiber, and (d) fabricated AFHFC plate.

FIGURE 3.1b TFFC fabrication movement diagram (a) tamarind tree (b) collected agro waste (c, d, e) soaking and extracted TFF, and (f) fabricated TFFC plate.

46 *Natural Polymers: Perspectives and Applications for a Green Approach*

TABLE 3.1 Unsaturated Polyester Properties at Liquid and Cured State

Liquid Resin	
Appearance	Yellow viscous liquid
Viscosity at 25 °C	200–300 cps
Specific gravity at 25 °C	1.24 ± 0.04
Volatile content	36 ± 1 wt.%
Acid value	23 ± 3 mg KOH/g
Cured Resin	
Tensile strength	35.0 ± 2 MPa
Tensile modulus	1.3 ± 0.2 GPa
Elongation at break	1.2 ± 0.25%
Flexural strength	43.70 ± 2.51 MPa
Flexural modulus	1.81 ± 0.21 GPa
Shear strength	4.5 ± 0.54 MPa
Impact strength	0.60 ± 0.03 J/cm^2
Rockwell hardness	62 ± 2 HRRW

3.2.3.2. *MICROSTRUCTURAL ANALYSIS OF AFHFC AND TFFC*

The broken surface of AFHFC and TFFC cracked tensile specimens is investigated using SEM analysis with a 15 kV electron beam accelerating potential. The AFHFC and TFFC samples are split into 10 mm pieces and placed on aluminum stubs using steel tapes to expose the fractured surface. Furthermore, a thin gold coating is applied to the surface to make it conductive.

3.2.3.3 *THERMAL ANALYSIS OF SELECTED FIBER AND COMPOSITE*

The thermal stability of AFHFC and TFFC is assessed using a Jupiter simultaneous TGA analyzer (Model STA 409 PC, Netzsch, Germany). About 10–20 mg of powdered fiber and composite samples were stored in an alumina crucible, which was meticulously instrumented with a thermocouple to record temperature variation over time using proteus thermal analysis software. The procedure is carried out in a nitrogen environment at a flow rate of 20 ml/min, with a temperature increase of 10°C/min from 32°C to 980°C.

Development of Bio-Composites From Industrial 47

3.2.3.4 THEORETICAL MODELLING OF AFHFC AND TFFC

The tensile strength of the polymer composite is predicted using theoretical models derived from fiber and matrix parameters. The tensile strength of randomly oriented short fiber composite materials is estimated using two models: series and Hirsch's. In order to develop a quantitative connection for the composite material from its base materials, the final results are compared to the actual experimental values within the linear range.

3.3 RESULTS AND DISCUSSION

3.3.1 INVESTIGATION OF SELECTED FRUIT FIBERS

Optical microscopy is used to examine the cross-section of AFHF and TFF to determine the fiber diameter, which is determined to be in the range of 396 μm to 476 μm and 564 μm to 789 μm, respectively. To determine the mass density, randomized fibers were pulverized and put in a known volume cylindrical container. The average density of AFHF was found to be 0.75 to 0.03 g/cm³, and TFF was found to be 1.02 to 0.14 g/cm³ correspondingly.

The chemical composition of AFHF and TFF influences their mechanical, morphological, and bonding properties, all of which have a direct impact on its use as a polymer composite reinforcement. Table 3.2 lists and compares the physicochemical and mechanical parameters of AFHF and TFF with those of other natural fibers. AFHF has a cellulose level of 57.35 wt.%, while TFF has a cellulose concentration of 72.8 wt.%, ensuring higher mechanical qualities [20, 21]. When utilized as a reinforcement for polymer composite production, lignin increases cell wall texture and aids in matrix bonding. Wax reduces the bonding properties of fiber with matrix, but happily, AFHF and TFF have an average wax content of 0.15 wt.%, which is lower than any other natural fibers (Table 3.2). Natural fiber mechanical qualities are influenced by species, growth conditions, and chemical makeup. The INSTRON (5500R) universal testing machine is used to evaluate 20 samples with gauge lengths ranging from 10 mm to 40 mm in steps of 10 mm (Table 3.2). Also, when employed as a reinforcement, the higher strain and low modulus of AFHF and TFF improve the composite's durability.

TABLE 3.2 AFHF and TFF Properties in Comparison with Other Natural and Synthetic Fibers

Fiber	Chemical Properties					Physical Properties			Mechanical Properties		
	Cellulose (wt%)	Hemi Cellulose (wt%)	Lignin (wt%)	Wax (wt%)	Moisture (wt%)	Diameter (µm)	Density (g/cm³)	Length (mm)	Tensile Strength (MPa)	Young's Modulus (GPa)	Elongation (%)
AFHF	57.3–58.2	13–15.4	23.1–24.1	0.1	7.3	396–476	0.7–0.8	10–60	147–322	1.1–3.1	10.2–13.1
TFF	72.8	11	15.3	0.2	6.3	564–789	1–1.2	10–160	1137–1360	11.2–20.7	6.5–10.1
Coconut	32–43.8	0.15–20	40–45	—	8	100–460	1.1–1.4	20–150	95–230	2.8–6	15–51.4
Palm	60–65	—	21–29	—	—	150–500	0.7–1.5	—	80–248	0.5–3.2	17–25
Cotton	82.7	5.7	—	—	—	10–20	1.5–1.6	15–56	287–800	5.5–12.6	7–8
Bagasse	55.2	16.8	25.3	—	8.8	10–34	1.25	0.8–2.8	290	17	1
Bamboo	26–43	30	21–31	—	8.9	14	0.6–1.1	2.7	140–230	11–17	—
Flax	71	18.6–20.6	2.2	1.5	7	10–25	1.5	10–65	345–1035	27.6	2.7–3.2
Kenaf	72	20.3	9	—	—	1.4–11	—	12–36	930	53	1.62
Jute	61–71	14–20	12–13	0.5	12	25–200	1.3	0.8–6	393–773	26.5	1.5–1.8
Hemp	68	15	10	0.8	9	25–35	1.41	5–55	690	70	1.6
Ramie	68.6–76.2	13–16	0.6–0.7	0.3	9.3	10–25	1.43	18–80	560	24.5	2.5
Abaca	56–63	20–25	7–9	3	15	—	1.52	—	400	12	3–10
Sisal	65	12	9.9	2	11	7–47	1.54	0.8–8	511–635	9.4–22	2–2.5
Banana	60–65	19	5–10	—	—	13–16	1.3	0.17	355	33.8	5.3
Pineapple	81	—	12.7	—	13	20–80	0.8–1.6	3–9	400–627	1.44	14.5
E-glass	—	—	—	—	—	8–15	2.5	—	2000–3500	70	2.5
Carbon	—	—	—	—	—	5–100	1.7	—	2400–4000	230–400	1.4–1.8
Aramid	—	—	—	—	—	15	1.4	—	3000–3150	63–67	3.3–3.7
S-glass	—	—	—	—	—	—	2.5	—	4570	86	2.8

Development of Bio-Composites From Industrial 49

3.3.2 INVESTIGATION OF SELECTED FRUIT FIBER COMPOSITE

3.3.2.1 MECHANICAL ANALYSIS OF AFHFC AND TFFC

Tensile, flexural, impact, and hardness values of the produced composites were determined according to ASTM standards and given as four subsections, respectively.

3.3.2.1.1 Tensile Properties of AFHFC and TFFC

Figures 3.2 and 3.3 show the influence of fiber content in the matrix on the tensile characteristics of AFHFC and TFFC, respectively. The two most convincing characteristics of a composite material's mechanical qualities are fiber alignment and the amount of reinforcement. If the fiber orientation is kept random, however, maximizing the fiber content for composite characteristics optimization is critical. The mechanical properties of various natural fiber reinforced polymer composites, such as jute, kenaf, banana, and *Sansevieria cylindrica*, were improved when the fiber content was increased to 40 wt.% by weight [1, 8, 18, 22, 23]. Similarly, at 40 wt.% fiber contents, AFHFC and TFFC displayed improved mechanical capabilities, and this is considered the ideal value.

In comparison to the hazardous synthetic E-glass fiber composites with the identical UPE matrix, Figures 3.2 and 3.3 show the tensile strength and modulus of AFHFC and TFFC with respect to the corresponding weight percentage of fiber. The strength of the UPE material is 37 MPa, and it increases with fiber addition, peaking at 40 wt.% fiber contents with a value of 68.2 MPa for AFHFC and 77.44 MPa for TFFC, respectively. At the same fiber weight content, the GFC (glass fiber composites) has a compressive strength of 72 MPa. However, after adding 40 wt.% fiber, the tensile strength of AFHFC and TFFC falls because of inadequate fiber-matrix bonding, which leads to matrix failure. The glass fiber composite is following a similar pattern. Table 3.3 shows the mechanical properties of AFHFC and TFFC, as well as additional fiber-reinforced components.

Except for Abaca and *Sansevieria cylindrica* reinforced polymer composites, the optimum value of AFHFC and TFFC at 40 wt.% fiber is higher than that of most other composites. UPE has a tensile modulus of 1.5 GPa, which is higher than AFHFC; however, it is extremely brittle. Glass fiber is stiffer than any natural fiber, and this is mirrored in the GFC material. In comparison to E-glass, the modulus of AFHF and TFF is in the range of

1.124–3.155 GPa and 11.2–20.7 GPa, respectively (70 GPa). Despite the fact that the modulus is low, the high specific strength of AFHFC and TFFC makes it a likely auxiliary for lightweight structural applications when stiffness is not a concern.

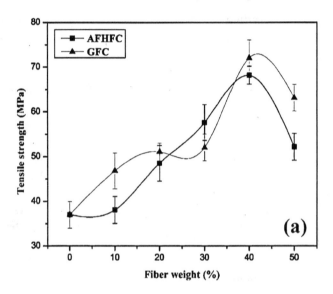

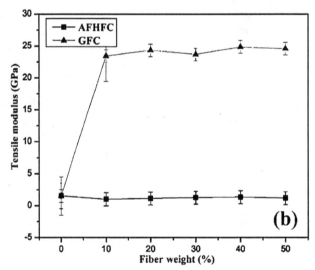

FIGURE 3.2 Tensile property of AFHFC (a) tensile strength vs. fiber weight and (b) tensile modulus vs fiber weight.

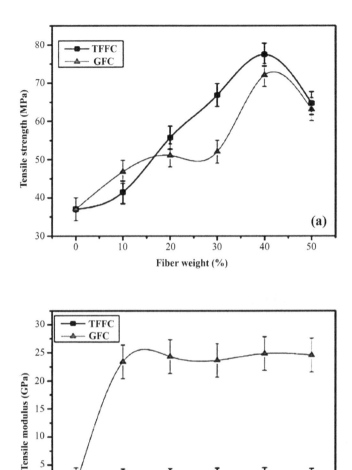

FIGURE 3.3 Tensile property of TFFC (a) tensile strength vs fiber weight and (b) tensile modulus.

3.3.2.1.2 Flexural Properties of AFHFC and TFFC

In terms of fiber composition, Figures 3.4 and 3.5 illustrate the flexural strength and modulus of AFHFC and TFFC in comparison to GFC. Flexural qualities are closely related to tensile properties due to unsystematic fiber

TABLE 3.3 AFHFC and TFFC Average Values in Comparison with GFC and Other Natural Fiber-Reinforced Polymer Composites

Sl No	Fiber/Matrix	Weight Percent (%)	Tensile Strength (MPa)	Tensile Modulus (GPa)	Tensile Strain (%)	Flexural Strength (MPa)	Flexural Modulus (GPa)	Flexural Strain (%)	Impact Strength (J/cm^2)	Hardness (HRRW)
1	Pure polyester	–	35	1.3	1.2	43.7	1.4	4.1	0.6	62
2	AFHF/polyester	40	68.2	1.3	5.1	73.9	1.6	4.5	6.8	76
3	TFF/polyester	40	77.4	1.4	5.2	88.5	1.5	5.5	7.3	90
4	GF/polyester	40	72.0	24.8	0.2	53.1	12.1	0.4	7.6	88
5	Banana/epoxy	16	16.3	0.6	2.5	57.5	8.9	0.6	–	–
6	Empty fruit bunch/PET	–	15.8	1.8	0.8	–	–	–	0.4	–
7	Hemp/UPE	–	58	9.3	0.6	110	13	0.8	–	–
8	Rice hull/HDPE	–	25	2.4	1.0	–	–	–	–	–
9	Coir/polypropylene	–	34	1.0	3.2	49.3	3.2	1.5	–	87
10	Coir/PLA	20	54	7	0.7	–	–	–	–	–
11	Banana/PCL	–	14.8	0.2	5.3	–	–	–	–	–
12	Sansevieria cylindrical/polyester	40	75	1.1	6.8	85	3	2.8	9.5	–

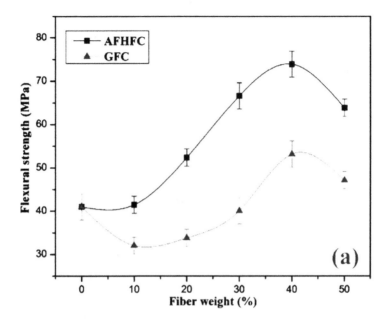

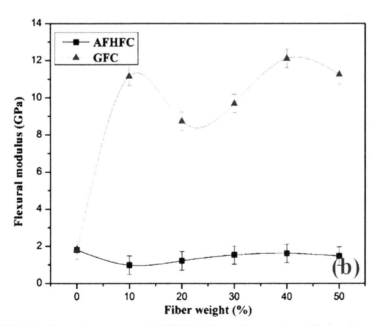

FIGURE 3.4 Flexural property of AFHFC (a) flexural strength vs fiber weight and (b) flexural modulus vs fiber weight.

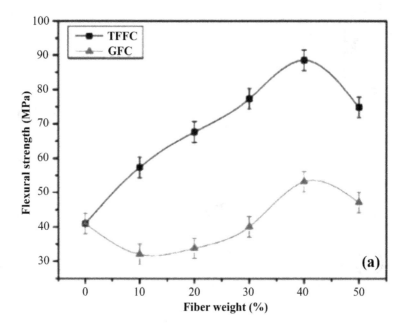

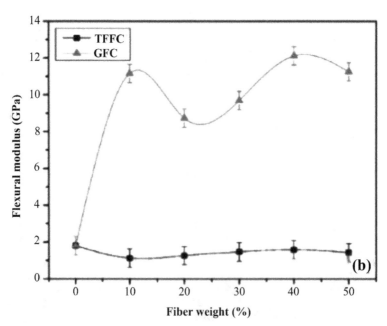

FIGURE 3.5 Flexural property of TFFC (a) flexural strength vs fiber weight and (b) flexural modulus vs fiber weight.

orientation. The addition of fiber content increases flexural strength significantly, reaching a maximum of 40 wt.% with a value of 73.9 MPa for AFHFC and 88.5 MPa for TFFC, which is more than 70 wt.% in comparison to the matrix material. The trend is similar for GFC, while AFHFC and TFFC have more possessions at low densities. In comparison to GFC's tensile qualities, the flexural modulus of AFHFC and TFFC is unremarkable, making it a general polymer composite material.

3.3.2.1.3 Impact Properties of AFHFC and TFFC

Impact tests are used to determine the materials' impact resistance and fracture toughness. At 40 wt.% fiber additions, the impact energy of AFHFC (Figure 3.6) and TFFC (Figure 3.7) increases with fiber concentration and is approximately identical to that of GFC. Despite having a greater modulus, the GFC has a stronger impact strength and does not fluctuate greatly in terms of fiber content, which is an important attribute of E-glass fiber. However, with the exception of *Sansevieria cylindrica* [18], the impact strength of AFHFC and TFFC is higher than that of most regularly used natural fiber composites.

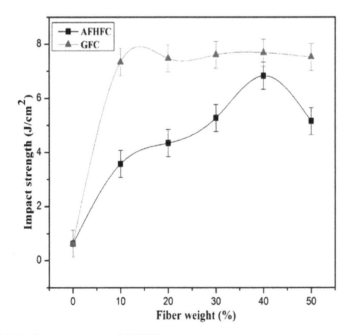

FIGURE 3.6 Impact property of AFHFC.

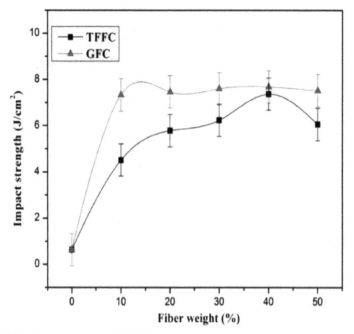

FIGURE 3.7 Impact property of TFFC.

3.3.2.1.4 Hardness Properties of AFHFC and TFFC

In general, the hardness of a composite is determined by the matrix's characteristics, as well as the distribution of filler/fiber in the matrix [18]. The testing of AFHFC (Figure 3.8), TFFC (Figure 3.9), and GFC material follow the same pattern. Due to the high modulus and toughness of glass fiber, GFC has a higher hardness. At the same time, AFHFC and TFFC have shown increased hardness as fiber content has increased, with an optimum value of 40 wt.% fiber content. Due to inadequate fiber-matrix bonding, increasing the fiber content beyond 40 wt.% significantly reduces most mechanical properties, which is one of the causes for the abrupt decline in hardness at 50 wt.% fiber concentrations.

3.3.2.2 MICROSTRUCTURAL ANALYSIS OF AFHFC AND TFFC

Figure 3.10 shows the surface morphology of an AFHFC fractured tensile sample. Figure 3.10(a) and (b) show a SEM picture of 10 wt.% AFHFC

Development of Bio-Composites From Industrial

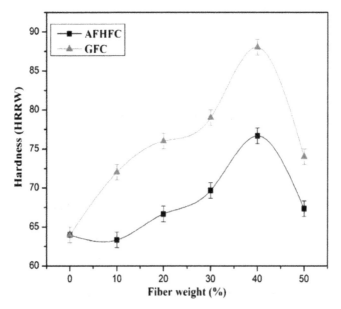

FIGURE 3.8 Hardness measurement of AFHFC.

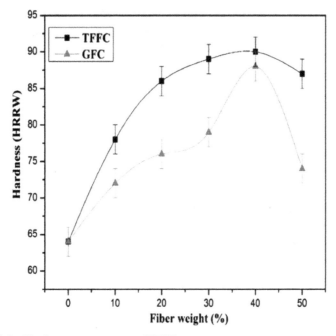

FIGURE 3.9 Hardness measurement of TFFC.

58   *Natural Polymers: Perspectives and Applications for a Green Approach*

weight, which shows a rough surface with numerous pores that enhances matrix bonding. The matrix fails to carry the load because the fiber content is low, and a fracture begins to form in the matrix, resulting in fiber-matrix debonding. The presence of microfibrils and trichomes, which are pierced deep structures on the fiber surface, improves mechanical bonding between the fiber and the matrix. The SEM picture of the fractured tensile specimen with 20% fiber content is shown in Figure 3.10(c) and (d). The good connection between the fiber and the matrix causes the fiber to shatter first, followed by the matrix, with no fiber pull out (Figure 3.10(c)). The broken tensile specimen reveals a honeycomb structure on the fiber surface at 30 wt.% fiber concentrations, indicating strong matrix bonding (Figure 3.10(f)).

At a fiber composition of 40%, the best mechanical qualities are achieved. The bonding properties deteriorate as the fiber content is increased to 50 wt.%. This is owing to the resins' limited ability to flow and conquer the space around the fiber during the production process. When subjected to tensile strains, Figure 3.10(i) and (j) exhibits debonding and withdrawal. This processing procedure has resulted in a dramatic loss in mechanical characteristics beyond 40 wt.% addition for both AFHF and GF in polymer composites due to inappropriate fiber-matrix bonding. The natural fiber surface treatment procedure [28] can achieve the incompatibility between the fiber and the matrix.

Figure 3.11 shows the fractography of TFFC at various fiber content levels. The porous fiber surface, which promotes stronger bonding, is visible in a fractured tensile specimen at 10% weight (Figure 3.11(a)). The break originates from the matrix at 20 wt.% fiber contents as well, resulting in TFFC failure (Figure 3.11(b)). The bonding properties are unaffected until the fiber content is increased to 40 wt.%, at which point the advantage of fiber strength contributes favorably to the TFFC's strength (Figure 3.11(c)). The increased mechanical characteristics at 40 wt.% fiber contents can be attributed to matrix penetration into the honeycomb fiber surface and a well-bonded fiber-matrix interface [29]. When the fiber content is increased to 50%, the bonding qualities become distorted due to insufficient matrix contact with the fiber and the creation of voids, resulting in worse mechanical properties (Figure 3.11(d)). This phenomenon is represented in the mechanical property evaluations of both TFFC and GFC.

3.3.2.3   *THERMAL ANALYSIS OF AFHF, TFF, AFHFC, AND TFFC*

The use of natural fibers from flora in a polymer matrix improves the composite's thermal resistance [30–32]. The TG-DTG curve of AFHF is

Development of Bio-Composites From Industrial 59

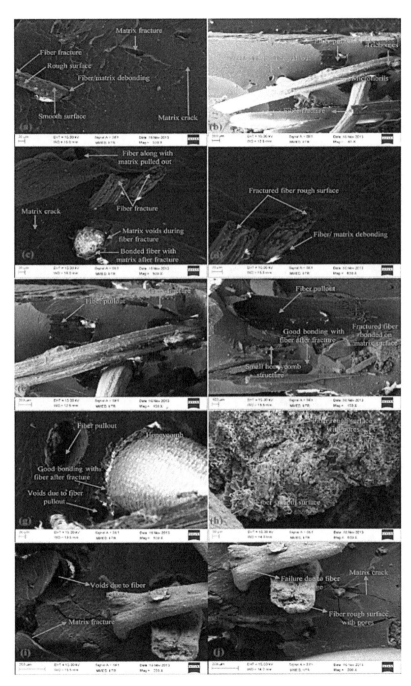

FIGURE 3.10 SEM micrograph of tensile fractured AFHFC specimen: (a, b) 10 wt.% fiber (c, d) 20 wt.% fiber (e, f) 30 wt.% fiber (g, h) 40 wt.% fiber, and (i, j) 50 wt.% fiber.

shown in Figure 3.12(a), while the TG-DTG curve of AFHFC is shown in Figure 3.12(c). The evaporation of moisture causes a small weight loss up to 100°C. The AFHF, as well as the composites formed with it in a polyester matrix, are thermally stable up to 230°C. Fiber burnout occurred at a temperature of roughly 325°C, which is a relatively high temperature for polymer manufacturing and AFHFC applications.

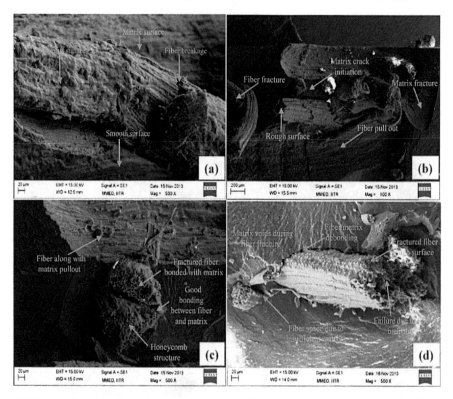

FIGURE 3.11 SEM image of tensile ruptured TFFC specimen: (a) 20 wt.% fiber (b) 30 wt.% fiber (c) 40 wt.% fiber, and (d) 50 wt.% fiber.

The TG-DTG analysis is used to assess the thermal consistency of TFF and TFFC, and the results are displayed in Figure 3.12(c) and (d). Due to the disappearance of moisture, minor weight loss occurred up to 100°C in both TFF and TFFC. The fiber's thermal stability was confirmed up to 230°C, and fiber burnout occurred at around 350°C (shown by the prominent peak in Figure 3.12(a) and (c), which is a high temperature for polymer composite processing and TFFC applications. TFF's ability to withstand a wide range of temperatures without degrading allows it to be used as a potential reinforcement for polymer composites.

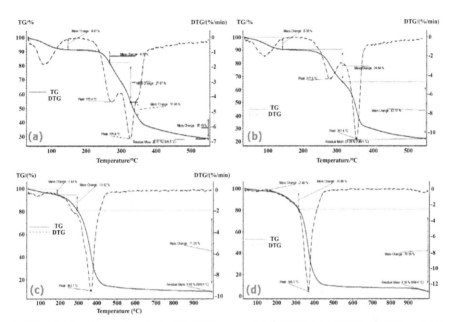

FIGURE 3.12 Thermal analysis: TG-DTG curve of (a) AFHF, (c) AFHFC, (b) TFF, and (d) TFFC.

3.3.2.4 THEORETICAL MODELLING OF AFHFC AND TFFC

In terms of many parameters, many theories are offered to anticipate the tensile properties of randomly oriented short fiber reinforced polymeric composites [18]. The following are the two models and their nomenclature for predicting the tensile properties of AFHFC and TFFC based on fiber and matrix properties:

i. **Series model:**

$$\sigma_c = \frac{\sigma_m \sigma_f}{\sigma_m V_f + \sigma_f V_m} \quad (3.1)$$

ii. **Hirsch's model:**

$$\sigma_c = \left[x\left(\sigma_m V_m + \sigma_f V_f\right)\right] + [(1-x) \times \frac{\sigma_f \sigma_m}{\sigma_m V_f + \sigma_f V_m}] \quad (3.2)$$

where; σ_c is the tensile strength of AFHFC& TFFC, N/m²; σ_m is the tensile strength of the polyester matrix, N/m²; σ_f is the tensile strength of AFHF& TFF, N/m²; V_f is the volume fraction of AFHF& TFF; V_m is the volume fraction of polyester matrix; and x is the parameter between 0 and 1.

The variation in volume percent of fibers is compared to the experimental and modeled tensile strength of AFHFC (Figure 3.13). The two models exhibit a similar tendency, with the composite's tensile strength increasing as the volume fraction of the fibers increases. Hirsch's model, on the other hand, achieves an excellent correlation between the experimental and theoretical tensile values, with an R2 value of 0.999. 'x' is a parameter in Hirsch's model that governs the stress transfer between fiber and matrix [33]. For AFHFC, the parameter 'x' = 0.4 results in a greater connection between experimental and theoretical tensile strength values. Below is the regression equation for the models that were used to forecast the tensile strength of AFHFC:

i. Series model:

$$\sigma_c = 5.1231 Vf + 34.749 \tag{3.3}$$

ii. HIRSCH'S model:

$$\sigma_c = 10.994 Vf + 35.649 \tag{3.4}$$

Due to inappropriate matrix penetration in between the fibers, tensile strength drops dramatically when the fiber fracture is greater than 40% (Figures 3.10 and 3.11(a)). The actual condition is not reflected in the model, and the models are only applicable to fiber content within the linear correlation limitations.

To validate the model, the projected tensile strength of TFFC (Figure 3.14) and GFC (Figure 3.15) in relation to the volume fraction of the fibers are compared to the actual experimental values. Up to the optimum range of fiber content, the model results correlated with the experimental data. The regression equations for the TFFC and GFC empirical and theoretical models are listed below:

iii. Experimental:

$$TFFC - \sigma_c = 166.69 Vf + 23.902 \tag{3.5}$$
$$GFC - \sigma_c = 195.81 Vf + 35.034 \tag{3.6}$$

iv. Series model:

$$TFFC - \sigma_c = 57.72 Vf + 34.803 \tag{3.7}$$
$$GFC - \sigma_c = 45.078 Vf + 36.61 \tag{3.8}$$

v. Hirsch's model:

$$\text{TFFC} - \sigma_c = 173.55 Vf + 35.022 \tag{3.9}$$

$$\text{GFC} - \sigma_c = 312.13 Vf + 36.645 \tag{3.10}$$

As a result of the above modeling, the experimental values of tensile strength of manufactured composites are closely connected with Hirch's model rather than the series model, as shown in the AFHFC (Figure 3.13), TFFC (Figure 3.14), and GFC (Figure 3.15).

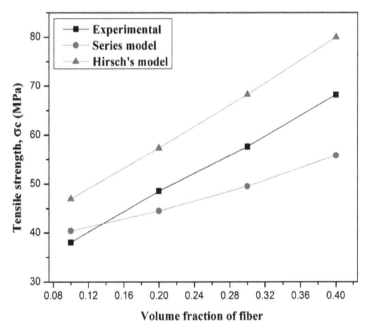

FIGURE 3.13 Model analysis of tensile strength of AFHFC.

3.4 CONCLUSION

In the current study, unused non-toxic AFHF and TFF were investigated for their possible usage as a reinforcement in polymer composites instead of harmful E-glass fibers. The high strength low-density characteristics of AFHF and TFF, along with improved bonding behavior in a polymer matrix, resulting in AFHFC and TFFC having substantially higher specific tensile strength than GFC. The effect of fiber content on mechanical qualities such as tensile, flexural, impact strength, and hardness was discovered, with an

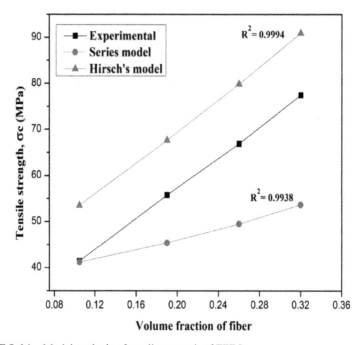

FIGURE 3.14 Model analysis of tensile strength of TFFC.

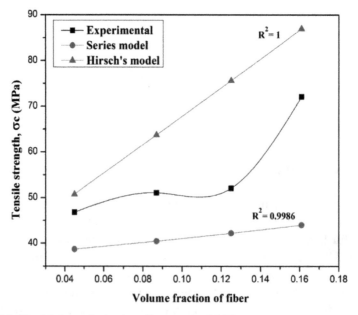

FIGURE 3.15 Model analysis of tensile strength of GFC.

increase in mechanical properties up to 40% fiber content and a subsequent drop due to improper bonding. When the fiber content was increased over 40% wt.%, the SEM picture of the fractured sample revealed a perforated fiber surface that allowed for good bonding; however, when the fiber content was increased above 40% wt.%, fiber pulls out, and voids were visible. AFHF and TFF have a higher degradation temperature (230°C), ensuring that the fiber can withstand the processing temperature. Below the crucial fiber content, Hirsch's model for predicting mechanical properties was closer to experimental results (40 wt.%). Furthermore, adequate surface treatment can improve the surface bonding between the fiber and matrix, resulting in improved mechanical properties.

ACKNOWLEDGMENT

The first author wishes to thank the Department of Science and Technology, Government of India, for awarding the INSPIRE fellowship for this research work. We are grateful to the Department of Metallurgical and Materials Engineering at the Indian Institute of Technology in Roorkee, India, for providing the facilities and services necessary to characterize some of the findings.

KEYWORDS

- **discarded fruit fiber**
- **bio-composites**
- **polymer matrix composites**
- **unsaturated polyester**
- **optimization**

REFERENCES

1. Venkateshwaran, N., Elaya, P. A., & Jagatheeshwaran, M. S., (2011). Effect of fiber length and fiber content on mechanical properties of banana fiber/epoxy composite. *J. Reinf. Plast. and Comp., 30*, 1621–1627.

2. Dhakal, H. N., Zhang, Z. Y., & Richardson, M. O. W., (2007). Effect of water absorption on the mechanical properties of hemp fiber reinforced unsaturated polyester composites. *Comp. Sci. and Tech., 67*, 1674–1683.

3. Behnaz, B., Mikael, S., & Lena, B., (2013). Manufacture and characterization of thermoplastic composites made from PLA/hemp co-wrapped hybrid yarn prepregs. *Comp. Part A, 50*, 93–101.

4. Boopathi, L., Sampath, P. S., & Mylsamy, K., (2012). Investigation of physical, chemical, and mechanical properties of raw and alkali-treated Borassus fruit fiber. *Comp. Part B, 43*, 3044–3052.

5. Chiachun, T., Ishak, A., & Muichin, H., (2011). Characterization of polyester composites from recycled polyethylene terephthalate reinforced with empty fruit bunch fibers. *Mate. and Des., 32*, 4493–4501.

6. Chuanwei, M., Wadood, Y. K., & Hamad, (2013). Cellulose reinforced polymer composites and nanocomposites: A critical review. *Cell, 20*, 2221–2262.

7. Venkateshwaran, N., Elaya, P. A., Alavudeen, A., & Thiruchitrambalam, M., (2011). Mechanical and water absorption behavior of banana/sisal reinforced hybrid composites. *Mate. and Des., 32*, 4017–4021.

8. Byoung-Ho, L., Hyun-Joong, K., & Woong-Ryeol, Y., (2009). Fabrication of long and discontinuous natural fiber reinforced polypropylene biocomposites and their mechanical properties. *Fib. and Poly., 10*, 83–90.

9. Paul, W., Jan, I., & Ignaas, V., (2003). Natural fibers: Can they replace glass in fiber-reinforced plastics? *Comp. Sci. and Tech., 63*, 1259–1264.

10. Senthil, K. K., Siva, I., Jeyaraj, P., Winowlin, J. J. T., Amico, S. C., & Rajini, N., (2014). Synergy of fiber length and content on free vibration and damping behavior of natural fiber reinforced polyester composite beams. *Mate. and Des., 56*, 379–386.

11. Ferreira, J. A. M., Capela, C., & Costa, J. D., (2010). A study of the mechanical properties of natural fiber-reinforced composites. *Fib. and Poly., 11*, 1181–1186.

12. Ku, H., Wang, H., Pattarachaiyakoop, N., & Trada, M., (2011). A review on the tensile properties of natural fiber-reinforced polymer composites. *Comp. Part B, 42*, 856–873.

13. Nazrul, I., Rezaur, R., Mominul, H., & Monimul, H., (2010). Physicomechanical properties of chemically treated coir reinforced polypropylene composites. *Comp. Part A, 41*, 192–198.

14. Youssef, H., Waleed, El-Zawawy, Maha, M., Ibrahim, & Alain, D., (2008). Processing and characterization of reinforced polyethylene composites made with lignocellulosic fibers from Egyptian agro-industrial residues. *Comp. Sci. and Tech., 68*, 1877–1885.

15. Indran, S., Edwin, R. R., & Sreenivasan, V. S., (2014). Characterization of new natural cellulosic fiber from Cissus quadrangularis root. *Carb. Poly., 110*, 423–429.

16. Chakrabarty, J., Masudul, H. M., & Mubarak, A. K., (2012). Effect of surface treatment on betel nut (*Areca catechu*) fiber in polypropylene composite. *J. Poly. Env., 20*, 501–506.

17. Dhanalakshmi, S., Ramadevi, P., Basavaraju, B., & Srinivasa, C. V., (2012). Effect of esterification on moisture absorption of single areca fiber. *Intnl. J. Agri. Sci., 4*, 227–229.

18. Sreenivasan, V. S., Ravindran, D., Manikandan, V., & Narayanasamy, R., (2011). Mechanical properties of randomly oriented short *Sansevieria cylindrica* fiber/polyester composites. *Mate. and Des., 32*, 2444–2455.

19. Uma, M. C., Obi, R. K., Muzenda, E., Shukla, M., & Varada, R. A., (2013). Mechanical properties and chemical resistance of short tamarind fiber/unsaturated polyester

composites: Influence of fiber modification and fiber content. *Intnl. J. Poly. Anal. and Charac., 18*, 520–533.

20. Yusriah, L., Sapuan, S. M., Zainudin, E. S., & Mariatti, M., (2014). Characterization of physical, mechanical, thermal, and morphological properties of agro-waste betel nut (*Areca catechu*). *Husk Fiber. J. Clea. Prod., 72*, 174–180.

21. Indran, S., & Edwin, R. R., (2015). Characterization of new natural cellulosic fiber from Cissus quadrangularis stem. *Carb. Poly., 117*, 392–399.

22. Omar, F., Andrzej, K. B., Hans-Peter, F., & Mohini, S., (2012). Biocomposites reinforced with natural fibers: 2000–2010. *Pro. in Poly. Sci., 37*, 1552–1596.

23. Moyeenuddin, A. S., Kim, L. P., & Alan, F., (2012). Analysis of mechanical properties of hemp fiber reinforced unsaturated polyester composites. *J. Comp. Mate., 47*, 1513–1525.

24. Haydaruzzaman, Khan, A. H., Hossain, M. A., Mubarak, A. K., & Ruhul, A. K., (2010). Mechanical properties of the coir fiber-reinforced polypropylene composites: Effect of the incorporation of jute fiber. *J. of Comp. Mate., 44*, 401–416.

25. Lai, W. L., Mariatti, M., & Mohamad, J. S., (2008). The properties of woven kenaf and betel palm (*Areca catechu*) reinforced unsaturated polyester composites. *Polym. Plast. Tech. and Eng., 47*, 1193–1199.

26. Ramanaiah, K., Ratna, P. A. V., & Hema, C. R. K., (2012). Thermal and mechanical properties of waste grass broom fiber-reinforced polyester composites. *Mate. and Des., 40*, 103–108.

27. Ratna, P. V., Atluri, K., Mohana, R. A. V. S., & Gupta, S. K. S., (2013). Experimental investigation of mechanical properties of golden cane fiber reinforced polyester composites. *Intnl. J. Poly. Anal. and Charac., 18*, 30–39.

28. Li, X., Tabil, L. G., & Panigrahi, S., (2007). Chemical treatments of natural fiber for use in natural fiber-reinforced composites: A review. *J. Poly. Env., 15*, 25–33.

29. Shinoj, S., Visvanathan, R., Panigrahi, S., & Kochubabu, M., (2011). Oil palm fiber (OPF) and its composites: A review. *Ind. Cro. and Pro., 33*, 7–22.

30. Binoj, J. S., Edwin, R. R., & Indran, S., (2018). Characterization of industrial discarded fruit wastes (*Tamarindus indica* L.) as potential alternate for man-made vitreous fiber in polymer composites. *Proc. Safe. and Envi. Prote., 116*, 527–534.

31. Binoj, J. S., Edwin, R. R., & Daniel, B. S. S., (2017). Comprehensive characterization of industrially discarded fruit fiber, *Tamarindus indica* L. as a potential eco-friendly bio-reinforcement for polymer composite. *J. Clea. Prod., 142*, 1321–1331.

32. Binoj, J. S., Edwin, R. R., Sreenivasan, V. S., & Rexin, T. G., (2016). Morphological, physical, mechanical, chemical, and thermal characterization of sustainable Indian areca fruit husk fibers (*Areca catechu* L.) as potential alternate for hazardous synthetic fibers. *J. Biol. Eng., 13*, 156–165.

33. Binoj, J. S., Edwin, R. R., Daniel, B. S. S., & Saravanakumar, S. S., (2016). Optimization of short Indian areca fruit husk fiber (*Areca catechu* L.)-reinforced polymer composites for maximizing mechanical properties. *Intnl. J. Poly. Anal. and Charac., 21*, 112–122.

34. Yusriah, L., Sauna, S. M., Zainudin, E. S., & Mariatti, M., (2014). Characterization of physical, mechanical, thermal, and morphological properties of agrowaste betel nut (*Areca catechu*) husk fiber. *J. Clean. Prod., 72*, 174–180.

35. Nesrine, S., Mahjoub, J., Foued, K., Najeh, T., Romdhani, Z., Wiem, I., Hamdaoui, M., & Bernard, D., (2016). *Fib. and Poly., 17*, 2095–2104.

36. Farias, M. A., Farina, M. Z., Pezzin, A. P. T., & Silva, D. A. K., (2009). Unsaturated polyester composites reinforced with fiber and powder of peach palm: Mechanical characterization and water absorption profile. *Mater. Sci. Eng. Part C., 29,* 510–513.
37. Seena, J. M. S., Sreekala, Z., Oommen, P., & Koshy, S., (2002). A comparison of the mechanical properties of phenol formaldehyde composites reinforced with banana fibers and glass fibers. *Comp. Sci. Tech., 62,* 1857–1868.

CHAPTER 4

Characteristics and Prospects of Controlled Release of Fertilizers Using Natural Polymers and Hydrogels

NEETHA JOHN

CIPET: IPT-Kochi, Udyogamandal P.O, Kochi, Kerala, India.
E-mail: neethajob@gmail.com.

ABSTRACT

Agriculture is the area where the whole world depends and needs much attention. Modified methods for agriculture can lead to the growth of nations. The role of polymers in the growth of agriculture is very much significant. The progress of the work is aiming for sustainable growth with regard to green polymers is discussed here. Usage of natural polymers does not make an impact on the quality of the crops and provides better routes for farming. Farming can be very well controlled by using various components such as fertilizers, pesticides, and insecticides. Controlled or modified release of fertilizers with the help of super absorbing natural polymers give greater scope as outlined in this chapter. Farmers will get help if the controlled release of fertilizers is used and it will very well reduce the burden due to climate changes. The controlled-release fertilizers are reducing the labor work of farmers and help them to get greater crop productivity without many changes in costs. The need of using super absorbent polymers and advanced material like hydrogels with the desirable characteristics in soil conditioning is discussed here.

4.1 INTRODUCTION

The world population is increasing at a faster rate and will be reaching seven billion by 2050 [1]. There is pressure on our environment, the reason for that could be the increasing global industrialization, problems for the sustainability of the environment, and lack of food security. It further created Global warming, accumulation of toxic chemicals, and contamination for the biological systems in soil [2]. The productivity per unit area has increased and the land used for crop production increased in order to supply food to the growing population in the 20th century. There is a greater thrust on agricultural development around the world. This can cause more exploitation of natural resources like water and plant and enormous usage of fertilizers and pesticides [3]. It could not be a sustainable method when considered for the future even though these practices created further crop yields in a small period. There observed depletion of natural water resources and get contaminated with toxic pesticides and chemical fertilizers. It further led to reduced productivity of the available lands which is threatening for the survival and well-being of all life forms of earth. The development in agriculture currently has converted to more sustainable methods to use land, water, and plant resources for agriculture and related applications. There is an intention of the current agricultural system is to maximize the usage of land and water, and not to make any threats to the environment and its resources.

Food supply is one of the main applications of agriculture. The trends presently indicate increased use for agriculture in comparison with available soil and water resources. Because of the global climatic anomalies, drought, lack of proper precipitation which is occurring in many parts of the world at different levels. Countries facing difficult situations related to drought may lead to starvation to a larger extend. It may also lead to problems in economy in the developing countries as their economy is depending greatly on agriculture only [4]. Agricultural polymers give a new way by focusing on natural/synthetic macromolecular materials like proteins, polyacrylates, polyacrylamides, and polysaccharides. Many natural as well as synthetic polymers are successfully used in various agriculture applications. With this respect, the current study takes an account of vast applications of natural polymer in agriculture, its modified types, controlled release mechanisms, etc. It is also presenting research trends, latest developments, and the future prospects.

4.2 NATURAL POLYMERS

Sustainable agriculture has to look into natural polymers to a greater extent. Natural polymers are taken directly from nature and make no harm by itself. It is very significant to think about the developments in agriculture using

Characteristics and Prospects of Controlled Release 71

natural polymers. The usage of natural polymers has increased over the last decade and many researches are going in this direction to make agriculture effective and environmentally sustainable.

Polymers taken naturally are termed as natural polymers. Majority of natural polymers are water-based as it contains major constituent water [5]. Large numbers of polymers come from nature. A human body is constituted of many polymers like proteins, enzymes DNA, RNA, etc. Polysaccharides, polypeptides natural polymers like keratin, silk, wool, hair, etc. One of the significant natural polymers is natural rubber. Plants consist of natural polymers; the most suitable example is cellulose with which the plants are structured [6]. Some of the natural polymers are significant components in the living organisms and part of life processes in living organisms as shown in Table 4.1, which illustrates the major types of natural polymers and their functions.

TABLE 4.1 Types and Examples of Natural Polymeric materials (Copyright © 2002 American Chemical Society)

Sl. No	Name	Structure and Functions
1	Proteins and polypeptides	They are natural polymers that constitute part of living organisms. Some of the proteins are called enzymes. One type of protein in our blood is hemoglobin which carries the oxygen from the lungs to the cells of a human body. Protein is a naturally occurring polyamide. This polymer contains an amide group present in the backbone chain [7].
2	Collagen	Collagen is one of the natural polymers and is a protein. It makes up the connective tissue present in the skin of human beings. It is a fiber that creates an elastic layer below the skin and thus helps in keeping it supple and smooth [7].
3	Latex	Latex is made of natural rubber, is a polymer. The natural form of latex is collected from the rubber trees and it is also found in a variety of plants which includes the milkweed. It can also be prepared artificially by the process of building up long chains of molecules of styrene [7].
4	Cellulose	Cellulose is one of the most abundant organic compounds found on the Earth and moreover, the purest form of natural cellulose is cotton. Paper is produced from trees and the supporting materials in leaves and plants consist of cellulose. Amylose is also a polymer which is made from glucose as the monomers [7].
5	Starch	Starch basically consists of glucose monomers, split into water molecules when combined chemically. Starch is also a member of basic food groups called the carbohydrates and it is found in grains, cereal, and potatoes. Starch is a polymer of monosaccharide glucose. The molecules of starch consist of two kinds of glucose polymers namely amylopectin and amylose which are the main component of starch in most plants [7].

72   *Natural Polymers: Perspectives and Applications for a Green Approach*

TABLE 4.1   *(Continued)*

Sl. No	Name	Structure and Functions
6	Chitin/ chitosan	Chitin is the second most important natural polymer in the world has sources of marine crustaceans, shrimp, and crabs. Chitosan, the most important derivative of chitin, which is soluble in acidic aqueous media, is used in many applications [7].
7	Carbohydrates	Carbohydrates, another group of polymers, form from glucose, just like cellulose. Sugar and starches, both forms of carbohydrates, serve as food for plants and animals. The glucose monomers connect differently in carbohydrates than in cellulose, though, bunching up instead of stretching out. One result of how these monomers connect is that carbohydrates dissolve in water. People can digest carbohydrates but not cellulose because carbohydrates dissolve in water but cellulose does not. Also, people lack the enzyme that will break the cellulose polymer [7].
8	DNA and RNA	Two nucleic acid polymers, deoxyribonucleic acid (DNA) and ribonucleic acid (RNA) form from monomer nucleotides. DNA contains the genetic code for an organism and RNA carries the genetic information from the DNA to the cytoplasm where proteins are then made. Like most natural polymers, nucleic acid polymers are condensation polymers [7].
9	Rubber	Natural rubber comes from the latex rubber trees. It has property like bouncing and stretching because of the flexibility in molecular chain. The monomers of a similar natural polymer called gutta-percha connect differently, resulting in a brittle rather than flexible material [7].
10	Xanthan Gum (XG)	This is a high molecular weight, water-soluble, anionic-bacterial heteropolysaccharide; it is a hydrophilic polymer, biocompatible, and inert and thus it provides time-dependent release kinetics [8]. It is used as a flow modifier and is produced from microbial fermentation of glucose from the bacterial coat of *Xanthomonas campestris*. Applications of XG in the CRFs industry are less common, findings prove that XG matrices exhibit quite consistent higher ability to retard drug release for controlled-release formulation. This calls for further investigations on its use in CRFs.
11	Carrageenan	This is a naturally occurring high molecular weight anionic gel-forming polysaccharide extracted from certain species of red seaweeds (Rhodophyceae) such as Chondruscrispus, Eucheuma, Gigartinastellata, and Iridaea [8].
12	Pectin	It is a methoxyester of pectic acid found in the higher plants cell walls. Certain fruits such as apple, quince, plum, gooseberry, grapes, cherries, and oranges also are known to contain pectin [8].
13	Tamarind seed polysaccharide (TSP)	TSP is a galactoxyloglucan (a monomer consists of three sugars-galactose, xylose, and glucose in a molar ratio of 1:2:3) separated from seed kernel of *Tamarindus indica* [8].

Characteristics and Prospects of Controlled Release   73

TABLE 4.1   *(Continued)*

Sl. No	Name	Structure and Functions
14	*Mimosa pudica* seed mucilage	Mimosa mucilage is known to act as a matrix forming agent for sustained delivery of formulations [8].
15	*Leucaena leucocephala* seed polysaccharide (LLSP)	LLSP is a galactoxyloglucan hydrophilic gum separated from seed kernel of *L. leucocephala*. In the controlled release art, LLSP has been used for controlled release of water-soluble plus water-insoluble drugs [8].
16	Guar gum	This is a nonionic naturally occurring, hydrophilic polysaccharide extracted from the seeds of *Cyamopsis tetragonolobus* and is used as binder and disintegrant [8]. It is a release-retarding polymer shows first-order release kinetic.
17	*Terminalia catappa* gum (TC)	It is a gum exudate obtained from *Terminalia catappa* Linn. It is a natural release retarding polymer. The retarding behavior of drug release of TC gum reported by Kumar et al. [8].
18	Mucuna gum	Mucuna gum is a biodegradable, amorphous polymer composed of mainly D-galactose along with D-mannose and D-glucose and isolated from the cotyledons of plant *Mucuna flagillepes* [8].
19	Gum copal (GC)	It is a naturally occurring hydrophobic resin isolated from the plant *Bursera bipinnata* and follows zero order release kinetics. There is a similar GC also exhibit release kinetics as in GC. GC is a naturally occurring hydrophobic gum obtained from plant *Shorea wiesneri* [8].
20	Karaya gum	It is a hydrophilic naturally occurring gum obtained from *Sterculia urens* and composed of galactose, rhamnose, and glucuronic acid. It swells in water and is thus used as release rate controlling polymers in different formulations [8].
21	Rosin	A clear, pale yellow to dark amber thermoplastic resin present in oleoresins of the tree Pinusroxburghii and Pinustaeda belonging to the family Pinaceae. Rosin is used in the development of controlled drug delivery acts as a hydrophobic matrix forming agent for systems [8].
22	Gum acacia	It is from stems of the Acacia Arabica tree and can be used as encapsulating agent. Locust bean gum provides excipient which gives sufficient muco-adhesive applications [8].

4.3   CONTROLLED RELEASE FERTILIZERS

Controlled release (CR) is the transfer with regulated permeation of an active component from larger storage to aimed components in order to control the required amount for a required time [9]. The comparative release mechanism of conventional and CR-assisted delivery is shown in Figure 4.1.

Figure 4.1 shows the comparison of conventional and controlled release of fertilizers. In a conventional release, all fertilizers will be lost in a short time. But in controlled release, soil will get the fertilizers in a controlled manner for a long time.

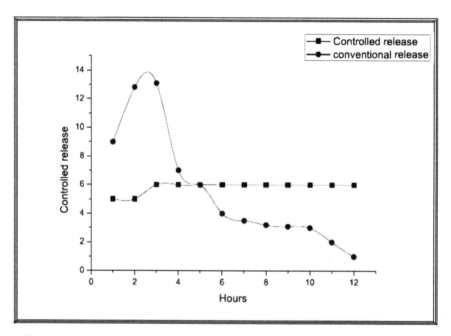

FIGURE 4.1 Comparison of controlled release and conventional release.

In conventional application very high initial dosage which rapidly falls below the effective level will be occurring. In the case of CR formulation, it can maintain the required quantity for a required and applicable time. Various applications are there for CR like in agricultural, biomedical, food, pharmaceutical industries, pesticides, herbicides, fertilizers, biomolecules, and drug release [10–18].

Traditional chemical fertilizer treatment has been improved by the use of controlled-release fertilizers (CRF). CRFs are types of systems that are prepared to improve the release kinetics of chemical fertilizers. New technologies are emerging that can improve the efficiency of CRFs. It gives additional functionality and reduces the cost to make CRFs a more viable alternative to conventional fertilizers.

Biodegradable polymers containing fertilizers give economic and environmental advantages as they release water and nutrients gradually to the environment and create no residues. They can reduce air and water pollution which are created from fertilizers. It enhances the capability of plants to access required nutrients, water retention will be improved, will increase the drought resistance, and reducing the amount of fertilizer to give maximum crop yields. Different strategies are being considered to solve such issues. For getting sustainable growing practices in agricultural industries one has to move toward more efficient practices. CRFs have the great potential to solve many problems and provide more profit in agriculture. CRFs are widely used in the plant nurseries for bigger trees as well as smaller shrubs. CRFs are also used as outdoor plants and greenhouse crops. Fertilizers that are soluble in water are easier to apply and the rates of application could be varied depending on the requirements of the crops, weather prevailing in that area, and the size of the plant. CRFs mainly are water-soluble fertilizers which are modified with a polymer coating. The polymer coating has a porosity in which water penetrates through the coating, solubilizes the fertilizer, and then the fertilizer slowly comes out from the coating by leaching through the coating. Nitrogen, phosphorous, and potassium detection of CRFs will be a different method and may not be considered with micronutrients. One can analyze the availability of fertilizers by the length of releases of the fertilizers and also how much time it will be available in the soil at ambient temperature.

Polymer coating thickness around the fertilizers will decide how long the release happens to the CRF. CRF release can be increased by enhancing the thickness of the polymer coating. When the release time of the fertilizer increases, the required application rates can also be increased. The process begins with moisture in CRF. When the growing medium dries out, reduces the amount of fertilizer released. Temperature influences to a greater extent on the release mechanism. At normal growing temperatures CRF works nicely; when the temperature is below, 10 °C fertilize cannot be released; and if it is above 32 °C, the release may happen to a greater extent and can be excessive.

CRF is applied as on the top of soil after planting or incorporated into the growing medium before the process of planting. CRF when placed on top of the growing medium surface after planting which is termed as top dressed. In case of top dress some disadvantages are that the fertilizers can fall out of the growing pots when there is any disturbances occur. When water is poured normally then it soaks into the growing medium, if the water is applied faster or water overruns then the fertilizers may float and can run out over the pot rim.

Pre-incorporation is the second application method in which the fertilizer gets mixed up with the CRF directly into the growing medium. This type of preincorporating has the advantage that the CRF is easily available for plants and it will remain in the growing medium. Supaporn Noppakundilograt et al. describe multilayer-coated nitrogen, phosphorous, and potassium (NPK) type compound fertilizer in hydrogel with controlled nutrient release and water absorbency [19]. There can be a three-layered controlled release coating for NPK fertilizer. Hydrogel can be prepared by dipping the NPK fertilizer granules poly(vinyl alcohol) (PVA) and chitosan (CS) solutions which were cross-linked.

Ahmed et al. prepared superabsorbent (PVP/CMC) based on polyvinyl pyrrolidone (PVP)/carboxymethyl cellulose (CMC) of different types of copolymer compositions by gamma radiation. The applicability of PVP/CMC in the agricultural fields shows larger growth rates on zea maize types plants. The growth rates of zea maize plant in soil mixed with PVP/CMC loaded fertilizers is greater in the case of untreated soil. PVP/CMC has very slow release rates of fertilizer, high swelling, and slow water retention behaviors. So that it can be used as better release systems for fertilizers and as a soil conditioner in agricultural applications [20]. Ding et al. was developed and reported the nutrients release from a gel-based low/CRF, which was produced by the combination of different types of natural, seminatural, and synthetic types of organic macromolecule materials and natural type inorganic mineral with conventional NPK fertilizers [21]. Souza et al. derived green composite based on polyhydroxybutyrate (PHB) and montmorillonite act as a carrier for CRFs like KNO_3 and NPK [22]. Rodrigues et al. describe application of potassium-containing microspheres based on CS and montmorillonite clay and the in situ soil release [23].

Characterization of the release of nutrients from a slow-release fertilizer (SRF)/CRF is used in assessing the efficiency of fertilizer. Trenkel and IFI Association [24] provide details of several methods of coated fertilizers to calculate the release of different SRFs/CRFs. Release characteristics may be affected by both physical effects and chemical effects. Physical effects are the reduced diffusion rates in the soil, moisture, fluctuation of temperature, etc. [25]. Chemical effects are pH changes, root excretion, action of the microorganisms on biodegradable materials, sulfur coating, waxes, etc. Evaluation of the properties of a typical CRF compound with some other parameters like water permeability, swelling ratio, and dissolution rate which account for release behavior, zeta potential, particle size, morphology, and thermal degradation properties [26].

Characteristics and Prospects of Controlled Release 77

4.3.1 BENEFITS AND LIMITATIONS OF CRFS

4.3.1.1 BENEFITS OF CRFS

1. Easy to use: CRFs plants will get the nutrients every time in spite of changing weather. Plants stay wet and do not need watering during the cool and cloudy weather. A planter should decide the correct amount of water and fertilizers in case of water-soluble fertilizer. Plants may get overwatered or may have nutrient deficiencies. In the case of CRF, it can release fertilizer continuously without the addition of water. It needs to be applied once, not many times as in the case of water-soluble fertilizers.
2. Low labor cost: For the planters, labor is always a matter of concern. CRF avoids the mixing of fertilizers and can reduce labor costs as in the case of water-soluble fertilizers, and no need of expensive injection equipment. Crop quality will be consistent as it is not affected by the improper mixing of water-soluble fertilizer in the stock solution.
3. Environmental friendly: The nutrients are released in a controlled rate and more ideally into the growing medium at the rate that plants needed them for their growth. Therefore no excess fertilizer of any lack of fertilizer situation arises. The amount of nutrients leached into the medium when watering is done is the minimum when compared to using water-soluble fertilizers.

4.3.1.2 LIMITATIONS OF CRFS

1. Growers flexibility is lacking: CRFs cannot be changed once the fertilizer has been incorporated. It is possible to adjust the fertilizer application rate and formulation as in case of water-soluble fertilizers, which is to meet the correct need of the fertilizer to the crop with the environmental conditions. Raining season may vary and late fall can happen or the temperatures can be unseasonably warm or cool. In such cases the fertilizer can be released very fast and due to root burn or dissolve very slowly, which may cause nutrient deficiencies.
2. Fixed rate of release: Plants require varying amounts of fertilizers and nutrients depending on the growing stage. Younger plants require less fertilizer than grown-up plants and more for the mature and actively

growing. The release rate of CRFs is fixed at a stable temperature and it may not provide the exact amount of nutrients at the time when they are required.

3. Limitations in storage: The nutrients are released generally based on temperature. They may start being released as soon as they are incorporated into the growing medium. It is of great concern that how long the medium can be stored before use. When a growing medium is using with an incorporated CRF, to avoid problems with high fertilizer release rates, it should be used within 7–10 days of preparation.

4.4 USAGE OF HYDROGELS AND SUPER ABSORBING POLYMERS IN THE FIELD OF AGRICULTURE

Polymeric hydrogels can work as an intelligent fertilizer also as superabsorbent and biosorbent. These materials can enhance the biodegradation processes in agriculture which are sustainable with respect to environmentally, technically, socially, and economically [27]. Agriculture is under abiotic stresses like drought, salinity, and temperature, chances to increase because of degradation of land, more and more urbanization, and drastic climate change. The world is looking forward to get a water-efficient agriculture system as getting water for irrigation is difficult. There are great challenges for food security as there is increasing food demand and declining water resources [28]. It needs more proper management techniques in order to preserve moisture and to increase water holding capacity of the soil. The yield of the crops may be lower when compared with normal conditions if moisture is not retained.

Hydrogel consists of polymer molecule, containing network chains in the gel which are linked to each other to form one larger size molecule on macroscopic scale. Hydrogel absorbs water also are cross-linked, absorb water from aqueous solutions using hydrogen bonding with water molecules. The gel is in a state which is not completely solid or liquid. It can show semiliquid-like and semisolid-like properties that are not found in any pure solid or a pure liquid state. It is very soft and has the capacity to store more and more water and thus unique [29]. Natural polymers based on proteins like collagen and gelatin and polysaccharides such as alginate and agarose.

In the agricultural sector, by the use of hydrogel, the growth increases considerably. It is applicable in both arid and semiarid regions of the country. It starts increasing the water retention of soil and improved the crop productivity [30]. These materials are multifunctional as they can be used in slow release of nutrients and made out of natural materials [31, 32]. Super absorbing polymers and hydrogels act as soil conditioners studied from 1980s where the application was published. Erickson's [33] patented method for the improvement of water retention and capacity of aeration of soil can be done by incorporating polymeric films containing water-swelling properties when placed in soil.

Hydrogels are water granules having retention properties as its wells to many times to their original dimensions when they come in contact with water. Water can be saved by the water holding capacity of hydrogels. Proper management practices needed to be implemented in order to maintain soil moisture. One of the best ways is to use super-absorbent polymers (SAPs) hydrogel which can be swollen and absorb a large volume of water or aqueous solution. Due to this property, there are lots of applications of this material. Chemical properties of the hydrogel are affected by molecular weight of hydrogel.

Three main types are in agricultural use: starch-graft copolymers cross-linked polyacrylate, cross-linked polyacrylamides, and acrylamide–acrylate copolymers. Potassium polyacrylate is one of the common materials used for making SAP and used as hydrogel for agricultural use as it has more retention and high efficiency in soil with no toxicity problems. These materials are prepared by polymerizing acrylic acid with a cross-linker. Cross-linked polymers can contain 400 times water and can release 95% to growing plants [34, 35]. Hydrogel can prevent leaching of water which can lead to increased water use efficiency and increasing frequency for irrigation. Plant can be under stress in summer months in semiarid regions due to lack of soil moisture. Hydrogels close to root area helps to increase growth and plant performance.

There are generally three types of hydrogels by their charges like negatively charged (anionic), positively charged (cationic), or with neutral charge. It occurs in both linear and cross-linked. Depending upon the charge, it can react with the soils and the solutes. Soil containing clay components has negative charges. Soil containing heavy metals has positive charges. Anionic hydrogel act as dispersants as it cannot directly bind but cationic can bind to clay components and act as flocculants. The affinity of the gel for other compounds depends on modified electrical charges, hydration levels, van der Waals forces, hydrogen bonding, etc. In the case of polyacrylamide polymer, it contains lots of positively charged, negatively charged, and neutral chain segments, all with varying affinities toward other molecules. The hydrogel has a greater ability to absorb water and gel in case of stronger attraction between the gel, surrounding solutes, and the soil particles which can create aggregate and stabilize the soil structure [36].

Hydrogel matrix has pores in it, which allows distribution of the fertilizers at a predesigned control system. The distribution of fertilizers can be controlled by changing the hydrogel composition to match the environment. This can be influenced by change in pH, temperature [37], and ionic strength [38]. The hydrophilicity of the hydrogel network is because of the presence of amide ($-CONH_2$) and sulfonic ($-SO_3H$) groups in the copolymer matrix. It can be synthesized by various methods with the help of thermal initiators such as azobisisobutyronitrile, photoinitiators such as benzoin, benzyl, and other sources such as gamma-ray. There will be polymer chain extension due to repulsion between charged carboxylate groups. This is because of high pH of the solutions, the $-CONH_2$ groups got hydrolyzed. There will be large space created between the molecules, due to this it can absorb greater amount of water molecules and increases by swelling [39]. Two phases are created when the solvent molecules penetrate into the polymer which causes an increase in volume of the polymer. Water reaches the interior of the hydrogel and swollen polymer of higher volume is created. The polymer acts as a matrix to hold water inside and this water in the hydrogel allows free diffusion of some solute molecules. As the number of ionic groups increases, the swelling capacity increases, as there is an increase in the number of counter ions inside the hydrogel. Sulfonic acid is a strong polyelectrolyte containing a high degree of ionization.

4.5 POLYMERIC SUPERABSORBENT POLYMER AND HYDROGELS AS SOIL CONDITIONERS

Hydrogels have great applications in conservation of agricultural lands. Various researches are going on different kinds of polymeric composites. These can increase the efficiency of SAPs for use as soil conditioners. Hydrogels and SAP have a great influence on the water holding capacity of soil in more sandy nature.

Callaghan et al. [40] tested superabsorbent polymers like polyacrylamide and polyvinyl alcohol in soils of Sudan as it is highly drought affected. It is proved that the field capacity of the soil can be increased by polyvinyl alcohol. It can increase the plant survival period and also provide evidence for the profitability in economic aspects of SAPs.

Woodhouse et al. [41] studied the effect of starch copolymer on the growth rate of barley and lettuce in coarse types of oil. Stahl et al. [42] studied first the rate of biodegradability of polyacrylamide and polyacrylate

Characteristics and Prospects of Controlled Release

superabsorbent copolymers and also pure polyacrylate in different types of soils. The biodegradability rate was estimated as CO_2 release from the materials. Shuterl et al. [43] declared that soil microorganisms naturally present were not able to biodegrade the conventional polymers.

Islam et al. [44] tried the study of polyacrylamide granules subjected to moderate and deficient irrigation conditions. SAPs can increase the irrigation's efficiency by more than 8.1% in cases of moderate conditions of drought, and 15.6% under the severe condition of drought.

It is the oxygen radical created during drought causes problems for the plants for its carbon fixation. This can be prevented by the presence of granular polyacrylamide as it can reduce the antioxidant activity of those enzymes.

The plant's growth rate was found proper in presence of SAP made of polyacrylamide even though the irrigation conditions are lower. The soils which are of sandy type have less water holding capacity. These SAPs are very well performing in the sandy soils even in disordered climactic conditions [45]. Johnson et al. [46] tried different cross-linked polyacrylamides along with sand to prepare a normal polymer material.

Hydrogels can make many changes to the soil properties like permeability, density, structure, texture, evaporation, and infiltration rates of water. Some of the properties are promoted like aeration and microbial activity, while irrigation frequency, compaction tendency, run-offs, etc. decreases.

In the case of water scarcity near the roots of the plants, there will be water stress experiences which will lead to leaf shedding and decreasing chlorophyll content. This also leads to lower yield of seeds, lower amount of fruits, and plant flower yield also diminishes. When proper hydrogels are used all such problems can be solved due to the improper irrigation. Hydrogels are water-retaining agents, can increase the cultivation period and can help in the increasing efficiency of irrigation in areas of arid and less arid regions.

Sharma et al. [47] studied the polyaspartic acid-based interpenetrating network type polymers which can biodegrade also. This is the case where the cross-linking can be improved using biochemicals in place of conventional cross-linking. This is materialized by the usage of charged functional groups which are hydrophilic in nature. These materials are further suitable for more bioabsorbancy and biodegradability. These materials are modified by hydrophilic functional groups and biochemical cross-linking also. Cellulose-based SAPs were used by Cannazza et al. instead of acrylic SAPs as they are more environmentally compatible. These cellulose-based materials are prepared from sodium salt of carboxymethyl cellulose and hydroxyl methylcellulose. The efficiency of swelling, water retention, and soil conditioning for the samples are similar to acrylate-based hydrogels. Biopolymeric materials like CS, alginate, starch, and other polysaccharides are studied with interpenetrating polymer networks (IPNs) by Dragan [49]. Chauhan et al. [46, 49] illustrate the applications of IPN-based biopolymer based on poly(methacrylamide) used for metal ion sorption from solutions. The cellulose components in the polymer network drastically increased the sorption capacity.

4.6 CHARACTERISTICS OF HYDROGEL FOR APPLICATIONS IN AGRICULTURE

Hydrogels have very special characteristics and can be used in agriculture. It can be listed as follows, lowest soluble content, low unreacted monomer content, lower price, higher durability, stability in the swelling environment, stability while in storage, steady, and gradual biodegradability without formation of any toxic components, pH neutrality after swelling in water, photostability, and rewetting capability. Some of them are illustrated below.

4.6.1 LESS AFFECTED BY SALTS

It can absorb water more than 400 times its dry weight and gradually release it and stable in soil for a minimum period of one year. It has low release rates in soil application for nursery horticultural crops. In cases of field crops, it reduces the leaching of herbicides and fertilizers. It can support the plants to resist moisture stress.

It also enhances the physical properties of soils. Many of the plant growing properties are improved by hydrogels like seed germination, seedling emergence rate, root growth, and density. It reduces the nursery establishment period, irrigation, and fertigation requirements of crops.

Hydrogel gives the advantages that it can make plants flowering in a dense manner and more fruiting on them. It also promotes tillering so that it can control more weeds. It can prevent plants from wilting earlier and enhances the growth of roots and thus water efficiency increases also the usage of nutrient in efficient dosages.

Abdurhman et al. [51] found that the swelling capacity of hydrogels in various salt solutions is affected by the reaction variables. The presence and

Characteristics and Prospects of Controlled Release

amount of salt in the medium will affect the values and rate of swelling. Presence of salt decreases the swelling capacity of gel prepared from cellulose [52]. There can be decrease in swelling of hydrogels which can control the conditions of hydrogel and can be favorable for the soil.

Liang et al. [53] studied the poly(acrylic acid co-acrylamide)/kaolin clay superabsorbent composite for the NPK adsorption by the entrapment onto the hydrogel method. Further by the addition of clay, cost can be reduced and they tried the details of its capacity of swelling in aqueous medium. The study shows that the solubility increases with temperature and the nutrient release is faster.

4.6.2 PHYSICAL–CHEMICAL CHARACTERISTICS

The most important properties of the structure of the hydrogel depend on the external conditions.

They are mainly characterized by the polymer–water interaction parameter. This will also be influenced by density and the presence of ionic groups and dissociation energy. There are other factors influencing the process of hydrogels like osmotic forces and elasticity. There is a zero swelling pressure for the hydrogel which the hydrogel gets at a point of equilibrium swelling [54]. Swelling is studied by the technique of geometric dimensions measured by optical methods. The samples are measured visually to get the correct cylinders and spheres. This is also tested for liquid remain after absorption by the registered volume [55].

4.6.3 WATER ABSORPTION WITH HYDROGEL

In the presence of water, hydrogels can expand to about 200–800 times its original hydrogel volume. Water reservoirs round roots are created by Hydrogels. It can be done by collecting rainwater and water supply for irrigation and allow to release slowly toward the crops. The process can be extended for a long time. The germinations of seed in the soil can be improved and also the permeability of the soil. The hydrogels are found to be compatible in all types of soils and enhance the performance. The plants and trees can be saved from loss by the attacks of insects when hydrogels are used. Hydrogels can thus be useful in retention of rainwater and prevent soil erosion by external means on sloped plains.

4.6.4 DROUGHT STRESS REDUCTION

Oxygen radicals may be created when there is drought stress which can increase lipid peroxidation. Lowering of leaf area, height, and matrix damages can be possible by that. Obviously, hydrogel can reduce the impact of drought on plants leading to lower stress and the formation of oxygen radicals. In the presence of hydrogels, there is a better scope for plant growth and greater productivity in case of bad climates also.

4.6.5 ENHANCED FERTILIZER EFFICIENCY

Agriculture and irrigation technology has lots of limitations in fertilizer herbicides and germicides application. The usage of synthetic fertilizers can be reduced when hydrogels are practiced. It is a good practice to use an appropriate agricultural process with all types of soil and all types of weather even though there are ecological constraints. There will be more safe and nontoxic methods which can prevent ecosystems pollution with the usage of potassium polyacrylate like materials.

4.6.6 POLYMERS AND BIODEGRADABILITY

Hydrogel can be decomposed in the soil by chemical or microbiological routes. Chemical steps may take place followed by microbiological breakdown [56]. As the chemical processes end quickly and the biological processes occur slowly hence it controls the decomposition rate. There have been great efforts that are going on over the past decade focusing on plastics disposal. Biopolymers have been used in innumerable applications with many studies for the improvement of their ultimate disposability. Biodegradation and biorecycling are attractive methods for the disposal of products as waste polymer incineration causes pollution. If the material degrades within a time of 3 years using natural biological processes into nontoxic carbonaceous soil, water, carbon dioxide, or methane it is to be certified as biodegradable. Such materials are the ability to degrade in the most common environments and are environmentally safe. The degradation process can be classified into many types as per their rate of degradation process.

Materials may be partially biodegradable if there is a minimal transformation in the physical nature of a material and leaving the molecule largely intact. Loss of any specific property of that substance which is due to the alteration of

Characteristics and Prospects of Controlled Release 85

a chemical structure and biological action may take place. It is not the required property as the intermediate chemicals formed can be more toxic than the original substrate. In this case, mineralization is the preferred method. The original products degrade into carbon dioxide and water in complete biodegradation. The molecular cleavage will be larger so that it may remove biological, toxicological, chemical, and physical properties of the material.

Readily biodegradable is a group of chemicals that have passed certain specified screening tests for ultimate biodegradability. Very stringent tests are conducted on compounds so that it will rapidly and completely biodegrade in aquatic environments under aerobic conditions. Inherently biodegradable materials are means a group of chemicals in which there is greater proof of biodegradation in any test of biodegradability.

Han et al. [57] studied the variations of morphology films and disintegration time. It is then recorded as a test for biodegradability. Studies have confirmed that hydrocarbons are very sensitive to the action of UV rays and degrade into oligomers containing smaller molecular weights. The polyacrylate can degrade at rates of 10%–15% per year into water convert into carbon dioxide and nitrogen-based compounds. The hydrogel molecules are too bigger in size to be absorbed into plant tissue and have zero bioaccumulation property [58]. Table 4.2 describes the dosages of different types of hydrogels in various types of soil.

TABLE 4.2   Application Rate of Hydrogel in soil [59] (Copyright © 2005, American Chemical Society)

Soil Type and Conditions	Hydrogel Dosage
Arid and semiarid regions	4–6 g/kg
Soil with all types water stresses and better irrigation period	2.25–3 g/kg
To delay permanent wilting point in sandy soil	0.2–0.4 g/kg or 0.8% of soil whichever is more
The irrigation water to reduce by 50% in loamy soil	2–4 g/kg
Water content and leaf water efficiency improved	0.5–2.0 g/pot
To reduce drought stress	0.2–0.4%
To prohibit drought stress totally	225–300 kg/ha
To decrease water stress	Cultivated area 3% by weight

4.7   CONCLUSIONS

Well-drained soils and conducive atmosphere is required for the better growth of roots and ultimately increases yield. Hydrogels can be used for improving soil properties, increasing water holding capacity of the soil, enhancement of water retention of the soil, enhancing the efficiency of irrigation, increasing the growth, and enhancement water intake of the crop. In agriculture, there are large numbers of application of polymers which can enhance productivity. The use of polymers in agriculture which are compatible environmentally, socially, and economically are sustainable practices. Biopolymers which are environmental friendly give greater biodegradability in the soil. The hydrogels and super absorbing polymers should maintain the required amount of swelling and should control the release of agriculture to the soil. These materials can make drastic changes to the agriculture technology and can have greater crop yield without making any damages with toxic chemicals.

KEYWORDS

- **hydrogels**
- **controlled release**
- **modified fertilizers**

REFERENCES

1. Ekebafe, L. O. D; Ogbeifun, E; Okieimen, F. E. Biochemistry, Preparation, Characterization, Properties' Evaluation and Agricultural Application of Starch Graft Copolymers. *Polym. Appl. Agric.*, **2011**, *23*(2), 81–89.
2. Yazdani, F; Allahdadi, I; AbasAkbari, G. Impact of Superabsorbent Polymer on Yield and Growth Analysis of Soybean (Glycine max L.) under Drought Stress Condition. *Pakistan J. Bio. Sci.* **2007**, *10*(23), 4190–4196.
3. Bhat, N.R; Suleiman, M.K; Abdal, M. Selection of Crops for Sustainable Utilization of Land and Water Resources in Kuwait. *World J. Agric. Sci.* **2009**, *5*(2), 201–206.
4. Okorie, F.C. Studies on Drought in the Sub-Saharan Region of Nigeria using Satellite Remote Sensing and Precipitation Data; Department of Geography of Lagos. Nigeria, 2003. *Atm. Climate Sci.* **2014**, *4*(4).
5. Kusum, K; RamBabu, S; Shweta, A. Natural Polymers and their Applications. *Int. J. Pharm. Sci. Rev. Res.* **2016**, *37*(2), Article No. 05, 30–36.
6. Girish, K; Jani, D. P; Shah, V; Prajapati, D; Vineet, C. J. Gums and Mucilages: Versatile Excipients for Pharmaceutical Formulations. *Asian J. Pharm. Sci.* **2009**, *4*(5), 309–332.
7. Nishiyama, Y; Langan, P; Chanzy, H. Crystal Structure and Hydrogen-Bonding System in Cellulose Iβ from Synchrotron X-ray and Neutron Fiber Diffraction. *J. Am. Chem. Soc.* **2002**, *124*(31), 9074–9082.

8. Khathuriya, R; Nayyar, T; Sabharwal, S; Jain, U. K; Taneja, R. *Int. J Pharm. Sci. Res.* **2015**, *6*(12), 4904–4919.
9. Milhou, A. P; Michaelkis, A; Krokos, F. D; Mazomenos, B. E; Couladouros, E. A. Controlled Pesticide Release from Biodegradable Polymers. *J. Appl. Entomol.* **2007**, *131*, 128.
10. Kharine, S.; Manohar, R., Kolkman, R. G. M.; Bolt, R. A; Steenberger, W.; Demull, F. F. Controlled Pesticide Release from Biodegradable Polymers. *Phys. Med. Biol.* **2003**, *48*, 357.
11. Peppas, N. A; Leobandung, W. Stimulus-Responsive Hydrogels: Theory, Modern Advances, and Applications. *J. Biomater. Sci. Polym.* **2004**, *15*, 125.
12. Kenawy, E.E; Sakran, M. Biologically Active Polymers. IV. Synthesis and Antimicrobial Activity of Polymers Containing 8-Hydroxyquinoline Moiety. *J. Appl. Polym. Sci.* **2001**, *82*, 1364–1374.
13. Davies, M. J. Genomics joint venture 1, *Trends Biotechnol.* **2001**, *19*, 489.
14. Tashima, S; Shimada. S; Ando, I; Matsumoto, K; Takeda, R;. Shiraishi, T. Effect of Physicochemical Properties in Water on Release. *J. Pest Sci.* **2000**, *25*, 128.
15. Liu, J; Lin, S; Liu, E; Li, L. Release of the Ophylline from Polymer Blend Hydrogels. *Int. J. Pharmaceut.* **2005**, *298*, 117.
16. Vidal, M. M; Filipe, O. M. S; Cruz Costa, M. C. Reducing the Use of Agrochemicals: A Simple Experiment. *J. Chem. Educ.* **2006**, *83*, 245.
17. Kulkarni, A. R; Soppimath, K. S; Aminabhavi, T. M; Dave, A. M; Mehta, M. H. Glutaraldehyde Crosslinked Sodium Alginate Beads Containing Liquid Pesticide for Soil Application. *J. Control. Rel.* **2000**, *63*, 97.
18. França, D.; Medina, Â. F; Messa, L. L.; Souza, C. F.; Faez, R. Chitosan Spray-Dried Microcapsule and Microsphere as Fertilizer Host for Swellable-Controlled Release Materials. *Carb. Polym.* **2018**, *196*, 47–55.
19. Milani, P.; Franca, D.; Balieiro, A. G.; Faez, R. Polymers and its Application in Agriculture. *Polimeros-Ciencia e Tecnologia.* **2017**, *1*, 1.
20. Supaporn, N; Natthaya P; Suda, K. Multilayer-Coated NPK Compound Fertilizer Hydrogel with Controlled Nutrient Release and Water Absorbency. *J. Appl. Polym. Sci.* **2015**, *132*(2), 41249.
21. Ahmed, M. E; Ghobashy M. O. Controlled Release Fertilizers using Superabsorbent Hydrogel Prepared by Gamma Radiation. *Radiochem. Acta.* **2017**, *105*(10), 865–876.
22. Souza, J. L.; Chiaregato, C.; GFaez, R. Green Composite Based on PHB and Montmorillonite for KNO_3 and NPK Delivery System. *J. Polym. Environ.* **2017**, *1*, 1–10.
23. Rodrigues, B.; Bacalhau, F. B.; Pereira, T. S.; Souza, C. F.; Faez, R.; Chitosan-Montmorillonite Microspheres: A Sustainable Fertilizer Delivery System. *Carbo. Polym.* **2015**, *127*, 340–346.
24. Ding, H; Zhang, Y. S; Li, W. H; Zheng, X. Z; Wang, M. K; Tang, L. N; Chen, D. L. *Appl. Environ. Soil Sci.* **2016**, Article ID 2013463, 13 pages.
25. Trenkel, M. E.; IFI Association. *Inter. Fert. Ind. Asso.* Paris, France, **1997**, 11.
26. Siafu, I.S; Hee, T. K; Egid, M; Askwar, H. Meticulous Overview on the Controlled Release Fertilizers. *Adv. Chem.* **2014**, Article ID 363071, 16 pages.
27. Gill, P; Moghadam, T. T; Ranjbar, B. *J. Biomol. Tech.* **2010**, *21*(4), 167–193.
28. Kreye, C; Bouman, B. A. M; Castaneda, A. R; Lampayan, R. M; Faronilo, J. E; Lactaoen, A. T; Fernandez, L. Possible Causes of Yield Failure in Tropical Aerobic Rice. *Field Crops Res.* **2009**, *111*, 197–206.

29. Shibayama, M; Tanaka, T. Volume Phase Transition and Related Phenomena of Polymer Gels. *Adv. Polym. Sci.* **1993**, *109*, 1–62.
30. Yangyuoru, M; Boateng, E; Adiku, S. G. K; Acquah, D; Adjadeh, T. A; Mawunya, F. Effects of Natural and Synthetic Soil Conditioners on Soil Moisture Retention and Maize Yield. *J. Appl. Ecol.* **2006**, *9*, 91–98.
31. Dar, S.B; Mishra, D; Zahida, R. Bull Impact of Hydrogel Polymer in Agricultural Sector. *Environ. Pharm. Life Sci.* **2017**, *6*(10), 129–135.
32. Dehkordi, K. D. Effect of Superabsorbent Polymer on Salt and Drought Resistance of Eucalyptus Globulus. *Appl. Ecol. Environ. Res.* **2017**, *15*(4), 1791–1802.
33. Erickson, R. E. Washington: U.S. Patent and Trademark Office, 1984, US Patent No. 4.424.247.
34. Johnson, M. S. The Effects of Gel-Forming Polyacrylamides on Moisture Storage in Sandy Soils. *J. Sci. Food Agric.* **1984**, *35*, 1063–1066.
35. Bowman, D. C; Evans, R. Y. Calcium Inhibition of Polyacrylamide Gel Hydration is Partially Reversible by Potassium. *Hort. Sci.* **1991**, *26*(8), 1063–1065.
36. Shahid, B. D; Dheeraj M; Zahida. R; Afshana B. B. Hydrogel: To Enhance Crop Productivity Per Unit Available Water Under Moisture Stress Agriculture. *BEPLS.* **2017**, *6*, 131.
37. Schmaljohann, D. The Impact of Water and Some Salt Solutions on Some Properties of Hydrophilic Acrylamide Copolymeric Hydrogels. *J. Mater. Sci.* **2006**, *58*(15), 1655–1670.
38. Chiu, H. C; Wu, A.T.; Lin, Y. F. Synthesis and Characterization of Acrylic Acid Containing Dextrane Release. Polymer. **2001**, *42*(4), 1471–1479.
39. Singh, B; Chauhan, G. S; Sharma, D. K; Chauhan, N. Synthesis and Characterization of Chemically Cross-Linked Acrylic Acid/Gelatin Hydrogels: Effect of pH and Composition on Swelling and Drug Release. *Int. J. Polym. Sci.*, **2007**, *67*(4), 559–565.
40. Callaghan, T. V; Abdelnour, H; Lindley, D. K. The Environmental Crisis in the Sudan: the Effect of Water-Absorbing Synthetic Polymers on Tree Germination and Early Survival. *J. Arid Environ.* **1988**, *14*(3), 301–317.
41. Woodhouse, J; Johnson, M. S. Effect of Superabsorbent Polymers on Survival and Growth of Crop Seedlings. *Agric. Water Manag.* **1991**, *20*(1), 63–70.
42. Stahl, J. D; Cameron, M. D; Haselbach, J; Aust, S. D. Biodegradation of Superabsorbent Polymers in Soil. *Environ. Sci. Pollut. Res. Int.* **2000**, *7*(2), 83–88.
43. Sutherland, G. R; Haselbach, J; Aust, S. D. Biodegradation of Crosslinked Acrylic Polymers by a White-Rot Fungus. *Environ. Sci. Pollut. Res. Inten.* **1997**, *4*(1), 16–20.
44. Islam, M. R; Xue, X; Mao, S; Ren, C; Eneji, A. E; Hu, Y. Effects of Water-Saving Superabsorbent Polymer on Antioxidant Enzyme Activities and Lipid Peroxidation in Oat (Avena sativa L.) under Drought Stress. *J. Sci. Food Agric.* **2011**, *91*(4), 680–686.
45. Ekebafe, L. O; Ogbeifun, D. E; Okieimen, F. E. Polymers and its Applications in Agriculture. *Am. J. Polym. Sci.* **2011**, *1*, 6–11.
46. Johnson, M.S. Effect of Soluble Salts on Water Absorption by Gel Forming Soil Conditioners. *J. Sci. Food Agric.* **1984**, *35*, 1196–1200.
47. Sharma, S; Dua, A; Malik, A. Polyaspartic Acid Based Superabsorbent Polymers. *Euro. Polym. J.* **2014**, *59*, 363–376.
48. Cannazza, G; Cataldo, A; De Benedetto, E; Demitri, C; Madaghiele, M; Sannino, A. Experimental Assessment of the Use of a Novel Superabsorbent Polymer (SAP) for the Optimization of Water Consumption in Agricultural Irrigation Process. *Water.* **2014**, *6*(7), 2056–2069.

49. Dragan, E. S. Design and Applications of Interpenetrating Polymer Network Hydrogels. A Review. *Chem. Eng. J.* **2014**, *243*, 572–590.
50. Chauhan, G. S; Mahajan, S. Use of Novel Hydrogels Based on Modified Cellulosics and Methacrylamide for Separation of Metal Ions from Water Systems. *J. Appl. Polym. Sci.* **2002**, *86*(3), 667–671.
51. Abdurhman, A; Abuabdalla, K; Sh-Hoob, E. M. *J. Mater. Sci.* **2017**, *1*, 7–15.
52. Grignon, J; Scallan, A.M. Effect of pH and Neutral Salts upon the Swelling of Cellulose Gels. *J. Appl. Polym. Sci.* **1980**, *25* (12), 2829–2843.
53. Liang, R; Liu, M; Wu, L. Controlled Release NPK Compound Fertilizer with the Function of Water Retention. *React. Funct. Polym.* **2007**, *67*(9), 769–779.
54. Chen, S. L; Zommorodi, M; Fritz, E; Wang, S; Huttermann, A. Using hydrogel and clay to improve the water status of seedlings for dryland restoration. *Struct. Funct.* **2004**, *18*, 175–183.
55. Geesing, D; Schmidhalter, U. Influence of Sodium Polyacrylate on the Water-Holding Capacity of Three Different Soils and Effects on Growth of Wheat. *Soil Use. Manag.* **2004**, *20*, 207–209.
56. Henderson, J. C; Hensley, D. L. Ammonium and Nitrate Retention by a Hydrophilic Gel. *Hort. Sci.* **1985**, *20*, 667–668.
57. Han, X; Chen, S; Hu, X. Controlled-Release Fertilizer Encapsulated by Starch/Polyvinyl Alcohol Coating. *Desalination.* **2009**, *240*(1–3), 21–26.
58. Neethu, T. M; Dubey, P.K; Kaswala, A. R. Prospects and Applications of Hydrogel Technology in Agriculture. *Inter. J. Curr. Microb. Appl. Sci.* **2018**, *7*(05), 1–9.
59. Zhan, F; Liu, M; Guo, M; Wu, L. Preparation of Superabsorbent Polymer with Slow-Release Phosphate Fertilizer. *J. Appl. Polym. Sci.* **2004**, *92*(5), 3417–3421.

CHAPTER 5

Chitosan Biopolymer for 3D Printing: A Comprehensive Review

DHILEEP KUMAR JAYASHANKAR[1], SACHIN SEAN GUPTA[1], RAJKUMAR VELU[2*], and ARUNKUMAR JAYAKUMAR[3]

[1]*Digital Manufacturing and Design Centre, Singapore University of Technology and Design, Singapore 487372*

[2]*Department of Mechanical Engineering, Indian Institute of Technology Jammu, J&K 1818221, India*

[3]*Department of Automobile Engineering, SRM Institute of Science and Technology, SRM Nagar, Kattankulathur 603 203, Kanchipuram, Tamil Nadu, Chennai, India*

*Corresponding author. E-mail: rajkumar7.v@gmail.com.

ABSTRACT

Additive manufacturing (AM) has the potential to completely redefine manufacturing in several areas. Manufacturers are now looking to 3D printing as a complement to existing and traditional manufacturing methods, particularly in the fabrication of biopolymer composites used for medical applications such as tissue engineering and artificial organ development. In light of the need for suitable materials to apply AM technologies to medicine, this chapter covers the benefits and methodology of using chitosan-based biopolymer and its composites with AM techniques. The fundamental extraction of chitin, and its derivative chitosan, as well as the latter's chemical, biological, and mechanical properties are reviewed and related to challenges with conventional manufacturing techniques such as compression molding, solution casting, and injection molding. Based on the critical review of the medical manufacturing literature, it is clear that increased functional requirements and complex geometries are required to produce higher-performing parts. Conventional techniques have many limitations towards

92 *Natural Polymers: Perspectives and Applications for a Green Approach*

these goals and thus the research trend to adopt 3D printing for chitosan-based biopolymer fabrication. This chapter should enable the reader to identify and relate the utilization of chitosan biopolymer in additive manufacturing and its process parameters for different applications, especially for largescale structural chitosan 3D printing and cell/organ 3D bioprinting.

5.1 INTRODUCTION

5.1.1 3D PRINTING

3D printing, or additive manufacturing (AM), in its broadest sense, is the building of a consolidated three-dimensional part in a layer-by-layer sequence with the use of computer-aided design [1]. 3D printing has rapidly developed over the past few decades and is a promising and emerging manufacturing technique for parts building. Furthermore, because 3D printing is an additive process that enables the placement of material only where it is required, material wastage is minimized; this contributes to sustainable manufacturing. According to recent Wohler's report on 3D printing [2], the revenue of 3D printing worldwide has significantly improved: in 2013, the total revenue was $3.07 billion, followed by $5.02 billion in 2015. Within three years after that, it increased to $12 billion and is expected to achieve $21 billion in 2020, as shown in Figure 5.1. This economic significance can be attributed to the fact that 3D printing is a flexible, powerful tool to develop and produce high-complexity part configurations and design geometries. It allows the user to take full control of design and manufacturability, with precision modeling to minimize zero error accommodation in the printed part. The first 3D printing technology, stereolithography, was invented by Charles Hull in 1987 [3]. However, it did not gain popularity until the end of the first decade of the 21st century, being previously inhibited by hurdles related to manufacturing cost, surface finish, and limited material availabilities. At that point of time, scientists, engineers, and industrialists recognized 3D printing as the next revolutionary manufacturing method with its untapped potential to uncover a variety of ways to produce a part; since then, the number of 3D printer manufacturers has rapidly increased. 3D printing has had an immense impact on society by opening new possibilities for the amateur user, empowering him or her to make a prototype with basic engineering insight and knowledge. Desktop 3D printers can be operated by anyone with any level of skill: the process of printing is automated, and the machine and its software are designed to be easy to operate. Furthermore, most desktop 3D printers do not require post-processing, meaning that users can directly use the parts that have been built by the machine. This democratization of technology is having profound consequences on the rate of discovery of new techniques, methods, and processes of making. This also applies to industrial and research contexts, where 3D printing provides geometric freedom and enables the creation of complex tooling and molds for various processing chains, such as automated fiber placement process on 3D printed molds [4, 5].

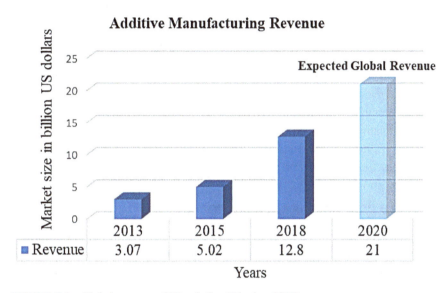

FIGURE 5.1 Global revenue of 3D printing (Kianian, 2017).

Different kinds of established 3D printing technologies are available today to produce a part with predefined purposes and requirements. The various technologies in AM are vat photopolymerization, powder bed fusion, material extrusion, material jetting, binder jetting, sheet lamination, and direct energy deposition method [6]. Every method has a unique way of producing a part, with the selected parameters mainly depending on the application or requirements for strength, surface finish, appearance and aesthetics, thermal resistance, and exposure to sunlight. Polymer printing technology especially is at the forefront of 3D printing research due to the diversity of polymers available. Polymer-based AM is typically realized as the technologies of vat polymerization, powder bed fusion, material extrusion, material jetting, binder jetting, and sheet lamination; therefore, direct energy deposition is currently the only 3D printing technology that is not applicable to polymers. Selection of the 3D printing technology is guided by the properties of the polymer printing material in question. For instance, thermoplastic materials are used in 3D printing technologies that depend on physical changes of the printing material, such as crystallization, glass transition, and drying, to facilitate processing and solidification. This generally includes processes of material extrusion such as fused deposition modeling (FDM), powder bed fusion such as selective laser sintering (SLS) [7], as well as various material jetting, binder jetting, and sheet lamination processes. Elastomers are typically mixed with thermoplastics in order to make them conducive for 3D printing, and therefore are used for selecting 3D printing processes such as FDM or SLS. On the other hand, thermoset materials, due to their irreversible phase change, are utilized in 3D printing technologies that depend on chemical reactions of the structure of the polymer. These include a wide range of vat photopolymerization, material extrusion, material jetting, binder jetting, and sheet lamination processes. Finally, the selection of an appropriate 3D printing technology for a class of polymer called natural polymers is a hot topic in the field that will be discussed below.

5.1.2 3D PRINTING OF NATURAL POLYMERS

Natural polymers are age-old inventions from nature and are categorized as polysaccharides, polypeptides, and rubbers. [8]. Each category has been broadly classified based on different sources as shown in Figure 5.2 [9]. Polypeptide-based natural polymers include functional materials like proteins, enzymes, and silk. Polysaccharide-based natural polymers are represented by structural polymers such as cellulose and chitin and storage materials like carbohydrates, DNA, and RNA. Cross-linked natural rubbers form the third category.

The breadth and depth of research in these natural polymers and the increasing adoption of 3D printing has led to the inevitable crossover of the two fields that have resulted in the application of natural polymers to 3D printing technologies. Interest and investment in this emerging field has a strong justification. Due to raised awareness of the ecological impacts of traditional manufacturing, it is logical to combine the waste minimization of 3D printing with green, nature-based polymers for environmentally conscious parts production. Natural polymers are highly abundant and stem from renewable resources, which potentially provide a durable feedstock for 3D printing. In addition, natural polymers are biocompatible and biodegradable, which makes them particularly beneficial for 3D bioprinting applications [10, 11]. However, the range of 3D printing processes applicable to natural polymers is currently limited, as seen by the challenges posed in Figure 5.3. Most natural polymers are neither photopolymerizable nor readily sinterable in their original form, barring them from several 3D printing processes suitable for synthetic polymers unless they undergo chemical modification. Furthermore, natural polymers can degrade in conditions like high-temperature characteristic of extrusion-based synthetic polymer printing processes such as FDM. In these cases, the natural polymer is usually limited to acting as particle or fiber reinforcement for a synthetic polymer matrix. However, in the creation of composite printing materials, the combination of natural polymers with synthetic materials can cause bonding or compatibility issues that must be resolved [12]. Despite of all these challenges, there exists a growing collection of research endeavors that have successfully integrated natural polymers into 3D printing processes and resulted in noteworthy advancements.

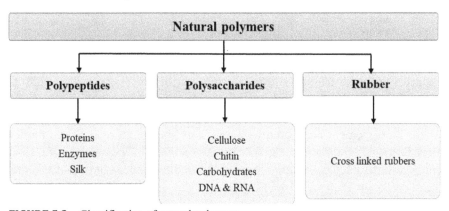

FIGURE 5.2 Classification of natural polymers.

A common usage of natural polymers in 3D printing is the replacement of synthetic fibers with natural fibers as a reinforcement phase in a thermoplastic matrix; such composites are termed as natural fiber-reinforced composites. Polysaccharide-based plant fibers offer better mechanical characteristics compared to polypeptide-based animal fibers [12]. While these properties are not as stellar as their synthetic counterparts like carbon fiber, glass fiber, and aramid fibers, plant fibers feature less embodied energy, biodegradability, renewability, extreme lightweight, and good impact resistance [13, 14]. A small selection of these natural fibers has been included in FDM processes as a reinforcement material in continuous form. For example, [15] have disclosed successful printing of continuous natural jute fibers in a polylactic acid (PLA) thermoplastic matrix by modifying a 3D printer head to accommodate both PLA filament and bundles of the jute fiber. They

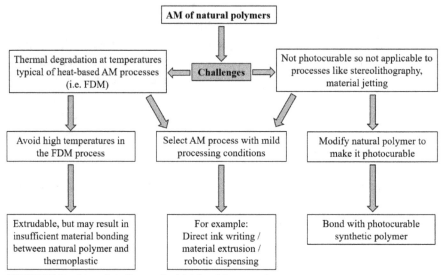

FIGURE 5.3 List of challenges in AM for chitosan polymers.

report a tensile modulus and tensile strength of 5.11 GPa and 57.1 MPa, respectively, for the jute-fiber-reinforced-plastic, corresponding to 157% and 134% of those shown by the PLA specimen; however, this is only a slight improvement compared to the carbon-fiber-reinforced PLA samples they tested (19.5 GPa and 185.2 MPa, respectively). They also reveal fiber pull-out during tensile testing, arguing that adhesion between the fibers and PLA could be improved, possibly due to the fact the fibers could not be heated along with the PLA to promote fusing in the printing nozzle. It is important to note here that while PLA is a bio-based thermoplastic, it cannot be considered a true natural polymer, as it is derived within a process chain from natural resources (i.e., starch) [16] rather than itself being naturally occurring. Milosevic et al. have accomplished the FDM printing of hemp- and harakeke-short-fiber-reinforced polypropylene composites by opting to prefabricate printing filaments through a twin-screw extrusion system. They measured the tensile properties of dog-bone samples printed using a low-temperature FDM process, finding that the harakeke-fiber-printed samples increased tensile strength by 49% but hemp-fiber-printed samples possessed an 18% reduction in tensile strength, compared to neat polypropylene. This is likely due to poor interlayer fusion in the nozzle during extrusion. These highlighted research efforts have circumvented the vulnerability of natural polymers to high temperatures to allow these fibers to be integrated in an FDM process, but as a result, natural fiber-reinforced thermoplastic composites clearly feature interfacial bonding issues.

Chitosan Biopolymer for 3D Printing: A Comprehensive Review

These issues of material compatibility in natural polymer composites could theoretically be resolved if it were possible to utilize a natural polymer matrix in conjunction with the natural fiber reinforcement. However, the susceptibility of natural polymers to high temperatures means that the printing of natural polymer composites could overall be better suited to 3D technique with milder processing conditions that do not require filament melting and subsequent solidification. One such process is direct ink writing (DIW), also known as material extrusion or robotic dispensing or deposition [17]. DIW involves the extrusion of viscous materials through nozzles and often depends on the rheological property of shear thinning, rather than resolidification, for the extruded filament to maintain its shape. In the realm of 3D bioprinting, tissue engineering is one field that makes use of DIW processes, as scaffolds for cell growth are required that feature soft material composition and complex morphology [18]. In this field, DIW is sometimes referred to as 3D plotting [19] and it extensively utilizes natural polymers to create highly-articulated biocompatible scaffolds for cell culture. This is because natural polymers, in addition to their shear-thinning abilities, can be very similar to macromolecular substances that can be perceived and metabolized in a biological environment, whereas synthetic polymers may either not be recognized by cells or be toxic to them [20]. Collagen is a polypeptide-based natural polymer that is one component of biological extracellular matrices and therefore has been investigated as a viable 3D printing material for artificial scaffold printing [10]. Kim et al. [20] have reported the 3D plotting of 95%-porosity collagen scaffolds for skin tissue repair with a maximum tensile strength of 2.8 MPa. Using this natural polymer scaffold, they achieved a keratinocyte/fibroblast co-culture that showed good cell migration and well-differentiated keratinocytes akin to human skin. A literature overview indicates that DIW/3D plotting clearly is the familiar method for tissue engineering applications, backed by a repertoire of successful endeavors [17].

However, research in the tissue engineering field is also underway to modify natural polymers to make them a feasible candidate for alternative AM processes. For example, gelatin, a hydrolyzed form of collagen, is another popular natural polymer that is similarly used for tissue engineering applications due to its capabilities in biocompatible hydrogel production and usefulness in vessels for controlled drug release [10]. Gauvin et al. [21] have managed to print gelatin methacrylate hydrogels scaffolds using a stereolithography vat photopolymerization process. The addition of methacrylamide moieties to side groups of gelatins made this type of printing process applicable to gelatin, with the result being the production of gelatin tissue scaffolds with complex architecture. Endothelial cells on the scaffolds grew with high density and excellent proliferation, indicating superior biological performance of this scaffold due to its composition and additively manufactured morphology. In addition to modification of the natural polymer, customization of the 3D printing process in question can enable the successful printing of natural polymers. Kim et al. [20] and Lam et al. [22] reported the binder jetting of mixtures of cornstarch, dextran, and gelatin powders to make porous composite scaffolds using only distilled water as a bonding agent. This specific use of water rendered this type of printing process applicable to natural polymers, eliminating the problems of a toxic fabrication environment and residual solvent. Lam and his team listed a compressive stiffness of 55.19 MPa and yield strength of 1.77 MPa for their starch-based scaffolds. Salmoria et al. [23] published research on the production of starch–cellulose and cellulose acetate scaffolds using direct SLS. In order to avoid thermal degradation and low mechanical strength of the natural polymers, the laser parameters of the printing process were optimized in terms of laser power and scan speed. The maximum elastic modulus and ultimate strength of their scaffold samples were 193.8 MPa and 3.729 MPa, respectively, and the specimens with small particle size exhibited satisfactory degrees of porosity. Understandably, chemical modification and customized 3D printing processes, as seen in Table 5.1, are active research areas that try to overcome the inherent challenges of manufacturing with natural polymers.

5.1.3   CHITOSAN BIOPOLYMER: A NEW 3D PRINTING MATERIAL?

As described above, primarily cellulose- and collagen-based natural polymers have been systematically investigated and applied copiously for a various range of 3D printing applications. The proposed chapter converges on recent research trends on another polysaccharide called chitin, and most notably its derivative, chitosan, a material exhibiting tremendous potential in natural polymer 3D printing applications. Chitin is the world's second most abundant biopolymer after cellulose, and composes the structural features of insects, fungi, and mollusks (Dutta et al., 2004). Chitosan, owing to its unique biochemical properties such as biocompatibility, biodegradability, nontoxicity, ability to form films, etc., has already found use in many promising biomedical applications [24]. Furthermore, chitosan has been demonstrated as a potential replacement for synthetic plastics in everyday products given its effective combination of processability and natural structural properties [37]. In order to fully understand and determine the

Chitosan Biopolymer for 3D Printing: A Comprehensive Review

viability of chitosan as a material to be used in 3D printing, the extraction and properties of chitosan as well as conventional manufacturing processes that utilize this natural material will first be reviewed.

TABLE 5.1 Additive Manufacturing Research in Chitosan

	Target Application	Description
Areas of Active Research: Additive Manufacturing of Chitosan	Structural parts	Exploit the mechanical stability and processability of chitosan for use as a building block for sustainable large-scale structures relevant to industries like product design, construction, automotive, and aerospace.
		AM process: Material extrusion/robotic dispensing
	Tissue engineering scaffolds	Leverage the inherent combination of mechanical stability, biodegradability, and biocompatibility of chitosan for use as robust scaffolds for cell culture.
		AM process: Direct ink writing/3D plotting

5.2 DERIVATION AND PROPERTIES OF CHITOSAN

Chitosan is a highly hygroscopic deacetylated form of chitin, which is the second most ubiquitous natural biopolymer on earth after cellulose. Chitin is obtained from arthropods, fungi, and mollusks. Henry Braconnot, a French chemist, professor and pharmacist, who first discovered chitin as an alkaline-insoluble fraction from mushroom in 1811, named it after "fungine" (the ancient name of chitin) [25]. Antoine Odier, in 1823, isolated those alkaline-insoluble fractions from insects, treated them with sodium hydroxide solution, and identified them as chitine (derived from Greek word Chiton: coat-of-mail shells, marine molluscs). Later, the English word came into existence as chitin. In 1859, Rouget treated chitin with concentrated sodium hydroxide and found the resulting substance was soluble in organic acid and so named it as "modified chitin." At last in 1894, Hoppe-Seyler discovered a method to dissolve chitin in dilute acetic acid and hydrochloric acid by treating the molluscs, crab, and shrimp shells with 180 °C-heated sodium hydroxide solution. Hoppe-Seyler named the resultant product as "chitosan" as seen in Figure 5.4. Meanwhile, other scientists like Ledderhosein 1878 claimed that chitin is composed of glucosamine and acetic acid, which was later confirmed by Gilson in 1894. In the subsequent years, chitin and chitosan were used interchangeably amongst other scientists and both are known as copolymers of N-acetyl-D-glucosamine and D-glucosamine units linked with b-(1–4)-glycosidic bonds.

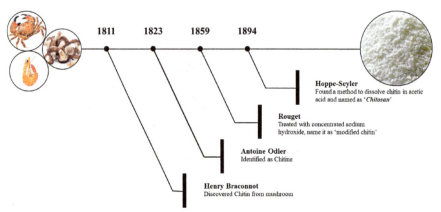

FIGURE 5.4 History of chitin and chitosan development (Khoushab & Yamabhai, 2010).

Between 1930 and 1940, the evolution of polymers had a major role in industrial revolution and evidently found more applications [26]. Due to the competition from synthetic polymers and lack of manufacturing facilities, natural polymers were suppressed for commercial development. In the late 1970s, with technological developments, scientists revealed keen interest on developing chitin and chitosan for research, industrial and commercial use. Nowadays, both the names of chitin and chitosan are generally categorized based on the degree of acetylation (DA, number of N-acetyl-D-glucosamine in the polymer) or degree of deacetylation (number of D-glucosamine in the polymer). Since the evolution of chitin and chitosan, they have been used all around the world, with more than 2000 numerous applications of chitin and its derivatives (Global Analysis).

5.2.1 EXTRACTION OF CHITIN AND CHITOSAN

Researchers identified that the fundamental sources of chitin are mostly from marine organisms like arthropod exoskeletons; the tendons and the linings of the respiratory, excretory and digestive systems of crustaceans and cephalopods; and other sources like mushroom and fungal mycelia [9]. Among all these sources, only shrimp and crab canning industries are considered two of the major sources for the extraction of chitin from waste [27]. Shrimp shells contain 30% chitin and are composed of complex network of proteins and calcium carbonate presented in the various layers. Steps involved in extraction of chitin are shown in Figure 5.5.

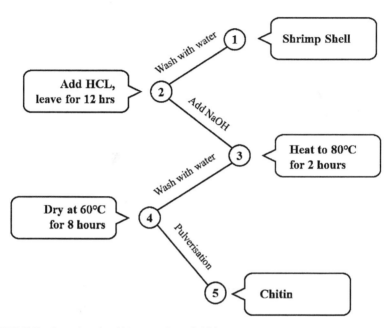

FIGURE 5.5 Steps involved in extraction of chitin.

The effective extraction of chitin involves the removal of proteins and inorganic calcium carbonate from shells by deproteinization and demineralization processes, respectively [28] but some pigments and lipids are also removed during these processes. The extracted chitins can further be biosynthesized and classified into polymorphic forms: α-chitin, β-chitin, and γ-chitin. An isolated chitin is a copolymer of 2-acetamido-2-deoxy-β-D-glucose and 2-amino-2-deoxy-β-D-glucose. As a point of distinction from other polysaccharides, chitin contains nitrogen. Chito-biose, O-(2-amino-2-deoxy-β-d-glucopyranosyl)-(1−4)-2-amino-2-deoxy-D-glucose, is the structural unit of local chitin. Thus, chitin is poly (β-(1-4)-N-acetyl-D-glucosamine) and can be dissolved up to 5% on the basis of dimethylacetamide containing 5%−9% LiCl (DMAc/LiCl) and N-methyl-2-pyrrolidinone/LiCl. It is expected that it may possess high concentration of polymer by mesomorphic property, based on its rigidness at room temperature.

Figure 5.6 represents the molecular formula, 2D and 3D chemical structure of chitin. Once the chitin is extracted from shrimp shells, an additional process is involved for producing chitosan from raw chitin. Moreover, chitin can also be directly incorporated for some end-use applications or it can be used to enhance the properties of the products by mixing it as an additive with other materials to attain the desired properties.

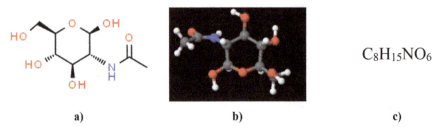

FIGURE 5.6 Molecular structure of chitin (a) 2D molecular structure (b) 3D molecular structure (c) Chemical formula [Image courtesy: Royal Society of Chemistry].

Chitosan is a deacylated form of chitin and at the basic elemental analysis has a degree of acetylation lower than 0.40 and nitrogen content higher than 7%. However, after the extraction of the chitosan from chitin with different processes, the availability in the market is typically based on low, medium, and high molecular weight and also called as practical grade for most general applications. In order to attain these types of chitosan, the extraction process requires the following steps. The extracted chitin from shrimp shells is processed by adding 40% of NaOH at 100 °C and kept in an agitator for 6 h–8 h to attain a complete chemical mixing process of NaOH and chitin. Once blended, NaOH is drained and rinsed several times to completely remove the acid from amalgam; furthermore to dry the amalgam at 60 °C for 8–10 h varying by quantity [27, 29]. Subsequently, the water particles are completely dried from the amalgam, and it is then pulverized and packed with different levels of molecular weight based on viscosity with 70%–85% deacetylated physical form for research purposes, as shown in Figure 5.7 and Table 5.2. The molecular formula, 2D and 3D chemical structure of chitosan is shown in Figure 5.8. Chitosan has the ability to bring the polymer into solution by salt formation with presence of number of 2-amino-2-deoxyglucose units. Also, it is a primary aliphatic amine with protonation constant pK as 6.3 that can be protonated by some selected acids like citrate, ascorbate, and acetate. Due to the presence of four essential elements: cationicity, amino group, derivatives of nitrogen, and capacity to form polyelectrolyte complexes, it is evident that chitosan is not considered as cellulose. The film-forming capacity of chitosan is another significant property that lacks with cellulose [30].

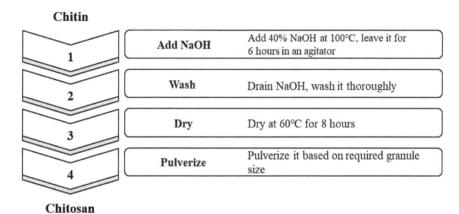

FIGURE 5.7 Production of chitosan from chitin (Islam et al., 2017), (Younes & Rinaudo, 2015).

TABLE 5.2 Availability of Chitosan with Different Molecular Weight [Sigma Aldrich]

Chitosan	Physical Form	Solubility	Molecular Weight
Low	75–85% deacetylated	Dilute aqueous: soluble	50,000–190,000 Da
Medium	75–85% deacetylated	Dilute aqueous: soluble	190,000–310,000 Da
High	≥75% deacetylated	H$_2$O, organic solvents: insoluble	310,000–375,000 Da
Practical grade	≥75% deacetylated	1 M acetic acid:10 mg/mL	190,000–375,000 Da

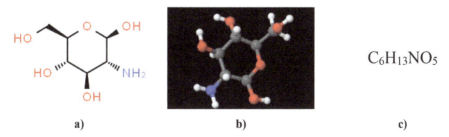

FIGURE 5.8 Molecular structure of chitosan (a) 2D molecular structure, (b) 3D molecular structure, and (c) chemical formula (image courtesy: Royal Society of Chemistry).

5.2.2 CHITOSAN PROPERTIES

Chitosan possesses inherent biological properties. To obtain biological and physicochemical properties such as solubility, crystallinity, dispersity, biocompatibility, biodegradability, etc., usually raw chitosan can be used. In most cases, the process of producing chitosan will have some variations; it is a challenging task to produce chitosan with same intensity (pigment), as manufacturers' methods may differ from one another. Due to the diversity of sources of natural chitin, the properties may vary in color, ratio of impurities, etc. The common differences between this chitosan are DA and molecular weight. Biodegradability and crystallinity are directly proportional to degree of *N*-acetylation and inversely proportional to molecular weight, whereas solubility, viscosity, and biocompatibility are inversely proportional to degree of *N*-acetylation. Other physicochemical properties for 3D printing of chitosan for engineering applications can be attained by doing different tests such as infrared spectra characterization, scanning electron microscopy analysis, differential scanning calorimetry (DSC) thermal analysis, X-ray diffraction study, thermogravimetric analysis, and thermally stimulated discharged currents study. As our focus also includes chitosan for 3D bioprinting, some unique properties are mentioned here, as these may apply for medical fields: antimicrobial activity, analgesic action, mucoadhesion, nontoxicity, cytocompatibility, anti-inflammatory action, hemostatic action, angiogenesis stimulation, granulation and scar formation, adsorption enhancer, and anticholesterolemic activity. Most of the biological properties are inversely proportional to degree of *N*-acetylation and molecular weight: antimicrobial, analgesic, antioxidant, hemostatic, and mucoadhesion, except antimicrobial. Yet, only few of them can be considered during material characterization based on different methods of 3D printing, for example, FDM for engineering applications, bioprinting, etc. Chitosan in its different forms can also be used as an additive or in mix with other materials for 3D printing. Some of the physical forms of chitosan are nanoparticles, microspheres, hydrogels, and fibers [31] that can directly be used in an AM process. Research on AM has increased all over the world and scientists have explored different 3D printing methods with chitosan such as binder jetting, stereolithography, and extrusion using the following combinations of materials: chitosan–raffinose in acetic acid, alginate–chitosan, hydroxy butyl chitosan, chitosan– polyethylene glycol (PEG) diacrylate, chitosan–pectin, chitosan–calcium phosphate, and chitosan–gelatin (Thomas kean, Maya Thanou).

5.3 CONVENTIONAL MANUFACTURING PROCESSES USING CHITOSAN

Currently, the manufacturing and processing of chitosan polymers into desired shapes for engineering purposes is an active research area that has been accomplished using diverse fabrication processes. Because these techniques have been time tested, there already exists a wealth of knowledge for producing polymers in the target form and with the desired properties. However, this knowledge gets questioned when having to consider the special characteristics of natural materials. The research involved in producing chitosan-based products with different methods such as compression molding, solution casting, and injection molding, as shown in Figure 5.9, will now be presented, as well as details about why the fabrication approach works for this particular material. Furthermore, in this section, a list of material properties of chitosan provided by various research endeavors will be presented to provide an overview of the range of properties afforded by these different fabrication processes. Finally, overarching limitations of the current methodologies and propositions for an alternative manufacturing approach to deal with the most pressing challenges in the field are reviewed.

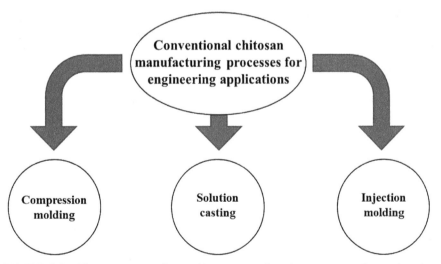

FIGURE 5.9 Three most prominent chitosan manufacturing processes for engineering applications.

Compression molding is a typical polymer manufacturing approach that offers short processing times, a high degree of automation, good repeatability,

106 *Natural Polymers: Perspectives and Applications for a Green Approach*

and excellent dimensional stability for both thermosets and thermoplastics [32]. Correlo et al. [33] investigated compression molding of blends of chitosan with synthetic aliphatic polyesters and to create porous scaffolds for tissue engineering applications. In this case, chitosan was selected for its high degree of biocompatibility. Powders of the natural and synthetic polymers were mixed with defined amounts of leachable salt particles and loaded onto a mold. Heat and pressure were applied that resulted in packing and melting of the polymers, fusing the polymer particles together to create a continuous polymeric network that provided mechanical stability. The molded specimen was then immersed in a solvent to dissolve the embedded salt particles, leaving functional pores.

Correlo et al. [33] tested these chitosan-based scaffolds for compressive properties, reporting a maximum compressive modulus and yield stress of 106 MPa and 9.9 MPa, respectively, for a composite consisting of 25% chitosan and 75% poly(butylene succinate) (PBS) by weight. This 25% chitosan–75% PBS also featured the highest melting temperature and heat of fusion of 102.3 °C and 52.9 J/g, respectively, as revealed by DSC. In the pursuit of creating biodegradable plastics, Galvis-Sánchez et al. [34] also explored compression molding as a means of creating chitosan films. They opted for a combination of choline chloride (ChCl) and citric acid (CA), which are eco-friendly materials, with the chitosan to create a eutectic system with good film-forming ability, flexibility, and mechanical resistance. Powders of chitosan, ChCl, and CA were ground and mixed into a paste with acetic acid as a solvent, and then the paste itself was compression-molded with different parameters tested such as compression load and time. The compression molding temperature was set at 120 °C because temperatures above or below would consistently lead to sample degradation or lack of film formation, respectively. DSC tests proved that the eutectic mixture of ChCl–CA was a good plasticizer for chitosan, given that the glass transition temperature of the composite dropped from about 170 °C for chitosan–CA mixtures to 150 °C for chitosan–ChCl–CA mixtures. The water absorption and water vapor permeability (WVP) of the films were measured, with the monolayer moisture content reported (0.002) much lower than colleagues (0.047–0.166), perhaps due to the method of fabrication, and WVP value (2.87 9 $\times$ 10^{-10} g m^{-1} s^{-1} Pa^{-1}) being in the range of other work (2.6–6.9 $\times$ 10^{-10} g m^{-1} s^{-1} Pa^{-1}). Finally, tensile modulus of the chitosan–ChCl–CA film was determined to be 20.5 $\times$ 10^2 MPa and the tensile strength 28 MPa, both of which were lower than that of chitosan–CA films (37.0 $\times$ 10^2 MPa, 63.8 MPa), confirming the eutectic mixture's role in plasticizing the ~0.04 mm-thick films.

Chitosan Biopolymer for 3D Printing: A Comprehensive Review

While compression molding is a less actively investigated method of producing films of chitosan, solution casting is a far more common technique. The fabrication of keratin–chitosan composite films via solution casting was explored by different researchers [35]. Keratin is a structural, yet flexible fibrous protein found in nature and thus has biomedical potential for use as a tissue scaffold in a similar manner to gelatine and collagen. In this work, chitosan was selected for its ability to reinforce the keratin. A keratin–chitosan mixture was poured into a polypropylene mold and dried to give a transparent film of desired thickness (0.01–0.02 mm). They disclosed that a film prepared from keratin mixed with 10 wt.% chitosan was flexible yet strong with an ultimate tensile strength 27 ± 8 MPa, ultimate elongation of $4 \pm 2\%$, and Young's modulus of 152 ± 76 MPa. They reported that further increase of chitosan content gave little increase of ultimate strength and elongation but significant increase of Young's modulus, highlighting the role of chitosan as a stiffener. Furthermore, the authors performed pH swelling tests on pure chitosan films, showing that at pH 5.3, chitosan film swelled 382% and even dissolved at pH 4.0. Ge et al. [36] took the opposite approach, using chitosan as the bulk material in a film and reinforcing it with another material, graphene. Graphene was formed into N-doped graphene sheets by the direct current arc-discharge method and dispersed in acetic acid aqueous solution. Chitosan was dissolved in this solution and then cast into a mold, dried, and vacuumed at 50 °C. Using the nanoindentation method for thin films, the mechanical properties of the composite films were measured, with Young's modulus of their composite chitosan film increasing by over ~200% to a maximum of ~7.5 GPa just with the addition of 0.3 wt.% graphene. In order to ensure that the addition of graphene did not upset the viability of the chitosan film to cell growth, 3-(4,5-dimethylthiazol-2-yl)-2,5-diphenyltetrazolium bromide (MTT) assays were conducted. Cells were exposed to grapheme–chitosan composites with a rising concentration of graphene from 0 to 0.6 wt.%, and MTT results showed that while the speed of cell attachment to the films decreases with increasing graphene concentration, no visible reduction in viability or change in morphology was observed. Therefore, these films maintained good biocompatibility within the tested range of graphene concentration.

These aforementioned examples validate the notion that solution casting is a simple yet effective way of producing thin, planar films. However, as a step forward towards more volumetric geometry at macroscopic scales, Fernandez and Ingber [37] have demonstrated their ability to produce chitosan objects of virtually any complex 3D shape via casting, as well as injection molding. By concentrating an initial dilute solution of chitosan in acetic acid, a pliable liquid crystal material was formed that conformed to the desired manufacturing process. While the chitosan underwent shrinking due to solvent evaporation, this was mitigated by the inclusion of particulates of wood flour. They showcased their capabilities by producing biodegradable versions of everyday objects like cups of different colors and chess pieces. Interestingly, they commented that chitosan has the ability to capture and retain small molecules, and they leveraged this property to modulate controlled release of colorants by varying pH. This contrasts to current methods of coloring synthetic plastics, which leaves the dye covalently bonded to the macropolymer, rendering plastics one of color unable to be recycled for another application requiring a different color. Finally, in terms of mechanical properties, they reported that the chitosan polymer could reach a tensile strength of up to 60 MPa and an ultimate tensile strain of 6.2% and emphasized that these values are similar to those of typical synthetic materials like polyethylene or polystyrene; by showcasing the ability to make commonplace functional objects entirely from chitosan, this work gives weighted plausibility to the idea of widespread manufacturing with nonoil-derived plastics. This concept of having freedom to form chitosan polymer into geometry of choice should not be undervalued. In the present literature on chitosan, it is clear that the pursuit of shaping chitosan into a desired construct, whether it is a porous scaffold, thin-film, or board game piece, is a path of research in itself that is piled upon the primary research goal for which chitosan is selected. Furthermore, as we have seen, natural polymers are not always complimentary with traditional manufacturing processes, and so fabrication parameters and setups must be adjusted accordingly in order to achieve the desired material fidelity. This requires time, resources, and effort that could be better applied to exploring and exploiting the properties of this natural polymer. These molding techniques for forming chitosan, despite their intricacies, have worked well until this point, but as functional requirements get more demanding, more complex geometries are sure to be required, as is already apparent in the fields of tissue engineering [18]. Producing elaborate molds is rarely a viable economic option for applications of such resolution. Thus, alternative fabrication approaches that provide the user with more freedom of materialization must be accommodated in order to fully catalyze the integration of environmentally sustainable, natural polymers such as chitosan and one such promising approach is 3D printing.

5.4   3D PRINTING OF CHITOSAN AND ITS COMPOSITES

Of the variety of natural polymers investigated, chitosan is a material showing tremendous potential for 3D printing applications that require good mechanical properties. As stated earlier, chitin is ubiquitous as a structural material in the biological world, and therefore the adoption of chitosan for 3D printing has leveraged these structural capabilities for producing stable printing materials. In particular, two applications in the use of chitosan polymer have attracted recent attention: for tissue scaffolds as with the other natural polymers listed above, and as a component material for additive manufacturing of large-scale (>1 m) structural parts. These two research areas are discussed more in-depth below.

Because it fulfills requirements for good mechanical properties, processability, surface characteristics, biocompatibility, and biodegradability, chitosan has been extensively explored as a material with significant potential for additively manufactured scaffolds for tissue engineering [10]. Chitosan is particularly known to facilitate cell attachment and maintenance of cell differentiation, which is crucial for tissue engineering applications [38]. 3D printing of chitosan for tissue engineering is characterized by a DIW process and many research endeavors have used 3D plotting processes to create complex chitosan-based artificial tissue architecture [39]. Notably, Yan et al. [40] developed an organ manufacturing technique to print a liver tissue construct made from a gelatin and chitosan hydrogel composite that replicated the function of the extracellular matrix. They emphasized the significance of the structural performance of the hydrogel matrix in supporting cell division, differentiation, and migration. In addition, chitosan has been combined with bioceramics to create natural polymer ceramic composites for hydrogel scaffolds of increased hardness and new inorganic functionalities, as produced by [19]. Further chitosan composite exploration could readily bring forth the necessary material developments for advancements in 3D printing of artificial tissues and organs.

A recurring question in 3D printing is its potential application to industries requiring large structural parts such as product design, construction, automotive, and aerospace. Coupled with an increasing desire to build with sustainable materials, this has directed investigations for natural materials with two important capabilities: firstly, sufficient manufacturability to be integrated into 3D printing processes, and secondly, good mechanical properties to withstand stresses at larger scales. chitosan, with its inherent strength and intrinsic ability to form composites in its natural form [41], has begun to be explored as a candidate for such a functional material component. Various researchers [42] have demonstrated the ability of solid cellulose–chitosan composite (fungal-like adhesive material [FLAM]) to be extruded as a paste using a DIW process mounted on an industrial robotic arm, reporting an optimal ratio between cellulose fibers and chitosan matrix resulting in a pliable printing material that could retain its shape after deposition. When dried, FLAM features mechanical attributes in the range of low-density woods and rigid foams, which are already used in construction applications. They demonstrated the natural polymer printing process with a prototype of a scaled wind turbine blade of 1.2 m in length. Similar such work of Mogas-Soldevila et al. [43] and Mogas-Soldevila and Oxman [44], also investigated a DIW process that deposited composite polymer pastes and hydrogels composed of chitosan, sodium alginate, cellulose, chitin, and other organic fillers in a variety of volumetric weight ratios. Chitosan served as the primary matrix material that readily bound with these natural fillers. They demonstrated their process by printing insect-wing-inspired self-supporting artifacts of up to 3 m in length. However, they stated that their printed objects are vulnerable to water and will degrade once exposed, and the application of this 3D process is currently geared towards producing temporary parts. In the context of these examples, it is evident that the state-of-the-art of chitosan AM for large-scale structural parts is based on extruding thick pastes that dry after deposition, which is primarily a result of challenges in processing the natural polymer without intense heat and pressure but still achieving formability.

5.5 RECOMMENDATION OF CHITOSAN COMPOSITES FOR 3D PRINTING AND THE EFFECT OF 3D-PRINTER OPERATING PARAMETERS

An additional characteristic of chitosan is its antibacterial characteristics; in addition to the antibacterial activities of these 3D-printed scaffolds, attention should also be paid to their biocompatibility and osteogenic activity [45]. Although chitosan composite has been well established for the 3D-printing technique [B, C] it involves few limitations [46]. The predominant challenges faced by tissue engineers are normally categorized into two; the first one being the research and development of novel bio-inks for diverse tissues or one universal bio-ink for all tissues. Preferably, a common bio-ink should be a blend of biomaterials that support native tissue viability, chemical cues, and growth factors for angiogenesis and channels for nerve innervation. These challenges can be addressed with availability of new technologies, such as additive manufacturing/3D printing, that enable fabrication of complex tissues with complex geometry. Thus 3D printing techniques have potential to build geometries that would be unimaginable through traditional manufacturing techniques [47]. In addition, vascularization is one of the most critical challenges in creating viable strategies to induce angiogenesis, including the addition of angiogenic growth factors (vascular endothelial growth factor), the addition of platelets, bone marrow clots, and using bioreactors [48]. The scope for fabricating symmetrically porous 3D scaffolds using chitosan hydrogels as the printing ink in a custom-designed nozzle extrusion-based 3D printer is a challenge which requires optimization of series of operating parameter such as laser spot size, scan radius, laser power, scan spacing, and the laser scanner parameters (scan spacing and scan parameters). The size of the laser spot and the scan radius are both machine-specific and may differ from one SLS machine to another of the AM machine [49]. The flow diagram in Figure 5.10 shows how the energy to the surface is a function of the operating parameters. The energy input at the surface is a function of the flux of the pulse, the duration of the pulse, and the delay between successive pulses. The design of the energy input is based on the laser output has a Gaussian intensity distribution. This assumption is valid for the CO_2 laser that is most commonly used in the SLS process [E].

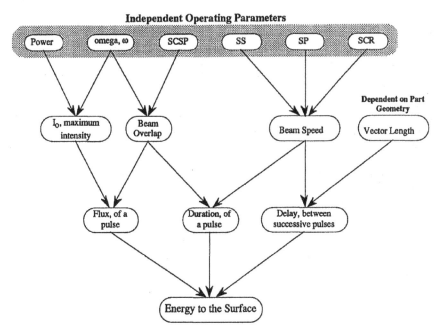

FIGURE 5.10 A flow diagram of the energy input at the surface as a function of the operating parameters.

The process parameters directly affect the amount of energy delivered to the surface of the thin layer and the energy density absorbed by the powders; therefore, the parameters decide the properties like physical and mechanical of the built parts, such as relative density, porosity, surface roughness, dimensional accuracy, strength, etc. [50]. Although porosity is the default phenomenon in the selective laser sintering-AM process, more investigation and research is still need to be performed based on the tissue and scaffold requirement so as to acquire desired porosity attributes of the scaffold. To attain such precise attributes, simulation tools can be used to model the behavior of a part under a range of operating conditions. Fine-tuning parameters cannot completely eliminate the porosity, but the density of the manufactured part is usually higher than the powder density.

5.6 CONCLUSION

Additive manufacturing/3D printing has been rapidly evolving as a reputed, economic, and sustainable manufacturing technique around the world for diverse applications in aerospace, automobile, construction, energy, medicine, etc. 3D printing of synthetic polymers are quite convincing as opposed to natural polymers due to certain difficulties such as thermal degradation, extrudability, and temperature which reduces the broader use of different AM methods in natural polymers and narrowed down to only few AM categories; Direct ink writing/3D plotting, material extrusion and robotic dispensing. However, research in 3D bioprinting using these technologies has found to be increasing in recent years and achieved few breakthroughs in the field of tissue engineering using gelatin, cornea 3D printing, etc. In light of using natural polymers in 3D bioprinting, chitosan has been highly proposed and researched next to cellulose which possesses tremendous mechanical properties along with biocompatibility and biodegradability. Chitosan were 3D printed with few combinations such as chitosan–raffinose in acetic acid, alginate–chitosan, hydroxy butyl chitosan, chitosan–PEG diacrylate, chitosan–pectin, chitosan–calcium phosphate, and chitosan–gelatin. Due to the chitosan's nature, the function for medical application is more critical owing to its inherent biological properties. The other targeted areas are printing large structural parts and in tissue engineering scaffolds. The properties of chitosan plays an important role in developing AM methods such as good mechanical properties, processability, surface characteristics for structural parts building, and cell attachment facilitation with cell differentiation is crucial for tissue engineering scaffolds through FDM. These aspects make 3D printing a versatile route for parts building and in the present study, its scope for biomedical application is well ascertained, specifically matching its compatibility with chitosan. Such a holistic assessment on the incorporation of chitosan biopolymer for 3D printing techniques may provide a comprehensive insight to researchers and biomedical engineers about their application to the life-saving medical industry.

ACKNOWLEDGMENTS

We would like to thank Murali Krishnan for his effort on data collection and referencing.

KEYWORDS

- **3D printing**
- **additive manufacturing**
- **chitosan biopolymer composites**
- **medical application**

REFERENCES

1. Gibson, I.; Rosen, D.; Stucker, B. *Additive Manufacturing Technologies: 3D Printing, Rapid Prototyping, and Direct Digital Manufacturing*, Second Edition; 2015. https://doi.org/10.1007/978-1-4939-2113-3.
2. Kianian, B. *Wohlers Report 2017: 3D Printing and Additive Manufacturing State of the Industry*, Annual Worldwide Progress Report: Chapters Titles: The Middle East, and Other Countries; 2017.
3. Terry Wohlers, T. G. *History of Additive Manufacturing*. 2017, 1–24. https://doi.org/10.4018/978-1-5225-2289-8.ch001.
4. Raspall, F.; Velu, R.; Vaheed, N. M. Fabrication of Complex 3D Composites by Fusing Automated Fiber Placement (AFP) and Additive Manufacturing (AM) Technologies. *Adv. Manuf. Polym. Compos. Sci.* **2019,** *5* (1), 6–16. https://doi.org/10.1080/20550340.2018.1557397.
5. Velu, R.; Vaheed, N.; Raspall, F. Design and Robotic Fabrication of 3D Printed Moulds for Composites. Proceedings of the 29th Annual International Solid Freeform

Fabrication Symposium—An Additive Manufacturing Conference, Austin, TX, USA, 2018, 1036–1046.

6. Velu, R.; Raspall, F.; Singamneni, S. Chapter 8 3D Printing Technologies and Composite Materials for Structural Applications. In *Green Composites for Automotive Applications; Koronis*, G., Silva, A., Eds.; Woodhead Publishing Series in Composites Science and Engineering; Woodhead Publishing, 2019; pp 171–196. https://doi.org/https://doi.org/10.1016/B978-0-08-102177-4.00008-2.

7. Velu, R.; Singamneni, S. Selective Laser Sintering of Polymer Biocomposites Based on Polymethyl Methacrylate. *J. Mater. Res.* **2014**, *29* (17), 1883–1892. https://doi.org/10.1557/jmr.2014.211.

8. Vishakha, K.; Kishor, B.; Sudha, R. Natural Polymers A Comprehensive Review. *Int. J. Res. Pharm. Biomed. Sci.* **2012**, *3* (4), 1597–1613.

9. Hitoshi Sashiwa, D. H. Advances in Marine Chitin and Chitosan II, 2017. *Adv. Mar. Chitin Chitosan II*, 2017 2018. https://doi.org/10.3390/books978-3-03842-678-3.

10. Bose, S.; Ke, D.; Sahasrabudhe, H.; Bandyopadhyay, A. Additive Manufacturing of Biomaterials. *Prog. Mater. Sci.* **2018**, *93*, 45–111. https://doi.org/10.1016/j.pmatsci.2017.08.003.

11. Velu, R.; Fernyhough, A.; Smith, D. A.; Joo Le Guen, M.; Singamneni, S. Selective Laser Sintering of Biocomposite Materials. *Lasers Eng.* **2006**, *35* (1), 173–186.

12. Ho, M. P.; Wang, H.; Lee, J. H.; Ho, C. K.; Lau, K. T.; Leng, J.; Hui, D. Critical Factors on Manufacturing Processes of Natural Fibre Composites. *Compos. Part B Eng.* **2012**, *43* (8), 3549–3562. https://doi.org/10.1016/j.compositesb.2011.10.001.

13. Tambyrajah, D. *Indulge & Explore Natural Fiber Composites Preface*. NFCDesign Platform, The Netherlands. 2015, 122.

14. Milosevic, M.; Stoof, D.; Pickering, K.L. Characterizing the Mechanical Properties of Fused Deposition Modelling Natural Fiber Recycled Polypropylene Composites. *J. Compos. Sci.* **2017,** *1* (1), 7. https://doi.org/10.3390/jcs1010007.

15. Matsuzaki, R.; Ueda, M.; Namiki, M.; Jeong, T. K.; Asahara, H.; Horiguchi, K.; Nakamura, T.; Todoroki, A.; Hirano, Y. Three-Dimensional Printing of Continuous-Fiber Composites by in-Nozzle Impregnation. *Sci. Rep.* **2016,** *6* (February), 1–7. https://doi.org/10.1038/srep23058.

16. Wijk, A. van; Wijk, I. van. 3D Printing with Biomaterials: Towards a Sustainable and Circular Economy; 2015. https://doi.org/10.3233/978-1-61499-486-2-i.

17. Lewis, J. A. Direct Ink Writing of 3D Functional Materials. *Adv. Funct. Mater.* **2006**, *16* (17), 2193–2204. https://doi.org/10.1002/adfm.200600434.

18. Melchels, F. P. W.; Domingos, M. A. N.; Klein, T. J.; Malda, J.; Bartolo, P. J.; Hutmacher, D. W. Additive Manufacturing of Tissues and Organs. *Prog. Polym. Sci.* **2012**, *37* (8), 1079–1104. https://doi.org/10.1016/j.progpolymsci.2011.11.007.

19. Ang, T. H.; Sultana, F. S. A.; Hutmacher, D. W.; Wong, Y. S.; Fuh, J. Y. H.; Mo, X. M.; Loh, H. T.; Burdet, E.; Teoh, S. H. Fabrication of 3D Chitosan-Hydroxyapatite Scaffolds Using a Robotic Dispensing System. *Mater. Sci. Eng. C* **2002**, *20* (1–2), 35–42. https://doi.org/10.1016/S0928-4931(02)00010-3.

20. Kim, G.; Ahn, S.; Yoon, H.; Kim, Y.; Chun, W. A Cryogenic Direct-Plotting System for Fabrication of 3D Collagen Scaffolds for Tissue Engineering. *J. Mater. Chem.* **2009**, *19* (46), 8817–8823. https://doi.org/10.1039/b914187a.

21. Gauvin, R.; Chen, Y. C.; Lee, J. W.; Soman, P.; Zorlutuna, P.; Nichol, J. W.; Bae, H.; Chen, S.; Khademhosseini, A. Microfabrication of Complex Porous Tissue Engineering Scaffolds Using 3D Projection Stereolithography. *Biomaterials* **2012**, *33* (15), 3824–3834. https://doi.org/10.1016/j.biomaterials.2012.01.048.

22. Lam, C. X. F; Mo, X. M.; Teoh, S. H.; Hutmacher, D. W. Scaffold Development Using 3D Printing with a Starch-Based Polymer. *Anesthesiol. Keywords Rev. Second Ed.* **2013,** *20,* 579.

23. Salmoria, G. V.; Klauss, P.; Paggi, R. A.; Kanis, L. A.; Lago, A. Structure and Mechanical Properties of Cellulose Based Scaffolds Fabricated by Selective Laser Sintering. *Polym. Test.* **2009,** *28* (6), 648–652. https://doi.org/10.1016/j.polymertesting.2009.05.008.

24. Aravamudhan, A.; Ramos, D. M.; Nada, A. A.; Kumbar, S. G. Chapter 4 Natural Polymers: Polysaccharides and Their Derivatives for Biomedical Applications. In *Natural and Synthetic Biomedical Polymers*; Kumbar, S. G., Laurencin, C. T., Deng, M., Eds.; Elsevier: Oxford, 2014; pp 67–89. https://doi.org/https://doi.org/10.1016/B978-0-12-396983-5.00004-1.

25. Tamura, H.; Furuike, T. Chitin and Chitosan. https://doi.org/10.1007/978-3-642-36199-9_322-1.

26. Khoushab, F.; Yamabhai, M. Chitin Research Revisited. *Mar. Drugs* **2010,** *8* (7), 1988–2012. https://doi.org/10.3390/md8071988.

27. Islam, S.; Khan, M.; Alam, A. N. Production of Chitin and Chitosan from Shrimp Shell Wastes. *J. Bangladesh Agric. Univ.* **2017,** *14* (2), 253–259. https://doi.org/10.3329/jbau.v14i2.32701.

28. Sivashankari, P. R.; Prabaharan, M. 5—Deacetylation Modification Techniques of Chitin and Chitosan. In *Chitosan Based Biomaterials Volume 1*; Jennings, J. A., Bumgardner, J. D., Eds.; Woodhead Publishing, 2017; pp 117–133. https://doi.org/https://doi.org/10.1016/B978-0-08-100230-8.00005-4.

29. Younes, I.; Rinaudo, M. Chitin and Chitosan Preparation from Marine Sources. Structure, Properties and Applications. *Mar. Drugs* **2015,** *13* (3), 1133–1174. https://doi.org/10.3390/md13031133.

30. Jayakumar, R.; Prabaharan, M.; Muzzarelli, R. A. A. *Adv. Polym. Sci.*; **1976,** 99. https://doi.org/10.1524/zpch.1976.99.4-6.310.

31. Kumar, M. N. V. R.; Muzzarelli, R. A. A.; Muzzarelli, C.; Sashiwa, H.; Domb, A. J. Chitosan Chemistry and Pharmaceutical Perspectives. *Chem. Rev.* **2004,** *104* (12), 6017–6084. https://doi.org/10.1021/cr030441b.

32. Mitschang, P.; Hildebrandt, K. Polymer and Composite Moulding Technologies for Automotive Applications. *Adv. Mater. Automot. Eng.* **2012,** 210–229. https://doi.org/10.1533/9780857095466.

33. Correlo, V. M.; Boesel, L. F.; Pinho, E.; Costa-Pinto, A. R.; Alves Da Silva, M. L.; Bhattacharya, M.; Mano, J. F.; Neves, N. M.; Reis, R. L. Melt-Based Compression-Molded Scaffolds from Chitosan-Polyester Blends and Composites: Morphology and Mechanical Properties. *J. Biomed. Mater. Res. A* **2009,** *91* (2), 489–504. https://doi.org/10.1002/jbm.a.32221.

34. Galvis-Sánchez, A. C.; Sousa, A. M. M.; Hilliou, L.; Gonçalves, M. P.; Souza, H. K. S. Thermo-Compression Molding of Chitosan with a Deep Eutectic Mixture for Biofilms Development. *Green Chem.* **2016,** *18* (6), 1571–1580. https://doi.org/10.1039/c5gc02231b.

35. Tanabe, T.; Okitsu, N.; Tachibana, A.; Yamauchi, K. Preparation and Characterization of Keratin-Chitosan Composite Film. *Biomaterials* **2002,** *23* (3), 817–825.

36. Ge, Z.; Jin, Z.; Fan, H.; Zhao, K.; Wang, L.; Shi, Z.; Li, N. Fabrication, Mechanical Properties, and Biocompatibility of Graphene-Reinforced Chitosan Composites. *Biomacromolecules* **2010,** *11* (9), 2345–2351. https://doi.org/10.1021/bm100470q.

37. Fernandez, J. G.; Ingber, D. E. Manufacturing of Large-Scale Functional Objects Using Biodegradable Chitosan Bioplastic. *Macromol. Mater. Eng.* **2014,** *299* (8), 932–938. https://doi.org/10.1002/mame.201300426.

38. Li, X.; Cui, R.; Sun, L.; Aifantis, K. E.; Fan, Y.; Feng, Q.; Cui, F.; Watari, F. 3D-Printed Biopolymers for Tissue Engineering Application. *Int. J. Polym. Sci.* **2014,** *2014,* 1–13. https://doi.org/10.1155/2014/829145.

39. Liu, I. H.; Chang, S. H.; Lin, H. Y. Chitosan-Based Hydrogel Tissue Scaffolds Made by 3D Plotting Promotes Osteoblast Proliferation and Mineralization. *Biomed. Mater.* **2015,** *10* (3). https://doi.org/10.1088/1748-6041/10/3/035004.

40. Yan, Y.; Wang, X.; Pan, Y.; Liu, H.; Cheng, J.; Xiong, Z.; Lin, F.; Wu, R.; Zhang, R.; Lu, Q. Fabrication of Viable Tissue-Engineered Constructs with 3D Cell-Assembly Technique. *Biomaterials* **2005,** *26* (29), 5864–5871. https://doi.org/10.1016/j.biomaterials.2005.02.027.

41. Ravi Kumar, M. N. V. A Review of Chitin and Chitosan Applications. *Eng. Sci. Fundam. 2015—Core Program. Area 2015 AIChE Annu. Meet.* **2015,** *1* (Figure 1), 4–6. https://doi.org/10.1016/S1381-5148(00)00038-9.

42. Sanandiya, N. D.; Vijay, Y.; Dimopoulou, M.; Dritsas, S.; Fernandez, J. G. Large-Scale Additive Manufacturing with Bioinspired Cellulosic Materials. *Sci. Rep.* **2018,** *8* (1), 8642. https://doi.org/10.1038/s41598-018-26985-2.

43. Mogas-Soldevila, L.; Duro-Royo, J.; Oxman, N. Water-Based Robotic Fabrication: Large-Scale Additive Manufacturing of Functionally Graded Hydrogel Composites via Multichamber Extrusion. 3D Print. *Addit. Manuf.* **2016,** *1* (3), 141–151. https://doi.org/10.1089/3dp.2014.0014.

44. Mogas-Soldevila, L.; Oxman, N. Water-Based Engineering & Fabrication: Large-Scale Additive Manufacturing of Biomaterials *Laia. Creat. Glob. Compet. Econ. 2020 Vis. Plan. Implementation, Vols 1–3* **2015,** *1795* (100), 27–32. https://doi.org/10.1557/opl.2015.

45. Rabea, E. I.; Badawy, M. E. T.; Stevens, C. V.; Smagghe, G.; Steurbaut, W. Chitosan as Antimicrobial Agent: Applications and Mode of Action. *Biomacromolecules* **2003,** *4* (6), 1457–1465. https://doi.org/10.1021/bm034130m.

46. Almeida, C. R.; Serra, T.; Oliveira, M. I.; Planell, J. A.; Barbosa, M. A.; Navarro, M. Impact of 3-D Printed PLA- and Chitosan-Based Scaffolds on Human Monocyte/Macrophage Responses: Unraveling the Effect of 3-D Structures on Inflammation. *Acta Biomater.* **2014,** *10* (2), 613–622. https://doi.org/10.1016/j.actbio.2013.10.035.

47. Jammalamadaka, U.; Tappa, K. Recent Advances in Biomaterials for 3D Printing and Tissue Engineering. *J. Funct. Biomater.* **2018,** *9* (1). https://doi.org/10.3390/jfb9010022.

48. Pandya, N. M.; Dhalla, N. S.; Santani, D. D. Angiogenesis—A New Target for Future Therapy. *Vascul. Pharmacol.* **2006,** *44* (5), 265–274. https://doi.org/10.1016/j.vph.2006.01.005.

49. Nelson, J. C.; Xue, S.; Barlow, J. W.; Beaman, J. J.; Marcus, H. L.; Bourell, D. L. Model of the Selective Laser Sintering of Bisphenol-A Polycarbonate. *Ind. Eng. Chem. Res.* **1993,** *32* (10), 2305–2317. https://doi.org/10.1021/ie00022a014.

50. Gu, D. D.; Meiners, W.; Wissenbach, K.; Poprawe, R. Laser Additive Manufacturing of Metallic Components. *Int. Mater. Rev.* **2013,** *57* (3), 133–164. https://doi.org/10.1179/1 743280411Y.0000000014.

51. Dutta, P. K.; Duta, J.; and Tripathi, V. S. 2004. "Chitin and Chitosan: Chemistry, Properties and Applications." *J. Sci. Ind. Res.* 63 (1): 20–31. https://doi.org/10.1002/chin.200727270

CHAPTER 6

Potential Applications of Chitosan-Based Sorbents in Nuclear Industry: A Review

ANUPKUMAR BHASKARAPILLAI[1,2]

[1]*Water and Steam Chemistry Division, Bhabha Atomic Research Centre Facilities, Kalpakkam, Tamil Nadu 603102, India*

[2]*Homi Bhabha National Institute, Anushakthi Nagar, Mumbai, Maharashtra 400394, India. E-mail: anup@igcar.gov.in.*

ABSTRACT

Chitosan is among the most abundant and studied biopolymers. This is due to the versatile applications of chitosan and the ease of modifications of its functional groups to impart desired binding and physical properties. There is exhaustive literature on chitosan covering its wide applications, varying from medical devices and pharma applications to pollution control, which is driven by the abundance and the special properties (such as biocompatibility) of chitosan. Nuclear industry, due to its necessity to deal with both radioactive and nonradioactive metal ions during various waste processing needs, is on constant lookout for efficient and selective metal ion removal agents. This chapter attempts to document the various reports that have been published on the potential applications of chitosan in the nuclear industry. Though limited, the published literature on the radioactive nuclide removal by chitosan in the last two to three decades is very encouraging and deserves stock-taking. The review gives special emphasis on the various studies our research group has carried out toward bringing chitosan closer to the application side for targeted applications in the nuclear industry. The general characteristics, metal ion binding properties, and relevant modifications attempted in modifying the physical and chemical properties of the chitosan are also discussed.

118 *Natural Polymers: Perspectives and Applications for a Green Approach*

6.1 INTRODUCTION

Chitosan (poly(β-(1→4)-2-amino-2-deoxy-D-glucose)), an *N*-deacetylated derivative of chitin, is a basic polysaccharide and is among the most widely available biopolymers next only to cellulose. Chitosan market size surpassed USD 1.5 billion in 2017 and is expected to witness a compound annual growth rate of more than 20% between 2018 and 2024 [1]. Current demand is fueled primarily by the cosmetics and water treatment industry, and in the future years, biotechnological applications are expected to be the major driving force in the chitosan market [1].

In the face of ever-increasing burden of nondegradable polymers being added into the environment, biopolymers such as chitosan and chitin deserve serious consideration as replacement, even if to a limited extent, for the conventional nondegradable synthetic polymers used in removing metal ions and other pollutants. The source of chitosan is the naturally abundant chitin. Chitin is a mucopolysaccharide extracted from the crustacean shells, and is only second to cellulose as the most abundant natural polymer. The degree of deacetylation (of chitin to yield chitosan) varies and can affect the sorption properties of the chitosan. Though there is not any rigid definition to differentiate between chitin and chitosan based on the degree of deacety-lation, in general, most literature reports use chitosan with a deacetylation of 70% or above. Clark et al. have carried out the first XRD investigation of chitin as early as in 1936 [2]. However, the definitive analysis of the XRD pattern was done about 60 years later in 1997 [3–5]. The first-generation chitosan, which was used in the earlier decades and is still in use to some extent, were not well defined in terms of purity and degree of deacetylation but were more or less supplied as mixtures of polymers of varying purity and varying composition and were unfit for development of specialized chitosan-based products. The currently available chitosans, however, are well defined in terms of the functional groups and purity and are well suited for well-defined and predictable modifications and development of products for targeted applications.

6.2 MODIFICATIONS OF CHITOSAN

The versatile nature of the functional groups in the chitosan makes it an effective metal ion sorbent. However, the solubility of chitosan in dilute organic and mineral acids (except H_2SO_4) [6] limits its applications. In particular, as most industrial effluents, including typical nuclear reactor decontamination

Potential Applications of Chitosan-Based Sorbents 119

formulations and backend waste processes, involve acidic to highly acidic solution conditions, this becomes a severe handicap in the application of chitosan. An effective way to overcome this limitation is incorporation of a crosslinked matrix. Through crosslinking of chitosan, its stability can be greatly improved leading to broadening of its applications. Crosslinking, however, can often bring in unwelcome modifications in the property such as reduction in capacity, swelling, kinetics, modified selectivity, etc. Hence, it becomes pertinent that the nature of crosslinker used and the mode of crosslinking are rationally chosen. The involvement of the metal ion binding functional groups of chitosan in the crosslinking reaction can be the major cause of reduced capacity. And the reduced capacity does not always result in increased selectivity. The extent of change in the capacity and selectivity would depend on the nature of crosslinker used and the chemistry involved in the crosslinking reaction. For example, epichlorohydrin and glutaraldehyde, the most commonly used chitosan crosslinkers, are reported to crosslink the chitosan in completely different ways leading to chitosan with different binding properties. Though epichlorohydrin can bind through both the amine and the hydroxyl groups, under alkaline conditions epichlorohydrin reacts with the nucleophilic hydroxyl groups of the chitosan leading to opening of the epoxide ring and the removal of chlorine atom. Thus, most often epichlorohydrin crosslinked chitosan have free amine groups while the hydroxyl groups are engaged in the crosslinking. However, the opening of the epichlorohydrin ring leads to formation of hydroxyl groups in the chitosan matrix. In case of glutaraldehyde, the crosslinking is done under acidic conditions. The crosslinking involves aldehyde groups of the glutaraldehyde forming covalent imine bonds with the amino groups of chitosan, due to the resonance established with adjacent double ethylenic bonds via a Schiff reaction [7, 8]. The crosslinking reactions are shown in Figure 6.1.

One of the biggest advantages of chitosan is the ease of chemical modifications it can be subjected to due to the presence of the reactive amino and hydroxyl groups. This has resulted in large number of reports on chemical modifications of chitosan to make it suitable for the targeted applications. Apart from modifying the chemical properties, better physical properties such as better stability under different conditions, better format, better polymer morphology, etc., can be introduced through modifica-tions. For example, water-soluble chitosan derivatives can be prepared through quaternization and *N*-carboxylation of chitosan derivatives [9]. Primarily, modifications in chitosan can be carried out through two ways: (1) introduction of new functional groups (grafting) and (2) crosslinking leading to strong interconnected chains in the polymer matrix. Chemical modifications through grafting introduce new functional groups onto the chitosan moieties leading to increased binding sites and faster and higher binding toward the targeted analytes. Crosslinking, in general, may lead to reduced binding sites due to the engagement of the active functional groups in the crosslinking reactions, and reduced kinetics due to the reduced accessibility to the binding sites embedded in the crosslinked matrix. However, as we have shown in a study, that is not always the case. While designing a chitosan-based sorbent for antimony, we have shown that the crosslinker, epichlorohydrin, in fact, enhance the antimony uptake capacity [7]. Apart from these two methods, property modifications can also be carried out through incorporation of chitosan as part of a composite with another material such as metal oxides [7]. The latter has opened excellent opportunities in bringing chitosan closer to real-life applications in the adsorption field.

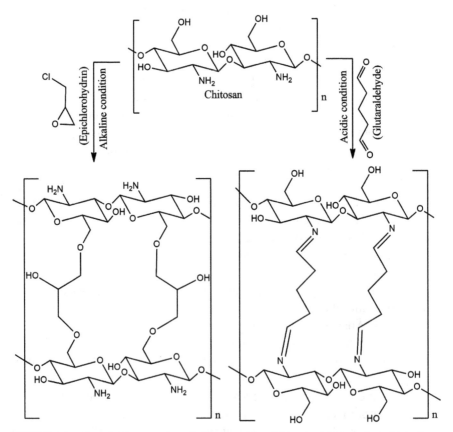

FIGURE 6.1 Crosslinking reaction of chitosan with epichlorohydrin and glutaraldehyde.

Incorporation of magnetite within a sorbent material for imparting magnetic properties to ease the separation of sorbent from the aqueous phase is used often with most types of sorbents. Researchers have adopted this methodology widely for chitosan as well, and have successfully prepared many chitosan-based sorbents with magnetic properties. A general methodology followed in the synthesis of such magnetic chitosan sorbent is shown in Scheme 6.1 [10].

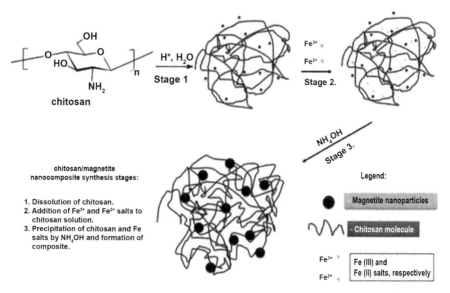

SCHEME 6.1 General methodology for the synthesis of magnetic chitosan (chitosan–magnetite composite) sorbent (reproduced from [10] (doi: https://doi.org/10.1186/s11671-016-1363-3) under Creative Common licence: http://creativecommons.org/licenses/by/4.0/).

Depending on the chitin source and the deacetylation reaction, the molecular weight of chitosan varies. High molecular weight chitosan exhibits high viscosity. The viscosity can be modified by altering the molecular weight. For example, the viscosity can be reduced through chitonolysis, which is depolymerization of chitosan which would lead to the formation of oligosaccharides and monomers. Oxidative processes using oxidants such as concentrated nitrous acid, hydrogen peroxide, etc. are used for the depolymerization of chitosan. The various methodologies used [11] for chitonolysis are shown in Figure 6.2.

The cationic nature of the binding sites of chitosan aids its use as a receptor for anionic species such as arsenic and antimony which are of concern as toxic pollutants in water sources. Further, antimony ions are of concern in the nuclear industry due to the presence of active isotopes of antimony. The sorption of these species can be enhanced by increasing the density of the cationic binding sites in the chitosan. Thus, chemical modifications that can enhance the cationic sites will be of high use. This can, for example, be done by reacting chitosan with an alkaline solution of dialkylaminoalkyl chlorides (Figure 6.3) [12].

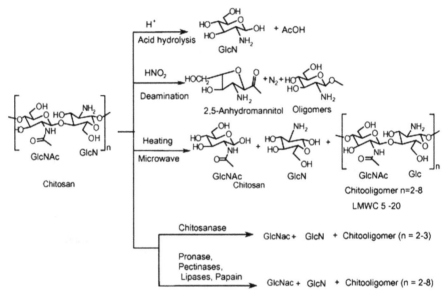

FIGURE 6.2 General methodologies used for chitonolysis (reproduced from [11], copyright (2008), with permission from Elsevier).

6.3 METAL ION BINDING PROPERTIES OF CHITOSAN

The conversion of chitin to chitosan is shown in Figure 6.4. The degree of deacetylation and the molecular weight have significant influence on the properties of the chitosan. For example, Kleine et al. [13] have synthesized chitosan–sodium tripolyphosphate composites using chitosan of varying degrees of deacetylation and molecular weight, and studied the properties of the composites. The results showed significant difference between the composites prepared with chitosan having different degrees of deacetylation. Also, high molecular weight chitosan was seen to incorporate better into the sodium tripolyphosphate nanoparticles. These differences were attributed to the differences in the polymer conformation and chain flexibility. Dung et

Potential Applications of Chitosan-Based Sorbents 123

al. [9] have introduced a methodology for almost complete deacetylation of chitin using 5% NaOH solution (unlike the general practice wherein 40–50% NaOH solutions are used) in presence of sodium borohydride (for preventing the polymer degradation).

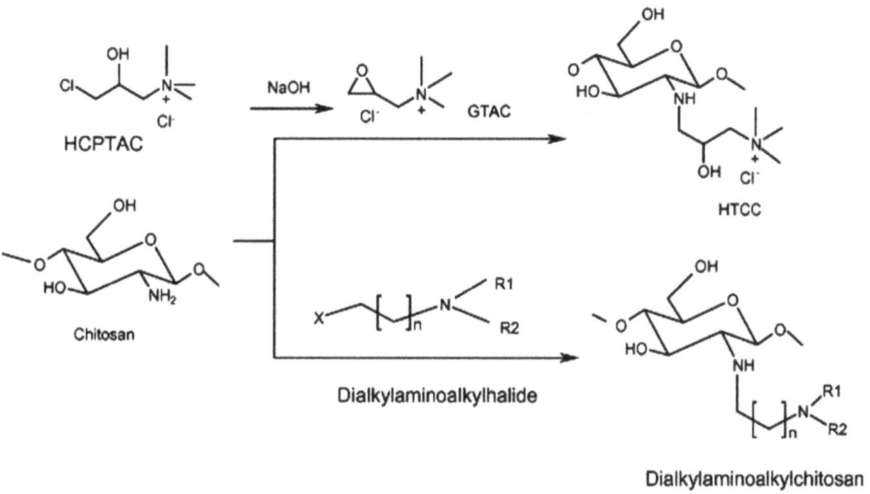

FIGURE 6.3 Methodologies for increasing the cation density on chitosan (Reproduced from [11], Copyright (2008), with permission from Elsevier).

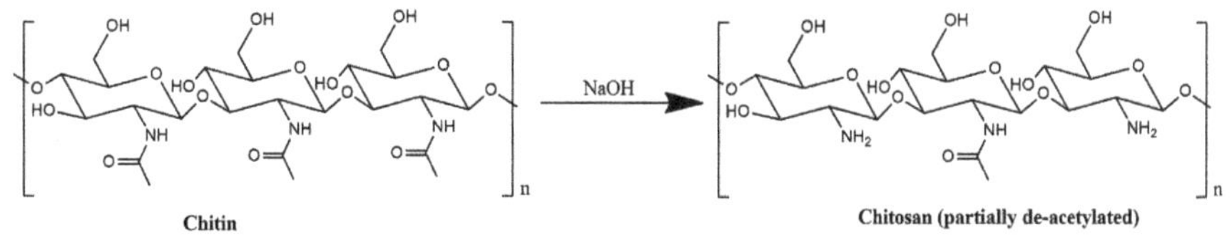

FIGURE 6.4 Deacetylation reaction of chitin to produce chitosan.

Chitosan, apart from the general properties of most biomaterials such as biodegradability, nontoxicity, affordability, etc, has excellent binding affinity toward various metal ions and organic compounds such as dyes [14–16]. Such versatile binding affinity combined with the advantages of high abundance and biodegradability has made chitosan and its derivatives highly popular amongst the researchers. The complexing nature of the chitosan is due to the reactive amino and hydroxyl functional groups present in the chitosan backbone, which make it an excellent receptor for a suite of metal ions. This property has been widely exploited in exploring chitosan for various applications involving removal of contaminants and in separation of targeted analytes. Thus, apart from its applications, in the field of medicine, cosmetics, and food industry [17–26], it has shown promise as a replacement for the synthetic polymer-based resins used for removal of metal ions including radioactive metal ions such as uranium and organic pollutants [27–30]. In general, chitosan does not show affinity toward the alkali metal and the alkali-earth metal ions. On the other hand, chitosan shows high affinity for the first transition series metal ions, in particular for their oxyanions. The binding of anions by chitosan is often not simple anion exchange instead it is seen to be through synergistic effect of various interactions between the chitosan functional groups and the metal ion species [31, 32]. First series transition metal ions that form strong chelating complexes (iron to zinc in the first series) bind chitosan through chelation. However, second-row elements do not behave as one would expect based on general chelation chemistry. For example, Muzarelli et al. have shown that, in the context of treating nuclear plant effluent with chitosan, Zr, Mo, and Ru behave in peculiar ways [33].

Binding by chitosan can occur through three different modes namely (a) electrostatic attraction, (b) chelation, and (c) amide bond formation depending on the solution conditions. In an acidic solution, when the amide groups are protonated, anionic species in the solution bind through electrostatic attraction with the chitosan. Transition metal ions such as iron, cobalt, etc., which form strong chelating complexes bind the amine groups through chelation. Covalent amide bond formation can happen between the amino group of chitosan and carboxylic compounds. For example, nitrilotriacetic acid (NTA), a complexant which is used in nuclear reactor decontaminations, can bind chitosan through amide bond formation (Figure 6.5). When a mixture of such complexants and various metal ions are present in a solution; the speciation in the solution, the extent of the amine group protonation, etc., which are in turn influenced by the solution pH decide the type of interaction that takes place. Thus, by judicious interplay of various parameters, it is possible to tune the selectivity of chitosan toward a particular analyte in a solution containing a mixture of metal ions and complexants *(vide infra)*.

The amine group of chitosan has a pKa value close to 6.5. Thence, in presence of chelating transition metal ions, chelation could be prominent at pH below 6.5. The point of zero charge pH_{pzc} determines the density of the cationic sites in the chitosan under given pH conditions. Degree of deacetylation, source of chitin, and the chemical modification performed to influence the pH_{pzc} values. Jha et al. [34] have reported point of zero charge as high as 8.6. Unuabonah et al. [35] have modified hybrid clay composite adsorbent (HyDCA) with chitosan using different ratios of HyDCA and

Potential Applications of Chitosan-Based Sorbents 125

chitosan to yield sorbents for Gram-negative bacteria. The report showed the dependence of the pH$_{pzc}$ on the sorbent composition (Figure 6.6).

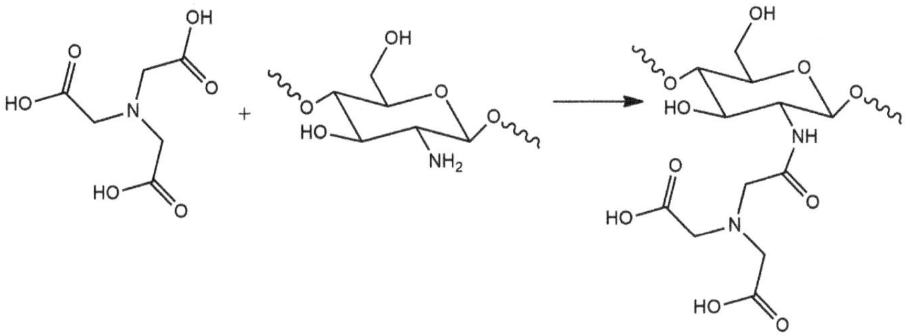

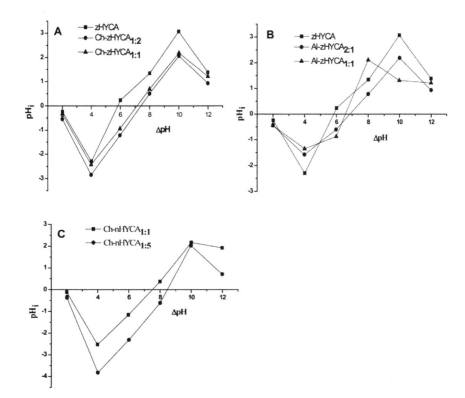

FIGURE 6.5 Reaction of the amine groups of chitosan with NTA to form amide groups.

FIGURE 6.6 Plots showing the pHpzc curves for the sorbents prepared with various ratios of hybrid clay adsorbent and chitosan (reproduced from [35] (doi: https://doi.org/10.1016/j.heliyon.2017.e00379) under Creative Commons license: https://creativecommons.org/licenses/by-nc-nd/4.0/).

Erosa et al. [36] have reported the binding of chitosan through chelation by the amino groups of chitosan. The report showed reduced binding of cadmium cations at the acidic pH and a maximum capacity of 150 mg/g at pH 6. The positive binding sites of chitosan make it highly efficient for the removal of metal ions such as arsenic, antimony, etc., that have high tendency to exist as anionic species in aqueous solutions. This explains the increasing number of reports wherein chitosan has been projected as a sorbent material for the removal of toxic arsenic, in particular from domestic water supplies [37–39]. Saleh et al. [40] modified vermiculite with chitosan to prepare As(III) sorbent. The adsorbent was synthesized by functionalizing the clay with oxalic acid and crosslinking with chitosan through the oxalic acid groups. This methodology ensured efficient and stable incorporation of chitosan. Kwok et al. [38, 41] have brought out the sorption–desorption equilibrium involved in the arsenic binding by chitosan, which is dictated by the solution pH. Lower pH was seen to significantly increase the arsenic uptake as the arsenic speciation is predominated by anions. Antimony, a nuclide of interest to the nuclear industry, has near similar speciation profile as arsenic. Hence, the former is also expected to follow similar binding behavior with these sorbents and hence of interest to the nuclear industry.

Some of the very early reports on chitosan have demonstrated the stability of chitosan against radiation [42]. The report showed gamma irradiation up to 50,000 krad dose did not make chitosan lose its binding affinity for metal ions of interest in the nuclear field, including cobalt. This was one of the earliest reports which showed potential for application of chitosan in the nuclear industry. However, till now none of the chitosan-based products have been seriously considered as an option by nuclear plant operators.

6.4 NEED FOR METAL REMOVING AGENTS IN THE NUCLEAR INDUSTRY

Nuclear power plants, like thermal power plants, must deal with the corrosion associated with interaction of water with the structural materials. Decontamination process, which is carried out for the periodic removal of these corrosion products which will have active nuclides such as cobalt, antimony, etc. trapped within, is an important maintenance procedure carried out through the operating life of nuclear reactors [43]. The dilute chemical decontamination involving removal of the metal oxides by dissolving them using a dilute mixture of complexants such as NTA, and further stripping the

Potential Applications of Chitosan-Based Sorbents 127

metal ions from solution on ion exchange resin beds is a common practice that is followed world over for the decontamination of coolant circuits. The stripping of the metal ions from the circulating decontamination formulation requires the use of ion exchange resins. Further, backend operations involving sequestration of active nuclides generated during waste treatment procedures, and general requirement of demineralization of water require large amounts of polymeric resins. Thus, as in other industries, there is constant need of sorbent materials for the removal of metal ions from aqueous solutions in the nuclear industry.

6.4.1 NEED FOR SELECTIVITY FOR THE RESINS IN THE NUCLEAR INDUSTRY

The metal ion-containing solutions generated during the decontamination process and other similar operations in nuclear power plants typically contain radioactive metal ions, often in significantly low physical quantity, and large excess of nonactive ions. Removal of active nuclides selectively, while excluding the nonactive ions, would lead to extensive savings in terms of the cost involved and easier radioactive waste management processes [44]. This requirement during the regular decontaminations is an important aspect considered by researchers involved in designing sorbents for nuclear applications. During decontamination procedures, which are done by recirculation of a mixture of complexing agents, the metal ions are brought out into the solution and sorbed by the ion-exchange resin beds that form a part of the clean-up circuit [44]. The main corroding surface in primary cooling water circuits of pressurized heavy water reactors is carbon steel. Hence, during chemical decontaminations of these reactors, large quantity of ferrous ion (large part of which is nonradioactive) is released into the decontamination formulation circulated through the coolant circuits as iron oxide is the major corrosion product. Major contribution toward the activity removed during decontamination is from cobalt and antimony nuclides in most nuclear reactors. The source of cobalt activity is the cobalt impurity present in some of the alloys used (nickel-containing alloys of steam generators and to some extent in carbon steel), and some parts that are made of cobalt-containing alloys (e.g., Stellite, which has 50% cobalt) [44]. The nonradioactive cobalt (Co^{59}) released from these parts form active cobalt nuclides (Co^{60}) due to neutron activation when they pass through the core. Similarly, other radioactive isotopes such as antimony are formed due to neutron activation of the nonactive antimony ions released from materials containing antimony such as pump seals. The physical quantity of these isotopes would be extremely less as compared to the large excess of nonactive metal ions that are present in the corrosion products. Thus, during decontamination, very small quantities of radioactive ions are brought out into the solution along with large concentration of ferrous ions.

The volume of resin beds to be used during decontamination is thus primarily dictated by the amount of ferrous ions as these are in large excess. Since the normal ion-exchange resins generally used in reactors lack high selectivity for the radioactive ions in the presence of ferrous ions, the small amount of radioactive ions that is brought out also spreads through all the resin beds employed. This leads to the generation of large volume of radioactive ion-exchange waste which requires costly and elaborate disposal procedures. If a resin exhibiting high selectivity toward radioactive ions such as cobalt and antimony ions in presence of ferrous ions was available, majority of the activity could be trapped within a small volume of the resin [44]. Hence, the volume of radioactive waste can be substantially reduced (Figure 6.7). Thus, metal ion sorbents which have selectivity toward radioactive ions of interest would be highly valued in the nuclear industry due to the benefits they would bring in.

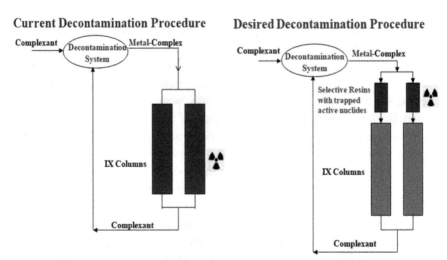

FIGURE 6.7 Regular decontamination process (left) where common IX resins are used, and the possible scenario if resins that are specific for the active nuclides are used in the decontamination process (right) (adapted with permission from [44]. Copyright 2009, American Chemical Society).

Potential Applications of Chitosan-Based Sorbents 129

6.5 CHITOSAN AND CHITOSAN-BASED SORBENTS FOR REMOVAL OF COBALT AND ANTIMONY

As discussed earlier, selective removal of radioactive ions from decontamination formulation can lead to large reduction in the volume of radioactive waste. Our group has studied in detail the metal-binding characteristics of chitosan to understand its utility in applications such as nuclear reactor decontamination where a mixture of complexing agents is used for dissolution of metal oxide deposits from the coolant surfaces. As the bulk of the activity generated during the regular decontamination campaigns come from cobalt and antimony, we have been working toward devising suitable receptors for these two active nuclides in particular. The work carried out by us and some of the reports from other groups on devising sorbents for these are discussed in the following sections.

6.5.1 CHITOSAN-BASED RECEPTORS FOR COBALT

In one of our earliest studies on chitosan, we have investigated the metal-binding properties of chitosan for cobalt with reference to copper binding properties as the latter is known to have strong binding with the chitosan functional groups [45]. The change in sorption observed with change in the solution pH was reflective of the influence of solution pH on the chitosan functional groups. Addition of the complexant NTA into the solution mixture further modified the binding characteristics.

The sorption results obtained clearly showed the operation of three modes of binding by chitosan namely, chelation, electrostatic attraction, and amide formation. For example, when Co(II) and NTA were present, it was seen that the uptake of Co(II) was more than that of NTA (Figure 6.8). This was attributed to sorption of metal ions happening through both chelation and electrostatic attraction (as the anionic metal ion-NTA complex) while NTA being bound only through electrostatic attraction as its anionic metal complex.

The results further showed the change in selectivity in the presence of complexing agent and revealed the feasibility of separation of copper and cobalt by simple change of pH in the presence of complexing agent. In presence of NTA, at an equilibrium pH of 2.9, the selectivity coefficient for Co^{2+} over Cu^{2+} was 2.06, while at pH 6.0, it was 0.072. This indicates a phenomenal change in selectivity which was driven by change in the solution speciation and the charge density on chitosan. This is an important observation as it could help in the separation of the radioactive cobalt in presence of copper. Thus, by making use of the fact that three modes of binding are operable in chitosan, by rationally modifying the chitosan functional groups and/or solution conditions, it may be feasible to obtain the desired selectivity.

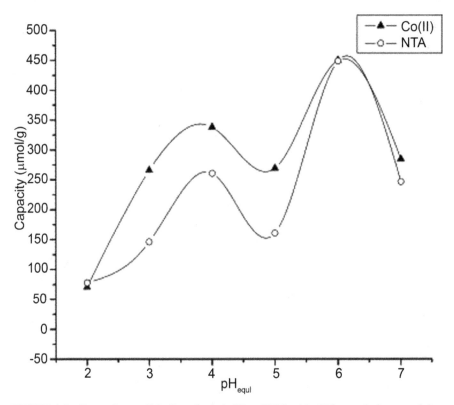

FIGURE 6.8 Dependence of binding of cobalt (II) and NTA with pH from solution containing both by chitosan (reproduced from [45], copyright (2011), with permission from Elsevier).

Chen and Wang [46] have prepared a xanthate modified magnetic chitosan sorbent for the removal of cobalt ions. The sorbent was prepared by crosslinking the xanthate modified magnetic chitosan with glutaraldehyde as the crosslinker. The report showed the influence of the thiol groups in enhancing the cobalt uptake. The authors claimed, based on the spectroscopic characterizations, the binding sites for Co(II) to be consisting of the nitrogen atoms of the amino groups in chitosan and the sulfur atoms of the attached xanthate moieties. The crosslinker's role in stabilizing the sorbent matrix was also brought out.

Potential Applications of Chitosan-Based Sorbents 131

Rebo et al. [47] made use of the strong chelation property of diethyl-enetriaminepentaacetic acid (DTPA) in preparing a chitosan-based sorbent for stripping Co(II) ions from its strong ethylenediaminetetraacetic acid (EDTA) complex. The modified chitosan was prepared by reacting chitosan with DTPA-anhydride in acetic acid–methanol solution. The sorption properties were compared with similar DTPA modified silica gel. The results showed comparable Co(II) removal properties from solutions containing the complexant and Co(II) ions. The chitosan-based sorbent was seen to be more effective than the modified silica gel in removing Co(II) ions at EDTA concentrations higher than that of Co(II) ions. This shows the better utility of chitosan in removing metal ions from complexant solutions through rational modifications of chitosan.

Zhu et al. [48] prepared polyvinyl alcohol-magnetic chitosan composite and showed its Co(II) removal properties. The IR spectra revealed the involvement of the amino and hydroxyl groups in complexing cobalt and a maximum capacity close to 15 mg/g was shown.

Zhuang et al. [49] reported maleic acid grafted chitosan as a sorbent for cobalt. The grafting was done by gamma irradiation. A 40% increase in the capacity for cobalt ions due to modification of chitosan was shown. The capacity obtained was moderate though.

6.5.1.1 COBALT ION IMPRINTING OF CHITOSAN

Encouraged by the results which showed that modification of selectivity of chitosan is feasible (vide supra), our group has further taken up efforts toward modifying the chitosan selectivity. While regular functional group modifications and solution conditions variations may be able to bring in change in selectivity, the order of selectivity needed in the nuclear industry needs better methodologies. Molecular imprinting is known to be a methodology that could lead to receptors containing binding sites with selectivity designed for a particular analyte [50]. In this technique, a functional monomer (or a group of functional monomers) is complexed with the chosen metal ion as the template and subjected to crosslinking polymerization with a crosslinker. Subsequently, the template metal ion is removed leaving binding sites that are expected to be selective for the metal ion used as the template. Bhaskarapillai et al. [44] have shown the extraordinary effect of metal ion imprinting in the synthesis of cobalt-specific polymeric sorbent which excluded excess ferrous ions completely under typical nuclear reactor decontamination conditions. Our group has attempted similar cobalt imprinting on chitosan [51]. The results showed a complete reversal of selectivity of chitosan on imprinting (Figure 6.9). The maximum sorption was seen at pH 4.8 (Figure 6.10a). An imprinting factor, which is a measure of selectivity obtained through imprinting, above 2.0 was shown (Figure 6.10b). The maximum sorption was seen at pH 4.8. This had for the first time showed that it was possible to reverse the selectivity of the biopolymer through the process of metal ion imprinting and encouraged more work on metal ion imprinting of chitosan.

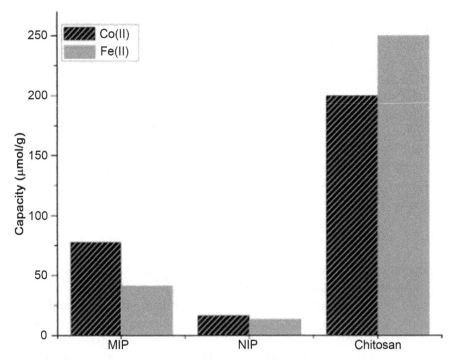

FIGURE 6.9 Reversal of selectivity of chitosan on Co(II) imprinting (reproduced from [51], copyright (2012), with permission from Elsevier).

As the crosslinker can bind through either the hydroxyl or the amino group, depending on the nature of the crosslinker used, the nature of crosslinker used is expected to have a strong influence on the selectivity of the imprinted polymer. For example, cobalt imprinted chitosan synthesized using two different types of crosslinkers show different orders of selectivity confirming the direct influence of the crosslinker on the selectivity.

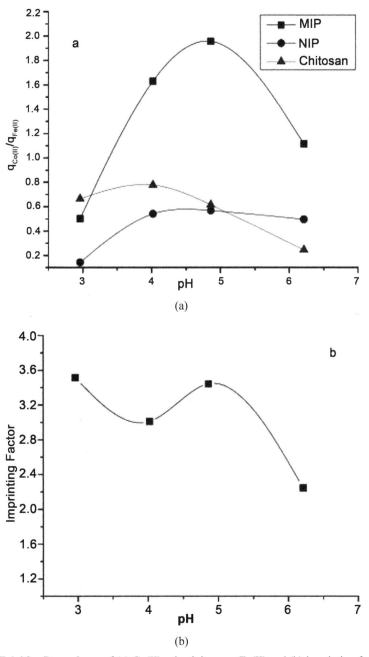

FIGURE 6.10 Dependence of (a) Co(II) selectivity over Fe(II) and (b) imprinting factor of Co(II) imprinted chitosan on pH (reproduced from [51], copyright (2012), with permission from Elsevier).

6.5.2 CHITOSAN COMPOSITE SORBENTS FOR REMOVAL OF ANTIMONY

Many nuclear reactors across the world grapple with the problem of removal of radioactive antimony from the reactor coolant surfaces [52]. The source of antimony activity is the antimony used in the antimony impregnated graphite pump seals. In many plants, this material was brought in as replacement for stellite, a cobalt-containing alloy due to the cobalt activity generated by the latter. The inactive antimony (^{121}Sb and ^{123}Sb) when released into the coolant through wear and tear gets activated to ^{122}Sb ($t_{1/2}$ 2.72 days) and ^{124}Sb ($t_{1/2}$ 60.2 days) under the neutron flux present inside the core. Complex speciation, low solubility, and inefficient removal by the regular resins make this a difficult problem to solve. Metal oxides are known to be able receptors for antimony due to the predominant presence of anionic species of antimony in aqueous solutions. For example, nanotitania was shown to be an excellent sorbent for antimony [53]. The format of the nanotitania, however, is not suitable for use in large scale applications. Zimmerman's group has [54] reported the preparation of chitosan–titania beads and evaluated the arsenic removal properties of the beads. These beads however suffered from instability at lower pH conditions due to the inherent instability of chitosan under such solution conditions. Reports [52, 55, 56] from our group revealed preparation of stable composite beads through judicious combination of titania and crosslinked chitosan (Scheme 6.2). The beads so obtained retained the antimony removal properties of titania and also exhibited stability under wide solution conditions.

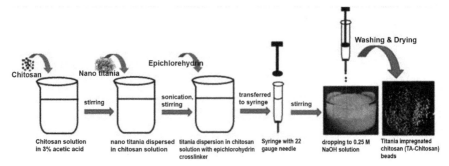

SCHEME 6.2 Methodology used for the synthesis of nanotitania impregnated crosslinked chitosan sorbent beads for antimony removal (reproduced from [52], copyright (2014), with permission from Elsevier).

Through various sorption experiments, it was shown that the nanotitania not only retained the antimony binding properties of titania but that the sorption capacity of the nanotitania was enhanced when present within crosslinked chitosan matrix (Figure 6.11). As much as 69% increase in antimony uptake by the nanotitania was seen on impregnating it within the crosslinked titania matrix.

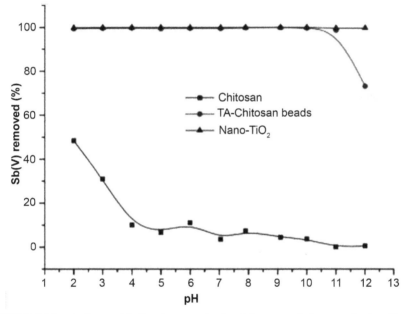

FIGURE 6.11 Antimony binding by nanotitania, chitosan, and the nanotitania chitosan composite beads (reproduced from [52], copyright (2014), with permission from Elsevier).

The epichlorohydrin crosslinked chitosan–titania composite beads were shown to perform well under flow conditions through column experiments (Figure 6.12).

It was also shown that the increase in the amount of epichlorohydrin used in the polymerization led to increase in the capacity. The ability of the beads to remove low-level antimony from typical decontamination formulation containing a mixture of complexing agents namely, NTA, ascorbic acid, and citric acid, was shown through column experiments. Further, it was also shown that the presence of large excess of iron, as to be expected during typical decontamination campaigns, did not negatively influence the antimony uptake. It was shown that the presence of iron actually enhanced the antimony sorption under column conditions. This could be attributed to the fact that iron keeps the complexant as its respective complexes in solution leaving more binding sites free for the antimony species. The radiation stability of the beads was also demonstrated making this a suitable material for application in the nuclear reactor decontaminations or in backend processes.

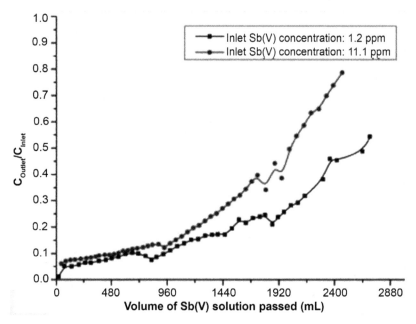

FIGURE 6.12 Removal of antimony on passing the antimony solution through column packed with chitosan-titania composite beads (reproduced from [52], copyright (2014), with permission from Elsevier).

Apart from being a major concern for the nuclear industry, antimony is also classified as a pollutant in water sources. Further to our work, some more reports have been published on the synthesis of chitosan-based sorbents for the removal of antimony. Most of these reports focused on this aspect of antimony removal. Nevertheless, there is high potential for exploring those sorbents as well for the nuclear applications. Lapo et al. [57] have reported chitosan modified with iron(III) beads as a sorbent for Sb(III) removal from aqueous solutions. Optimum pH for sorption was found to be 6.0. It was shown that the incorporation of iron into the chitosan matrix improved the Sb(III) uptake by 47.9%, as compared to neat chitosan. The report did not

Potential Applications of Chitosan-Based Sorbents   137

mention about the uptake of Sb(V) ions. Another work reported the application of chitosan-modified pumice for antimony adsorption from aqueous solutions [58], which also dealt with Sb(III) alone.

6.6 CHITOSAN-BASED SORBENTS FOR RADIOACTIVE NUCLIDES OTHER THAN COBALT AND ANTIMONY

Many fission products are also to be dealt with during the waste processing activities in the nuclear industries. Strontium-90 ($t_{1/2}$ = 28.7 years) is one of the major fission products seen in radioactive waste and hence its removal/separation is of interest in the treatment of radioactive waste generated from nuclear power plants. Chen and Wang reported [59] the synthesis and strontium removal properties of magnetic chitosan beads. These were synthesized by incorporation of noncrosslinked chitosan in magnetite. The beads were shown to exhibit a maximum sorption capacity of 11.6 mg/g and the binding was found to be predominantly through the NH_2 groups of the chitosan. Yin et al. [60] reported an interesting chitosan-based sorbent for the removal of strontium. The sorbent was synthesized by immobilizing *Saccharomyces cerevisiae*, a bacterial species, in magnetic chitosan microspheres. As many related bacilli have been reported to be possessing strontium binding properties, the authors attempted this combination for the synthesis of strontium sorbent which could be considered a completely natural material. The sorbent was shown to exhibit the strontium removal properties associated with the microorganism immobilized in the chitosan matrix. The report showed grid-like structure for the magnetic chitosan with the immobilized *S. cerevisiae* appearing largely shrunk. These reports show the effectiveness of chitosan in acting as a matrix for immobilizing a variety of metal-binding materials.

Xu et al. [61] have synthesized magnetic chitosan resin and further modified the resin with triethylene–tetramine. The chitosan in the magnetic chitosan was reacted with epichlorohydrin before carrying out the modification with triethylene–tetramine. Triethylene–tetramine binds to the chitosan in the resin by replacing the chloride of the epichlorohydrin moiety attached to the chitosan amine groups. This leads to formation of long pendant groups containing multiple amino groups that are conducive for metal ion complexation. The sorption experiments revealed the Th(IV) binding nature of the resin. Sorption carried out with simulated nuclear industry waste solution showed selectivity for Th(IV). A relative selectivity coefficient of above one was obtained for Th(IV) against the common competitor ions present in the waste solution.

138 *Natural Polymers: Perspectives and Applications for a Green Approach*

Wang et al. [62] used epichlorohydrin crosslinked chitosan for removal of uranyl ions from aqueous solutions. Mahfouz et al. [63] reported magnetic chitosan nanoparticles grafted with diethylenetriamine. The sorbent was shown to bind uranyl ions from sulfuric acid solutions, and acidic urea was shown to be an efficient desorbing agent.

Humelnicu et al. [64] prepared a composite of chitosan and clinoptilolite (a natural zeolite) as a sorbent for uranyl and Th(IV) ions. The selectivity obtained for the uranyl and thorium ions was seen to follow the order: $Cu^{2+} > UO_2^{2+} > Fe^{2+} > Al^{3+}$, and $Cu^{2+} > Th^{4+} > Fe^{2+} > Al^{3+}$, respectively. This report, like some of the earlier reports, showed the importance of crosslinking of the chitosan matrix for better stability under varying solution conditions. Also, hydrophilic hydroxyl groups generated by the crosslinking reaction of epichlorohydrin was claimed to enhance the binding properties. The resin showed higher sorption capacity for UO_2^{2+} as compared to Th(IV), which was attributed to relatively higher hydration radius of the thorium ions. Another chitosan composite with natural clay material namely bentonite was reported by Anirudhan and Rijith [65]. The composite—poly(methacrylic acid)-grafted chitosan/bentonite—was prepared through graft copolymerization of methacrylic acid and chitosan in the presence of bentonite, and N,N'-methylenebisacrylamide as a crosslinker. The uranyl binding ability of the composite was demonstrated using simulated nuclear waste solution. The use of inorganic materials such as clays and zeolites to prepare chitosan composites is advantageous as they can enhance the radiation stability. The proven uranyl binding nature of these composites makes them a potential material for use in the nuclear industry provided more studies are undertaken.

Elwakeel et al. [66] modified magnetic chitosan with tetraethylenepentamine (TEPA) to prepare a uranyl sorbent. In this synthesis, glutaraldehyde was used for the crosslinking. The authors proposed that the binding of uranyl ions by the resin was through chelation of the uranyl ions by TEPA groups in the modified chitosan. The chelation was suggested to be involving the formation of eight (five-membered) chelating rings through the interactions between the uranyl ions with the NH_2 and OH groups of TEPA. The high stability of the complex formed was claimed to be the reason for the high uranyl removal efficiency of the sorbent. The report also showed the applicability of the resin under column conditions. The maximum sorption capacity of the column at slower flow rates was seen to be close to the capacity obtained in the batch sorption experiments.

Schiff bases, which are formed by a nucleophilic addition reaction between amino and carbonyl compounds, are known to have metal ion

Potential Applications of Chitosan-Based Sorbents 139

binding properties. The metal ion coordination involves the azomethine nitrogen of the Schiff bases [67–69]. Thus, by the incorporation of suitable Schiff bases in solid sorbent matrices, metal ion binding properties of the adsorbents can be modified. This approach has been attempted in modifying the sorption properties of chitosan as well. Elwakeel and Aita [70] have prepared chitosan modified with magnetite and Schiff's base of thiourea and glutaraldehyde. The former, an amino compound, and the latter a carbonyl compound and a well-known crosslinker for chitosan, formed the Schiff base which was then reacted with chitosan–magnetite composite. The chitosan amino group formed Schiff base and yielded a sorbent containing Schiff base moieties in the chitosan matrix (Scheme 6.3).

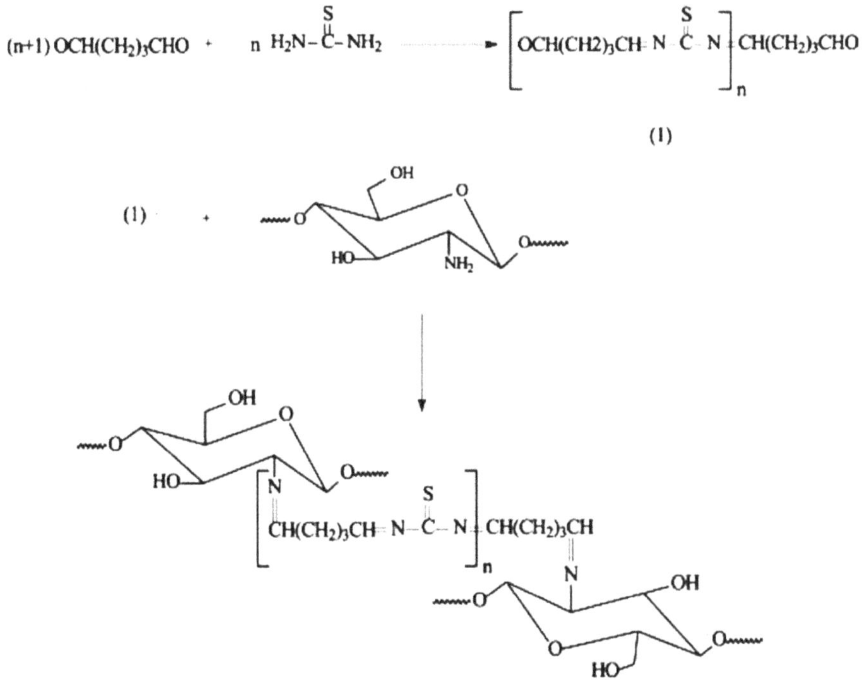

SCHEME 6.3 Scheme for introduction of Schiff base moities on the chitosan matrix (reproduced from [66], Copyright (2014), with permission from Elsevier).

The resultant sorbent was shown to exhibit uranyl binding properties. The pH effect was however seen to reflect the typical behavior of chitosan containing free amino groups.

Sureshkumar et al. [71] have shown that sodium tripolyphosphate cross-linked chitosan beads exhibit good uranyl removal properties. Beads with two different crosslinking densities were prepared by carrying out the cross-linking reaction at different pH values. It was shown that higher crosslinking led to increased uranyl binding and that the binding occurred through the phosphate groups. The beads were prepared through a procedure reported by Lee et al. for ionic crosslinking of chitosan with sodium tripolyphosphate, which involved the addition of acetic acid solution of chitosan to alkaline solution of tripolyphosphate (Figure 6.13) [72].

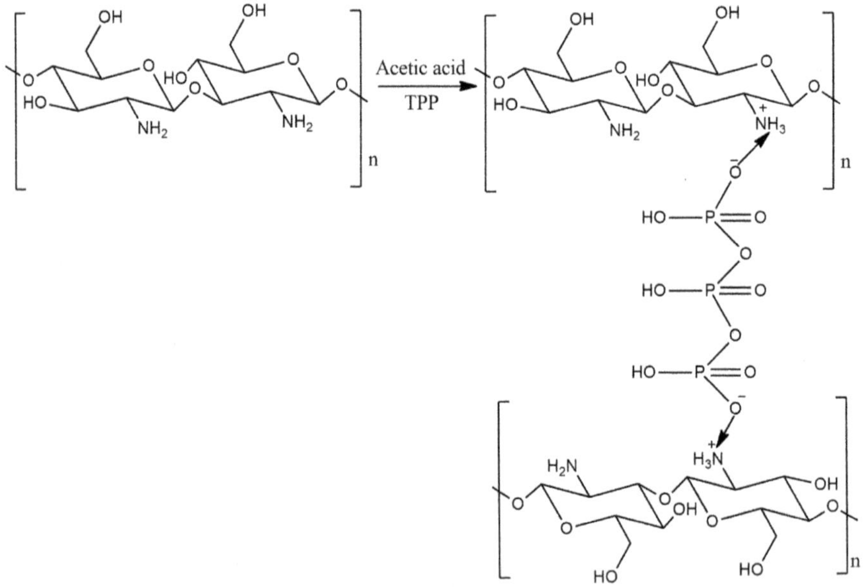

FIGURE 6.13 Crosslinking of chitosan with sodium tripolyphosphate (adapted from [71]).

Chitosan also has been explored as an additive to other cheap natural sorbents such as coal to improve the stability and sorption properties toward radioactive ions Cs, Co, and Eu [73]. The report showed that the influence of chitosan on modifying the binding properties of coal was pH-dependent. Using metal ion imprinting as a methodology to improve the selectivity of chitosan toward uranyl ions has been reported, which showed moderate selectivity against thorium and a few transition metal ions.

Lupa et al. [74] successfully impregnated chitosan with an ionic liquid that has binding affinity toward Cs and Sr. The resultant sorbent was shown to prefer cesium over strontium. The ionic liquid impregnation was done by

Potential Applications of Chitosan-Based Sorbents

ultrasonication. There is vast literature on the use of ionic liquids for separation of metal ions [75] of interest to nuclear industry. Ionic liquid is efficient binders of other nuclides of interest such as iodine. For example, a very recent report demonstrated the very high capacity of ionic liquid derived polymer for different species of iodine [76]. Thus, this methodology offers promise in expanding the use of chitosan as a matrix for devising sorbents of interest to the nuclear industry.

A recent report [77] revealed the preparation of a chitosan-based ternary composite sorbent which was prepared by caging copper nickel cobalt hexacyanoferrate (an analog of the well-known cesium removal agent Prussian blue) caged in chitosan surface-decorated carbon nanotubes. The sorbent was shown to remove cesium and strontium. The interference from other cations was shown to follow the order $K^+ > Ca^{2+} > Mg^{2+} > Na^+$ for adsorption of cesium, and $Mg^{2+} > K^+ > Na^+ > Ca^{2+}$ for adsorption of strontium. Though the advantage offered by incorporation of chitosan into a well-known cesium sorbent was not brought out clearly in the report, it nevertheless has detailed an interesting option wherein the salient properties of carbon nanotubes such as high surface area and robust physical stability was made use of in preparing a chitosan-based sorbent. The key advantage of this material is the near pH-independent uptake of Cs and Sr, which was over 200 mg/g. Table 6.1 summarizes some of the chitosan-based sorbents reported for the removal of various metal ions of interest to the nuclear industry.

6.7   CONCLUSION

Chitosan, due primarily to its amenability to various chemical modifications and ease of forming composites with a variety of materials shows enormous potential for application in the nuclear industry. Though there have been multitude of reports on the synthesis and evaluation of chitosan-based sorbents for active nuclides, there are very few reports that address the real-life situations faced in the nuclear industry. For example, it is very important that the sorbent is stable to irradiation, have faster kinetics, and high selectivity (preferably specificity). Even if incorporating these properties involve an increase in cost, due to the nature of the waste that is to be handled, any nominal increase should be acceptable to the nuclear industry. Further, being able to mass-produce the sorbent in a format suitable for use in column mode would enormously increase the utility. Many reported studies do not take the investigation toward that direction and instead stop with batch studies and evolving sorption isotherms. Hence, the need of the hour is to take

TABLE 6.1 List of Some of the Chitosan-Based Sorbents Reported for the Removal of Metal Ions of Interest to Nuclear Industry

Metal Ion	Sorbent	Remarks	Ref.
UO_2^{2+}	Tetraethylenepentamine modified magnetic chitosan	Capacity: 1.8 mmol/g;	[66]
		Desorption medium (desorption from uranyl loaded column): 0.2 M HCl	
		Sorption shown under batch and flow conditions	
	Diethylenetriamine functionalized magnetic chitosan	Capacity: 0.75 mmol/g	[63]
		Desorption medium: acidified urea	
	N-[2-(1,2- dihydroxyethyl)tetrahydrofuryl] chitosan (ascorbic acid derivative of chitosan	Capacity: 3.4 mmol/g	[78]
		Desorption medium: not studied	
		Prepared by reducing ascorbic acid solution of chitosan with sodium cyanoborohydride	
	Chitosan–tripolyphosphate (CTPP) beads	Capacity: 1 mmol/g;	[71]
		Synthesized by crosslinking chitosan with sodium tripolyphosphate.	
		Binding was shown to be through the phosphate groups of the crosslinker	
	Uranyl imprinted magnetic chitosan	Capacity: 0.68 mmol/g	[79]
		Desorption medium: 0.5 M HNO_3	
		Prepared through metal ion imprinting method; Shown to be selective against Th(IV); Fe(III); Cu(II); Co(II), Ni(II), and Zn(II);	
	Chitosan–attapulgite composite	Capacity: 0.05 mmol/g	[80]
		Desorption medium: humic substances (reversible uptake in presence of humic acid) Emphasis on removal of uranium from contaminated natural water sources containing humic and fulvic acids.	
Cs^+	Porous magnetic bentonite–chitosan beads	Capacity: 0.43 mmol/g	[81]
		Desorption medium : 0.1 M Mg^{2+} solution	
		Binding through chelation with the chitosan amine groups and cation exchange with bentonite.	

TABLE 6.1 *(Continued)*

Metal Ion	Sorbent	Remarks	Ref.
	Chitosan–grafted magnetic bentonite	Capacity: 1.21 mmol/g Desorption medium: not studied Argon plasma was used for grafting chitosan onto magnetic bentonite. Interference from alkali and alkaline earth metal ions were seen	[82]
Sr^{2+}	Saccharomyces cerevisiae immobilized in magnetic chitosan microspheres	Capacity: 0.94 mmol/g Desorption medium: not studied The binding was primarily attributed to the suite of functional groups present on the cell walls of the immobilized microbe. The report has shown the suitability of chitosan as a matrix for immobilization of microbes with metal-binding properties.	[60]
	Chitosan iron oxide composite	Capacity : 0.62 mmol/g Desorption medium: not studied Higher oxide content in the composite was shown to destabilize the sorbent. Chitosan was seen to provide the stability while retaining the Sr removal properties of the iron oxide in the composite.	[83]
	Chitosan modified graphene oxide	Capacity: 2.1 mmol/g Desorption medium: not studied A composite of graphene oxide, known to be an excellent receptor for Sr, and chitosan	[84]
Co^{2+}	Chitosan–montmorillonite	Capacity: 2.5 mmol/g. Desorption medium: sorption was reported to be irreversible making regeneration difficult. Composite was synthesized as 100 mesh size particles. Chitosan to Montmorillonite ratio of 0.25 was shown to have the highest uptake	[85]

TABLE 6.1 *(Continued)*

Metal Ion	Sorbent	Remarks	Ref.
	Chitosan-coated perlite beads	Capacity: 1.0 mmol/g	[86]
		Desorption medium: 0.1 M NaOH	
	Porous carboxymethyl chitosan beads	Capacity: 0.76 mmol/g	[87]
		Desorption medium: 0.1 M HCl	
		Carboxymethyl derivative of chitosan was used. PEG was used as a porogen, and calcium chloride and glutaraldehyde were used as physical and chemical crosslinkers, respectively.	
	Chitosan	Capacity: 0.40 mmol/g	[45]
		Desorption medium: not studied	
		The report focused on the binding mechanism involved in cobalt uptake and the influence of a complexing agent (NTA) on the uptake. It was shown that, in presence of NTA, selectivity between Co^{2+} and Cu^{2+} could be modified by the control of solution pH.	
	Cobalt imprinted chitosan	Capacity: 0.075 mmol/g (in citrate medium)	[51]
		Desorption medium: 0.25 M H_2SO_4	
		The selectivity of chitosan was reversed through metal ion imprinting in favor of cobalt over ferrous ions. First report of selectivity reversal of chitosan through metal ion imprinting.	
Ru	Polyethylenimine coated chitosan-bacterial biomass composite	Capacity: 1.1 mmol/g	[88]
		Desorption medium: not studied	
		Acetic acid waste solution containing ruthenium was used as a model waste solution. Oxidation state of the Ru in the solution was not determined. The ruthenium uptake was shown to be 7 times higher in the composite than with the raw bacterial mass, and 17 times higher than a commercial strong base anion resin. Batch and column studies were done.	

TABLE 6.1 *(Continued)*

Metal Ion	Sorbent	Remarks	Ref.
	Ru(III) imprinted beads based on 2-pyridylthiourea modified chitosan	Capacity: 2.5 mmol/g Desorption medium: 0.5 M HCl/thiourea solution Clear effect of imprinting on enhancing the selectivity for Ru(III) was shown. Selectivity coefficient values between 5.5 and 12.4 were reported.	[89]
Pd(II)	Glycine-modified crosslinked chitosan resin	Capacity: 1.1 mmol/g Desorption medium: 0.7 M thiourea + 2 M HCl solution The resin removed Pt(IV) and Au(III) as well	[90]
	L-lysine modified crosslinked chitosan resin	Capacity: 1.03 mmol/g Desorption medium: 0.7 M thiourea+2 M HCl solution Showed uptake of Pt(IV) and Au(III) as well.	[91]
	Chitosan hydrogel	Capacity: 1.8 mmol/g; Desorption medium: 0.25 M thiourea solution Hydrogels were prepared by ionotropic gelation in polyphosphate. The sorption mechanism was suggested to involve electrostatic attraction of the chloro-anionic species onto protonated amine groups	[92]
Sb(V); Sb(III)	Nanotitania-crosslinked chitosan composite beads	Capacity: up to 0.49 mmol/g (depended on bead composition and solution conditions) Desorption medium: 5 M HCl Various parameters controlling the antimony removal have been delineated. Column experiments under typical nuclear reactor decontamination conditions demonstrated the suitability of the beads for use during nuclear reactor decontaminations for antimony removal.	[52] [55] [56]

146   Natural Polymers: Perspectives and Applications for a Green Approach

into account the requirements and the operating conditions in the nuclear industry and design the sorbents accordingly. As has been discussed above, considering the ease of chemical modifications and formation of composite beads with chitosan, concerted efforts would yield ideal sorbents for nuclear industry. Strategies that lead to nil or reduced use of organic crosslinkers for the synthesis of chitosan sorbents could further increase the appeal of the sorbents. Judicious choice of inorganic oxides and other inorganic materials to derive chitosan composites could prove to be a useful strategy. With the projected rapid growth in the nuclear industry in countries like India, this may prove to be an effective strategy to decrease the use of large amounts of synthetic organic chemicals used in the synthesis of organic resins.

KEYWORDS

- chitosan
- biocompatibility
- nuclear industry
- radioactive

REFERENCES

1. Global Market Insights. https://www.gminsights.com/industry-analysis/chitosan-market (accessed May 01 2019).
2. Clark, G.L.; Smith, A. F. X-ray Diffraction Studies of Chitin, Chitosan, and Derivatives. *J. Phys. Chem.* **1996**, *40*, 863–879.
3. Ogawa, K.; Yui, T.; Okuyama, K. Three D Structures of Chitosan. *Int. J. Biol. Macromol.* **2004**, *34*, 1–8.
4. Okuyama, K.; Noguchi, K.; Miyazawa, T.; Yui, T.; Ogawa, K. Molecular and Crystal Structure of Hydrated Chitosan. *Macromolecules* **1997**, *30*, 5849–5855.
5. Barbara, B; Ilenia, D. A; Sabrina, S; Amelia, G; Attilio, C. "The Good, the Bad and the Ugly" of Chitosans. *Mar. Drugs* [Online] **2016**, *14(5)*, 99 https://www.mdpi.com/1660-3397/14/5/99 (accessed May 01 2019).
6. Pillai, C. K. S; Paul, W.; Sharma, C. P. Chitin and Chitosan Polymers: Chemistry, Solubility and Fiber Formation. *Prog. Polym. Sci.* **2009**, *34*, 641–678.
7. Nishad, P. A.; Bhaskarapillai, A.; Velmurugan, S. Enhancing the Antimony Sorption Properties of Nano Titania-Chitosan Beads using Epichlorohydrin as the Crosslinker. *J. Hazard. Mater.* **2017**, *334*, 160–167.

Potential Applications of Chitosan-Based Sorbents

8. Filipkowska, U.; Jóźwiak, T. Application of Chemically-Cross-Linked Chitosan for the Removal of Reactive Black 5 and Reactive Yellow 84 Dyes from Aqueous Solutions. *J. Polym. Eng.* **2013**, *33*, 735–747.

9. Dung, P.; Milas, M.; Rinaudo, M.; Desbrières, J. Water Soluble Derivatives Obtained by Controlled Chemical Modifications of Chitosan. *Carbohydr. Polym.* **1994**, *24(3)*, 209–214.

10. Pylypchuk, L. V.; Kołodyńska, D.; Kozioł, M.; Gorbyk, P. P. Gd-DTPA Adsorption on Chitosan/Magnetite Nanocomposites. *Nanoscale Res. Lett.* [Online] **2016**, *11*, 168 https://nanoscalereslett.springeropen.com/articles/10.1186/s11671-016-1363-3 (accessed May 01 2019).

11. Maurya, V. K.; Inamdar, N. N. Chitosan-Modifications and Applications: Opportunities Galore. *React. Funct. Polym.* **2008**, *68(6)*, 1013–1051.

12. Jain, A.; Gulbake, A.; Shilpi, S.; Jain, A.; Hurkat, P.; Jain, S. A New Horizon in Modifications of Chitosan: Syntheses and Applications. *Crit. Rev. Ther. Drug Carrier Syst.* **2013**, *30*, 91–181.

13. Kleine-Brueggeney, H.; Zorzi, G.K.; Fecker, T.; El Gueddari, N. E.; Moerschbacher, B. M.; Goycoolea, F. M. A Rational Approach Towards the Design of Chitosan-Based Nanoparticles Obtained by Ionotropic Gelation. *Colloids Surf. B.* **2015**, *135*, 99–108.

14. George Z. Kyzas, G. Z.; Bikiaris, D. N. Recent Modifications of Chitosan for Adsorption Applications: A Critical and Systematic Review. *Mar. Drugs* **2015**, *13(1)*, 312–337.

15. Wan Ngah, W. S.; Teonga, L. C.; Hanafiahab, M. A. K. M. Adsorption of Dyes and Heavy Metal Ions by Chitosan Composites: A Review. *Carbohydr. Polym.* **2011**, *83* 1446–1456.

16. Bhatnagar, A.; Sillanpaa, M. Applications of Chitin-and Chitosan-Derivatives for the Detoxification of Water and Wastewater—A Short Review. *Adv. Colloid Interface Sci.* **2009**, *152(1–2)*, 26–38.

17. Zhang, J.; Xia, W.; Liu, P.; Cheng, Q.; Tahirou, T.; Gu, W.; Li, B. Chitosan Modification and Pharmaceutical/Biomedical Applications. *Mar. Drugs* **2010**, *8*, 1962–1987.

18. Ahsan, S.M.; Thomas, M.; Reddy, K.K.; Sooraparaju, S.G.; Asthana, A.; Bhatnagar, I. Chitosan as Biomaterial in Drug Delivery and Tissue Engineering. *Int. J. Biol. Macromol.* **2018** *110*, 97–109.

19. Wen, Z. S.; Xu, Y. L.; Zou, X. T.; Xu, Z. R. Chitosan Nanoparticles Act as an Adjuvant to Promote Both Th1 and Th2 Immune Responses Induced by Ovalbumin in Mice. *Mar. Drugs* **2011**, *9*, 1038–1055.

20. Venkatesan, J.; Kim, S. K. Chitosan Composites for Bone Tissue Engineering—An Overview. *Mar. Drugs* **2010**, *8*, 2252–2266.

21. Jimtaisong, A.; Saewan, N. Utilization of Carboxymethyl Chitosan in Cosmetics. *Int. J. Cosmet. Sci.* **2014**, *36*, 12–21.

22. Libio, I. C.; Demori, R.; Ferrão, M. F.; Lionzo, M. I. Z.; da Silveira, N. P. Films Based on Neutralized Chitosan Citrate as Innovative Composition for Cosmetic Application. *Mater. Sci. Eng. C Mater. Biol. Appl.* **2016**, *67*, 115–124.

23. Xu, J.; Strandman, S.; Zhu, J. X. X.; Barralet, J.; Cerruti, M. Genipin-Crosslinked Catechol-Chitosan Mucoadhesive Hydrogels for Buccal Drug Delivery. *Biomaterials* **2015**, *37*, 395–404.

24. Wang, H.; Qian, J.; Ding, F. Emerging Chitosan-Based Films for Food Packaging Applications. *J. Agric. Food Chem.* **2018**, *66(2)*, 395–413.

25. Chen, J. K.; Yeh, C. H.; Wang, L. C.; Liou, T. H.; Shen, C. R.; Liu, C. L. Chitosan, the Marine Functional Food, is a Potent Adsorbent of Humic Acid. *Mar. Drugs* **2011**, *9*, 2488–2498.

26. Cheng, M.; Gao, X.; Wang, Y.; Chen, H.; He, B.; Xu, H.; Li, Y.; Han, J.; Zhang, Z. Synthesis of Glycyrrhetinic Acid-Modified Chitosan 5-Fluorouracil Nanoparticles and its Inhibition of Liver Cancer Characteristics In Vitro and In Vivo. *Mar. Drugs* **2013**, *11*, 3517–3536.

27. Muzzarelli, R. A. A. Potential of Chitin/Chitosan-Bearing Materials for Uranium Recovery: An Interdisciplinary Review. *Carbohydr. Polym.* **2011**, *84(1)*, 54–63.

28. Varma, A. J.; Deshpande, S. V.; Kennedy, J. F. Metal Complexation by Chitosan and its Derivatives: A Review. *Carbohydr. Polym.* **2004**, *55(1)*, 77–93

29. Zhang, L.; Zeng, Y.; Cheng, Z. Removal of Heavy Metal Ions Using Chitosan and Modified Chitosan: A Review. *J. Mol. Liq.* **2016**, *214*, 175–191.

30. Vakili, M.; Rafatullah, M.; Salamatinia, B.; Abdullah, A. Z.; M. H.; Tan, K. B.; Gholami, Z.; Amouzgar, P. Application of Chitosan and its Derivatives as Adsorbents for Dye Removal from Water and Wastewater: A Review. *Carbohydr. Polym.* **2014**, *113*, 115–130.

31. Muzzarelli, R. R. A.; Rocchetti. R. The Use of Chitosan Columns for the Removal of Mercury from Waters. *J. Chromatogr.* **1974**, *96*, 115–121.

32. Muzzarelli, R. R. A.; Rocchetti. R. The Determination of Vanadium in Sea Water by Hot Graphite Atomic Absorption Spectrometry on Chitosan after Separation from Salt. *Anal. Chim. Acta* **1974**, *70*, 283–289.

33. Muzzarelli, R. A. A.; Rocchetti, R.; Marangio, G. *J.* Separation of Zirconium, Niobium, Cerium and Ruthenium on Chitin and Chitosan Columns for the Determination of Cesium in Nuclear Fuel Solutions. *Radioanal. Nucl. Chem.* **1972**, *10(1)*, 17–25.

34. Jha, I.N.; Iyengar, L.; Prabhakara Rao, A. V. S. Removal of Cadmium Using Chitosan. *J. Environ. Eng.* **1988**, *114*, 962–974.

35. Unuabonah, E. I.; Adewuyi, A.; Kolawole, M. O.; Omorogie, M. O.; Olatunde, O. C.; Fayemi, S. O.; Günter, C.; Okoli, C. P.; Agunbiade, F. O.; Taubert, A. *Heliyon* [Online], **2017**, *3*, Article e00379. https://www.heliyon.com/article/e00379 (accessed May 01 2019).

36. Erosa, M. S. D.; Medina, T. I. S.; Mendoza, R. N.; Rodriguez, M. A.; Guibal, E. Cadmium Sorption on Chitosan Sorbents: Kinetic and Equilibrium Studies. *Hydrometallurgy* **2001**, *61(3)*, 157–167.

37. Sacco, L. D.; Masotti, A. Chitin and Chitosan as Multipurpose Natural Polymers for Groundwater Arsenic Removal and As2O3 Delivery in Tumor Therapy. *Mar. Drugs,* **2010**, *8(5)*, 1518–1525.

38. Kwok, K. C. M.; Koong, L. F.; Chen, G.; McKay, G. Mechanism of Arsenic Removal Using Chitosan and Nanochitosan. *J. Colloid Interface Sci.* **2014**, *416*, 1–10.

39. Elson, C. M., Davies, D. H.; Hayes, E. R. Removal of arsenic from Contaminated Drinking Water by a Chitosan-Chitin Mixture. *Water Res.* **1980**, *14(9)*, 1307–1312.

40. Saleh, A. T.; Sari, A.; Tuzen, M. Chitosan-Modified Vermiculite for As(III) Adsorption from Aqueous Solution: Equilibrium, Thermodynamic and Kinetic Studies. *J. Mol. Liq.* **2016**, *219*, 937–945.

41. Kwok, K. C. M.; Koong, L. F.; Al Ansari, T.; McKay, G. Adsorption/Desorption of Arsenite and Arsenate on Chitosan and Nanochitosan. *Environ. Sci. Pollut. Res. Int.* **2018**, *25(15)*, 14734–14742.

42. Muzzarelli, R. A. A.; Tubertini, O. *J. Radioanal. Nucl. Chem.,* **1972**, *12(1)*, 431.

43. Sathyaseelan, V. S.; Rufus, A. L.; Subramanian, H.; Bhaskarapillai, A.; Wilson, S.; Narasimhan, S. V.; Velmurugan, S. High Temperature Dissolution of Oxides in Complexing Media. *J. Nucl. Mater.* **2011**, *419 (1–3)*, 39–45.

44. Bhaskarapillai, A.; Sevilimedu, N. V.; Sellergren, B. Synthesis and Characterization of Imprinted Polymers for Radioactive Waste Reduction. *Ind. Eng. Chem. Res.* **2009**, *48 (8)*, 3730–3737.

45. Padala, A. N.; Bhaskarapillai, A.; Velmurugan, S.; Narasimhan, S. V. Sorption Behaviour of Co (II) and Cu (II) on Chitosan in Presence of Nitrilotriacetic Acid. *J. Hazard. Mater.* **2011**, *191(1–3)*, 110–117.

46. Chen, Y.; Wang, J. The Characteristics and Mechanism of Co (II) Removal from Aqueous Solution by a Novel Xanthate-Modified Magnetic Chitosan. *Nucl. Eng. Des.* **2012**, *242*, 452–457.

47. Repo E.; Malinen, L.; Koivula, R.; Harjula, R.; Sillanpaa, M. Capture of Co (II) from its Aqueous EDTA-Chelate by DTPA-Modified Silica Gel and Chitosan. *J. Hazard. Mater.* **2011**, *187*, 122–132.

48. Zhu, Y.; Hu, J.; Wang, J. Removal of Co^{2+} from Radioactive Wastewater by Polyvinyl Alcohol (PVA)/Chitosan Magnetic Composite. *Prog. Nucl. Energy.* **2014**, *71*, 172–178.

49. Zhuang, S.; Yin, Y.; Wang, J. Removal of cobalt ions from aqueous solution using chitosan grafted with maleic acid by gamma radiation. *Nucl. Eng. Technol.* **2018**, *50*, 211–215.

50. Bhaskarapillai, A; Narasimhan, S. V. A Comparative Investigation of Copper and Cobalt Imprinted Polymers: Evidence for Retention of the Solution-State Metal Ion–Ligand Complex Stoichiometry in the Imprinted Cavities. *RSC Adv.* **2013**, *3 (32)*, 13178–13182.

51. Nishad, P. A.; Bhaskarapillai, A.; Velmurugan, S.; Narasimhan, S. V. Cobalt (II) Imprinted Chitosan for Selective Removal of Cobalt During Nuclear Reactor Decontamination. *Carbohydr. Polym.* **2012**, 87(4), 2690–2696.

52. Nishad, P. A.; Bhaskarapillai, A.; Velmurugan, S. Nano-Titania-Crosslinked Chitosan Composite as a Superior Sorbent for Antimony(III) and (V). *Carbohydr. Polym.* **2014**, *108*, 169–175.

53. Nishad, P. A.; Bhaskarapillai, A.; Velmurugan, S. Towards Finding an Efficient Sorbent for Antimony: Comparative Investigations on Antimony Removal Properties of Potential Antimony Sorbents. *Int. J. Environ. Sci.* **2017**, 14(4), 777–784.

54. Miller, S. M.; Spaulding, M. L.; Zimmerman, J. B. Optimization of Capacity and Kinetics for a Novel Bio-Based Arsenic Sorbent, TiO_2-Impregnated Chitosan Bead. *Water Res.* **2011**, *45(17)*, 5745–5754.

55. Nishad, P. A.; Bhaskarapillai, A.; Velmurugan, S. Enhancing the Antimony Sorption Properties of Nano Titania-Chitosan Beads Using Epichlorohydrin as the Crosslinker. *J. Hazard. Mater.* **2017**, *334*, 160–167.

56. Nishad, P. A.; Bhaskarapillai, A.; Velmurugan, S. Removal of Antimony over Nano Titania–Impregnated Epichlorohydrin-Crosslinked Chitosan Beads from a Typical Decontamination Formulation. *Nucl. Technol.* **2017**, *197(1)*, 88–98.

57. Lapo, B.; Demey, H.; Carchi, T.; Sastre, A. M. Antimony Removal from Water by a Chitosan-Iron(III)[ChiFer(III)] Biocomposite. *Polymers* [Online] **2019**, *11*, 351 https://www.mdpi.com/2073-4360/11/2/351 (accessed May 05 2019).

58. Sari, A.; Tuzen, M.; Kocal, I. Application of Chitosan-Modified Pumice for Antimony Adsorption from Aqueous Solution. *Environ. Prog. Sustain Energy.* **2017**, *36(*6), 1587–1596.

59. Chen Y.; Wang, J. Removal of Radionuclide Sr^{2+} Ions from Aqueous Solution using Synthesized Magnetic Chitosan Beads. *Nucl. Eng. Des.* **2012**, *242*, 445–451.

60. Yin, Y.; Wang, J.; Yang, X.; Li, W. Removal of Strontium Ions by Immobilized Saccharomyces Cerevisiae in Magnetic Chitosan Microspheres. *Nucl. Eng. Technol.* **2017**, *49*, 172–177.

61. Xu, J.; Zhou, L.; Jia, Y.; Liu, Z.; Adesina, A. Adsorption of Thorium(IV) Ions from Aqueous Solution by Magnetic Chitosan Resins Modified with Triethylene-Tetramine. *J. Radioanal. Nucl. Chem.* **2015**, *303*, 347–356.

62. Wang, G.; Liu, J.; Wang, X.; Xie, Z.; Deng, N. Adsorption of Uranium (VI) from Aqueous Solution onto Cross-Linked Chitosan. *J. Hazard. Mater.* **2009**, *168(2–3)*, 1053–1058.

63. Mahfouz, M. G.; Galhoum, A. A.; Gomaa, N. A.; Abdel-Rehem, S. S.; Atia, A. A.; Vincent, T.; Guibal, E. Uranium Extraction Using Magnetic Nano-Based Particles of Diethylenetriamine-Functionalized Chitosan: Equilibrium and Kinetic Studies. *Chem. Eng. J.* **2015**, *262*, 198–209.

64. Humelnicu, D.; Dinu, M. V.; Dragan, E. S. Adsorption Characteristics of UO22+ and Th4+ Ions from Simulated Radioactive Solutions onto Chitosan/Clinoptilolite Sorbents. *J. Hazard. Mater.* **2011**, *185*, 447–455.

65. Anirudhan, T. S.; Rijith, S. Synthesis and Characterization of Carboxyl Terminated Poly(Methacrylic Acid) Grafted Chitosan/Bentonite Composite and its Application for the Recovery of Uranium(VI) from Aqueous Media. *J. Environ. Radioact.* **2012**, *106*, 8–19.

66. Elwakeel, K. Z.; Atia, A. A; Guibal, E Uptake of U (VI) from Aqueous Media by Magnetic Schiff's Base Chitosan Composite. *J. Clean. Prod.* **2014**, *70*, 292–302.

67. Yousif, E.; Majeed, A.; Al-Sammarrae, K.; Salih, N.; Salimon, J.; Abdullah, B. Metal Complexes of Schiff Base: Preparation, Characterization and Antibacterial Activity. *Arab. J. Chem.* **2017**, *10*, S1639–S1644.

68. Amiri, A.; Saadati-Moshtaghin, H. R.; Abdar, A.; Zonoz, F. M. Magnetic Solid-Phase Extraction Using Schiff Base Ligand Supported on Magnetic Nanoparticles as Sorbent Combined with Dispersive Liquid-Liquid Microextraction for the Extraction of Phenols from Water Samples. *Int. J. Environ. Anal. Chem.* **2018**, *98(11)*, 1017–1029.

69. Ben-Saber, S. M.: Maihub, A. A.; Hudere, S. S.; El-Ajaily, M. M. Complexation Behavior of Schiff Base Toward Transition Metal Ions. *Microchem. J.* **2005**, *81(2)*, 191–194.

70. Elwakeel, K. Z.; Atia, A. A; Guibal, E. Fast Removal of Uranium from Aqueous Solutions using Tetraethylenepentamine Modified Magnetic Chitosan Resin. *Bioresour. Technol.* **2014**, *160*, 107–114.

71. Sureshkumar M. K., Das, D.; Mallia, M. B.; Gupta, P. C. Adsorption of Uranium from Aqueous Solution Using Chitosan-Tripolyphosphate (CTPP) Beads. *J. Hazard. Mater.* **2010**, *184(1–3)*, 65–72.

72. Lee, S. T.; Mi, F. L.; Shen, Y. J.; Shyu, S. S. Equilibrium and Kinetic Studies of Copper (II) Ion Uptake by Chitosan-Tripolyphosphate Chelating Resin. *Polymer*, **2001**, *42*, 1879–1892.

73. Mizera, J.; Mizerova, G.; Machovic, V.; Borecka, L. Sorption of Cesium, Cobalt and Europium on Low-Rank Coal and Chitosan. *Water Res.* **2007**, *41(3)*, 620–626.

74. Lupa, L.; Voda, R.; Popa, A. Adsorption Behavior of Cesium and Strontium onto Chitosan Impregnated with Ionic Liquid. *Sep. Sci. Technol.* **2018**, *53(7)*, 1107–1115.

75. Sun, X.; Luo, H.; Dai, S. Ionic Liquids-Based Extraction: A Promising Strategy for the Advanced Nuclear Fuel Cycle. *Chem. Rev.* **2012**, *112(4)*, 2100–2128.

76. Bhaskarapillai, A; Thangaraj, V; Srinivasan, M. P.; Velmurugan, S. Crosslinked Poly (1-Butyl-3-Vinylimidazolium Bromide): A Super Efficient Receptor for the Removal and Storage of Iodine from Solution and Vapour Phases. *New J. Chem.* **2019**, *43 (3)*, 1117–1121.

77. Li, T.; He, F.; Dai, Y. Prussian Blue Analog Caged in Chitosan Surface-Decorated Carbon Nanotubes for Removal Cesium and Strontium. *J. Radioanal. Nucl. Chem.* **2016**, *310*, 1139–1145.

Potential Applications of Chitosan-Based Sorbents 151

78. Muzzarelli, R. A. A. Chelating Derivatives of Chitosan Obtained by Reaction with Ascorbic Acid. *Carbohydr. Polym.* **1984**, *4(2)*, 137–151.
79. Zhou L.; Shang, C.; Liu, Z.; Huang, G.; Adesina, A. A. Selective Adsorption of Uranium (VI) from Aqueous Solutions Using the Ion-Imprinted Magnetic Chitosan Resins. *J. Colloid Interface Sci.* **2012**, *366(1)*, 165–172.
80. Pan, D.; Fan, Q.; Fan, F.; Tang, Y.; Zhang, Y.; Wu, W. Removal of Uranium Contaminant From Aqueous Solution by Chitosan@ Attapulgite Composite. *Sep. Purif. Technol.* **2017**, *177*, 86–93.
81. Wang, K.; Ma, H.; Pu, S.; Yan, C.; Wang, M.; Yu, J.; Wang, X.; Zinchenko, A.; Wei, C.; Anatoly, Z. Hybrid Porous Magnetic Bentonite-Chitosan Beads for Selective Removal of Radioactive Cesium in Water. *J. Hazard. Mater.* **2019**, *362*, 160–169.
82. Yang, S.; Okada, N.; Nagatsu, M. The Highly Effective Removal of Cs+ by Low Turbidity Chitosan-Grafted Magnetic Bentonite. *J. Hazard. Mater.* **2016**, *301*, 8–16.
83. Zemskova, L.; Egorin, A.; Tokar, E.; Ivanov, V.; Bratskaya, S. New Chitosan/Iron Oxide Composites: Fabrication and Application for Removal of Sr^{2+} Radionuclide from Aqueous Solutions. *Biomimetics* [Online] **2018**, *3(4)*, 39 https://www.mdpi.com/2313-7673/3/4/39/htm (accessed 06 May 2019).
84. El Rouby, W. M. A.; Farghali, A. A.; Sadek, M. A.; Khalil, W. F. Fast Removal of Sr(II) From Water by Graphene Oxide and Chitosan Modified Graphene Oxide. *J. Inorg. Organomet. Polym.* **2018**, *28*, 2336–2349.
85. Wang H., Tang H., Liu Z., Zhang X., Hao Z., Liu Z. Removal of Cobalt(II) Ion from Aqueous Solution by Chitosan-Montmorillonite. *J. Environ. Sci. (China)* **2014**, *26(9)*, 1879–1884.
86. Kalyani, S.; Krishnaiah, A.; Boddu, V. M. Adsorption of Divalent Cobalt from Aqueous Solution onto Chitosan–Coated Perlite Beads as Biosorbent. *Sep. Sci. Technol.* **2007**, *42(12)*, 2767–2786.
87. Luo, W.; Bai, Z.; Zhu, Y. Fast Removal of Co(II) from Aqueous Solution using Porous Carboxymethyl Chitosan Beads and its Adsorption Mechanism. *RSC Adv.* **2018**, *8*, 13370–13387.
88. Kwak, I. S.; Won, S. W.; Chung, Y. S.; Yun, Y. S. Ruthenium Recovery from Acetic Acid Waste Water Through Sorption with Bacterial Biosorbent Fibers. *Bioresour. Technol.* **2013**, *128*, 30–35.
89. Monier, M.; Abdel-Latif, D. A.; Youssef, I. Preparation of Ruthenium(III) Ion-Imprinted Beads Based on 2-Pyridylthiourea Modified Chitosan. *J. Colloid Interface Sci.* **2018**, *513*, 266–278.
90. Ramesh, A; Hasegawa, H; Sugimoto, W; Maki, T; Ueda, K. Adsorption of Gold(III), Platinum(IV) and Palladium(II) onto Glycine Modified Crosslinked Chitosan Resin. *Bioresour. Technol.* **2008**, *99*, 3801–3809.
91. Fujiwara, K; Ramesh, A; Maki, T; Hasegawa, H; Ueda, K. Adsorption of Platinum (IV), Palladium(II) and Gold(III) from Aqueous Solutions onto l-Lysine Modified Crosslinked Chitosan Resin. *J. Hazard. Mater.* **2007**, *146*, 39–50.
92. Sicupira, D; Campos, K; Vincent, T; Leao, V; Guibal, E. Palladium and Platinum Sorption Using Chitosan-Based Hydrogels. *Adsorption* **2010**, *16*, 127–139.

CHAPTER 7

Spray-Dried Chitosan and Alginate Microparticles for Application in Hemorrhage Control

SRIKANT SUMAN, SANDARBH KUMAR, RITVESH GUPTA, AKHIL BHIMIREDDY, and DEVENDRA VERMA*

Biotechnology and Medical Engineering Department, National Institute of Technology Rourkela, Rourkela, Odisha 769008, India

Corresponding author. E-mail: dev.rivan@gmail.com

ABSTRACT

Significant blood loss due to injury, trauma, or high blood pressure is termed as hemorrhage which is one of the leading causes of death worldwide. The efficient way to overcome the condition is using a material which can quickly clot the blood. Hemostatic agents like polysaccharides, bone wax, microfibrillar collagen, gelatin foams, etc. are available in the market; however, their clotting efficiency is low. Chitosan is a well-known and widely used component of many hemostatic agents available commercially. An interdisciplinary approach that utilizes nanotechnology and the materialistic properties of chitosan and alginate has shown tremendous potential as an efficient hemostatic agent. Our work on the polyelectrolyte complexes (PECs) of chitosan and alginate has shown that it can clot the blood at least 6 times faster than the commercially available Celox. Both chitosan and alginate have been shown to exhibit healing and antibacterial properties. Chitosan mainly causes hemostasis by activating the intrinsic pathway of the coagulation cascade. However, there are a few disadvantages of chitosan such as low strength that can be taken care of by the application of alginate with it. Microparticles of chitosan and alginate were prepared using the spray-drying technique and mixed in different ratios. When exposed to blood or aqueous condition, they interacted to form a PEC. These particles easily dissolve in blood, and large surface area to volume ratio helps to adhere more red blood cells to them which ultimately helps the blood to clot faster. Our experiments also suggest that a high proportion of chitosan helps to adhere more platelets to them. About 83% and 70% platelet adhesion was observed for 1:1.5 and 2:1 Alginate-Chitosan samples, respectively. The swelling ratio increased tremendously for the first 5 min after keeping it in phosphate buffer saline. Activated partial thromboplastin time and the prothrombin time was found to be 34 and 14 s for 2:1 alginate–chitosan sample, whereas for Celox time was about 155 and 28 s. Detailed studies suggest that PECs of chitosan and alginate microbeads have potential application in hemorrhage control.

7.1 INTRODUCTION

Application of pressure is the foremost action taken to stop blood loss after a trauma or physical injury. In 1615, William Harvey observed that blood runs via a closed circulatory system [1]. Since then, a lot has been done in the field of hemostasis. The idea opened doors for inventions like tourniquet [2] that are still in use. What took it so long to get the idea that the application of pressure can help in hemostasis is still a matter of curiosity. Although there were ambiguities about the mode of blood circulation, it is interesting to know that Galen, a Greek physician claimed that the liver produces blood that is then distributed to the body in a centrifugal manner. He succeeded to interpret the connection of veins and arteries with lungs but the system contained an endpoint of veins and arteries [3, 4]. The idea thrived for the 15th century until Harvey published his book "On the Motion of Heart and Blood in Animals" in 1628 where he explained blood circulation in a closed system [4]. Greeks and Romans had discovered some methods like the application of some vegetables and mineral styptics for the healing of wounds in 332 BCE. Later the study of Egyptian mummification practice opened many hidden mysteries, not only regarding the hemostasis but various other great medical field advances [5]. Topical hemostatic agents that are used in the modern era only came in the early 20th century. In 1909, Bergel first introduced fibrin as a hemostatic agent. Fibrin sealants containing bovine thrombin with human plasma were used for the acceleration of blood clotting [6]. With time, the technology evolved and several new hemostatic devices were invented with improvement in understanding of the blood-clotting mechanism and the factors affecting the clotting process [7–9].

Spray-Dried Chitosan and Alginate Microparticles 155

Clotting-time became an important factor which produced the need for advanced hemorrhage control methods, devices and materials that could clot the blood faster. Today a number of hemostatic and hemorrhage control devices are available that target certain specific steps in the blood-clotting mechanisms. The blood clotting occurs in three steps: vasoconstriction, platelet plug formation, and the series of cascade reactions [10]. These steps lead to the formation of a mesh of fibers that help in hemostasis. A material that could help in accelerating any of the steps of hemostasis can accelerate the process. After hemostasis vasodilation occurs that assists the healing of the wound by increasing the flux of neutrophils, macrophages, and lymphocytes to reach the site [11]. Some of the commercially available products also claim to have wound healing along with the hemostatic properties.

Wound-specific devices are also present in the market. Broadly speaking, there are two prominent types of wounds: penetrating and nonpenetrating [12]. Penetrating wounds are those which break or tear through the layers of the skin reaching the underlying tissues and organs. The penetrating object may go inside the body, may come out by the path it entered, or pass through the tissues and come out through some another exit. It can be caused by pieces of bone or by a foreign object. They are classified as stab wounds, skin cuts, surgical wounds, and gunshot wounds. Nonpenetrating wounds, on the other hand, do not break or tear the skin. These result due to friction with other surfaces. Some of the nonpenetrating wounds are also caused due to blunt trauma. These wounds are mainly classified into abrasions, lacerations, puncture wounds, contusions, and concussions.

Depending on the severity and the type of wounds a variety of options are available. Many a time we have to deal with a number of types of wounds simultaneously. In a battlefield scenario, a soldier may be subjected to a myriad of situations such as stabbing, gunshot, burning, loss of limb, and various other types of injuries [13–15]. This makes the need for a multitasking hemostatic agent even more crucial. Further effective clotting time and sustainability become crucial factors in the battlefield. Major blood loss is a primary reason for death; not only in the battlefield but in road accidents also. QuickClot and Celox are two commercially available products with the claim of having faster blood-clotting actions. Their small granular size and hence high surface area help the material to interact with more number of blood cells and swelling properties help in platelet plug formations by thickening the blood.

The idea of small size particles is governed by top-down and bottom-up approaches. Both have their advantages and disadvantages. The top-down technology utilizes heavy machinery and has a limited size range but the control is easy. Bottom-up approaches, on the other hand, take the advantage of the self-assembly of molecules that may or may not involve chemical reactions and are cheaper than the top-down approach. However, top-down approaches have less control over the size of the particles.

As said earlier we need to deal with many factors at the same time for efficient hemostatic action. We are looking for an "all in one" product. The interdisciplinary approach can help us better in this regard. Recent advances in the field of nanotechnology and biomaterials have made mind-blowing advancements in the field of hemostasis [16, 17]. Miracles happen at nano- and microlevels of the particles. Compacting the boundaries of the atoms and molecular systems of macroparticles gives rise to nanosystems with altered physical properties. As the size approaches nanoscale, the percentage of atoms at the surface becomes significant [18]. Increased surface area to volume ratio plays a lead role in the interaction characteristics with various molecules. Charge, on the other hand, has also significant effects on the interactions between the particle and the molecules or components with which we are dealing [in our case it is the red blood cells (RBCs)]. But which one is more dominating—charge or size?

Polyelectrolyte complexes PEC are the arrangements of nanoparticles or nanofibers of polyanions and polycations in a fashion to create a nano-complex. Oppositely charged polymers bind together under the forces of attraction and form viscous solutions. It is quite a fact that they contain the properties of both salts and polymers. In short, they are large ionic polymers with useful physical properties. There are many methods by which PEC can be formed. Out of these the most simple and popular method for the synthesis of the PECs is the PEC titration method or drop-by-drop method [19]. In this type of PEC formation, one of the poly electric salt or polyelectrolyte is added dropwise in a beaker consisting of polyelectrolyte of the opposite charge. Disadvantages of the method include dilution of the solution by the presence of a solvent in the titrant and consumption of the polyelectrolyte in the process. This results in the overall reduction of the amount of poly-electrolyte. So to eradicate such problems generally a higher concentration of the sample is taken in a beaker. The concentration of the polyelectrolyte in the beaker is 5–10 times more compared to the titrated polyelectrolyte. The idea of PEC formation is based on the interaction of oppositely charged particles. The other way to do it, is preparing the charged particles first and then allowing them to form PEC in the blood itself. Blood already has enough amount of water to form PEC. In the later one, both "synthesis of PEC" and "hemostatic action" occur simultaneously. This chapter utilizes the same modus operandi for the analysis of hemostasis.

Spray-Dried Chitosan and Alginate Microparticles 157

A summary of the chapter would help you find your ways through the chapter and to gain the most from it. Section 7.3 gives you a brief explanation of the choice of material. Among the wide variety of biomaterials available in the market, we chose chitosan and alginate for our purpose. The choice is not random. Apart from having exceptional qualities, there is a lot of scope of modifications in the chitosan-based hemorrhage control devices, which has been given under this section. As we proceed, the chemistry of chitosan with alginate, starts getting easier to understand.

Section 7.4 interprets the significance of the combination (chitosan with alginate) as PECs. A brief understanding of the significance of scale—"why nano or micro?" and the effect of size and charge of the PECs on blood have been presented. Both size and charge do not affect the interactions of the PECs with other molecules, equally. This significantly changes the scenario of hemorrhage control.

The blood-clotting assay is the soul of this chapter. The outcomes associated with the assay prove the efficacy of the material. From blood collection to the clotting assay, their results and expected explanation have been explained as per the knowledge and understandings of the writer in Section 7.5. Celox, a commercially available product has been used as a control. The section also includes prothrombin time (PT), activated partial thromboplastin time (APTT), and hemocompatibility of the material. Roles of swelling and its effects in the context of PECs for hemorrhage control have been thoroughly explained in Section 7.6. Readers should also keep in mind that the complete chapter is solely based on the experiments done by our team on "spray-dried chitosan and alginate powder for the application in hemorrhage control" and hence everything is discussed in this context. Again, it also does not restrict the explanations, inferences, and facts to be true only for this particular study.

Furthermore, the other applications and scope of the material have been discussed in Section 7.7. If commercialized, the market potential and the subsequent strategies to be followed have also been discussed at the end of the chapter.

7.2 HEMOSTASIS—AN OVERVIEW

Hemostasis is the phenomenon by the virtue of which the body repairs ruptured/damaged blood vessels, in an attempt to minimize blood loss. As iterated at the beginning of the chapter, the degree of hemostatic activation

158   *Natural Polymers: Perspectives and Applications for a Green Approach*

and need prominently depends upon the type of wound and subsequent chemical activation. This is known as the blood-clotting phase. In this phase, platelets come to the site of injury and change their shape to amorphous to help them to bind with the wound. These platelets release chemical signals that activate fibrin. Fibrin forms a mesh-like structure that acts as a sticky glue that binds the platelets with each other. This results in the formation of the clot that prevents further blood loss.

Conceptually, hemostasis involves three basic steps that are as follows:

1. *Vasoconstriction (or vascular spasm):* It refers to a narrowing of the blood vessels in response to physical damage. This signal transmission is usually calcium-based. The smooth muscle cells of the vessels contract and reduce blood loss. In addition to this, body heat is retained while there is an increase in vascular resistance.
2. *Platelet plug formation:* It occurs when the person is injured. This is a crucial step in hemorrhage control. In regular conditions, there will be no platelet plug formations. It consists of mainly three steps that include platelet adhesion, platelet activation, and platelet aggregation. If the wound is minor, the platelet plug might be sufficient to stop the bleeding without the coagulation cascade.
3. *Blood coagulation*: The platelet plugs thus formed are not tight enough to maintain the bloodstream. This attachment is framed as blood coagulation, referred to as a clot. The coagulation changes the dissolvable fibrinogen proteins into insoluble fibrin that forms a mesh like over the injury. The general blood coagulation process is exceptionally complicated and comprises a number of reactions including the initiation of many factors with the assistance of different co-elements. The process consists of the following steps (Figure 7.1).

The initial step of blood coagulation is the activation of prothrombinase (or prothrombin activator) that results in the conversion of prothrombin into thrombin. There are two basic mechanisms that describe the activation of prothrombinase. These two mechanisms are described as pathways that are known as intrinsic and extrinsic pathways:

Intrinsic pathway: This pathway is also called the contact *activation pathway* and is followed when there is an injury to platelets or the introduction of blood to collagen. This pathway begins with the activation of factor XII (Hageman factor). This factor gets activated in the presence of negatively charged particles that are given by the inorganic phosphate released by the activated platelets. Other than inorganic phosphate, collagen can also activate the factor XII. When factor XII gets initiated, it activates factor XI in the presence of high molecular weight kininogen. This activated factor XI activates factor IX in the presence of Ca^{2+} ions. Activated factor IX ties with activated factor VIII and the presence of Ca^{2+} ions initiate factor X that at last ties with activated factor V and the presence of calcium initiates the prothrombinase or prothrombin activator.

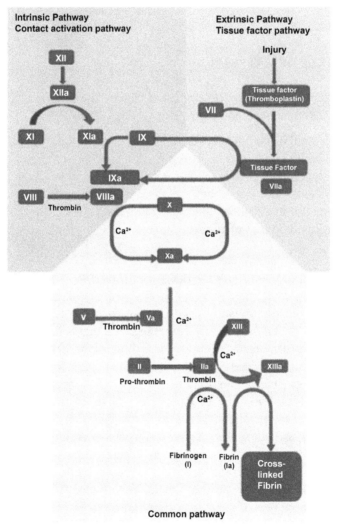

FIGURE 7.1 Schematic representation of the coagulation pathway (the intrinsic and extrinsic pathways merge into a common pathway).

Extrinsic pathway: This pathway is also called a *tissue factor pathway*. It begins with the release of factor III (tissue thromboplastin), which is presented to the blood when there is harm. Amid damage, different tissue elements are released by the tissue that thus activates factor VII. Factor VII alongside calcium and tissue elements results in the activation of factor X. When factor X is activated, it takes an indistinguishable pathway followed in the inherent pathway where factor X forms prothrombinase or prothrombin activator.

7.3 CHOICE OF MATERIAL

While selecting an appropriate hemostatic agent, the following key aspects should be taken into consideration [20]:

1. It should be simple and convenient in preparation.
2. It should be easily administered.
3. It should cover the injured area effectively.
4. It should not cause any adverse reactions, infection, or allergy response in the patient during or after use.
5. It should be cost-effective.

7.3.1 *CHITOSAN—A POTENT POLYSACCHARIDE*

Chitosan has been abundantly used in pure form for topical hemorrhage control [21]. It is obtained by deacetylating chitin that is found in abundance in exoskeletons of crustaceans, in insects, fungi, and others. It is a natural linear polysaccharide that consists of alternate units of *N*-glucosamine, and *N*-acetyl-D-glucosamine. Chitosan molecules mimic native ECM glycosaminoglycan structure that provides enhanced tissue compatibility, cell attachment, and proliferation.

Chitosan is insoluble at neutral and basic pH values but the protonation of the amine groups allows it to solubilize in an acidic medium. Its polycationic nature supplements it with mucoadhesion, hemocompatibility, and antifungal, as well as antibacterial properties. The positively charged molecules attach readily to negatively charged mucus membrane.

When chitosan comes in contact with whole blood, it forms a coagulum due to its polycationic nature and nonspecific binding to cell membranes [22]. Because of its positive charge, chitosan also interacts with the negative cell

Spray-Dried Chitosan and Alginate Microparticles

membrane of various microbes. This results in the opening of the cellular gap junctions and the microbes lose their cellular contents which ultimately kills them [23].

Blood protein adsorption on chitosan layers has also been investigated. It has been observed that chitosan improved wound healing. When incubated in serum, large amounts of chitosan depositions were observed. However, it was also validated that only weak transient activation of the complement system took place, whereas there was no effect on the intrinsic pathway. In addition to this, when acetylated strong activation of the alternative complement pathway was also observed [24, 25].

Benesch et al. also found that acetylated chitosan was a strong coagulation activator but did not bind fibrinogen or other plasma proteins as deacetylated chitosan did [23].

Chitosan hemocompatibility results from the attraction of anionic RBCs by positively charged chitosan molecules. Hence, the rapidly adhering cells to the bleeding site form the plug [26]. The underlying specific mechanism of hemostatic activity of chitosan is not known yet but the experimental data suggest three aspects of discussion:

1. Plasma sorption
2. Erythrocytes coagulation
3. Platelet adhesion, aggregation, and activation.

It has been experimentally validated with chitosan that the primary plasma sorption leads to the concentration of blood cells at the injury site. Chitosan can absorb about 50%–300% of its primary weight and this sorption rate is directly affected by changes in molecular weight or/and degree of deacetylation. Also, enhanced erythrocyte agglutination was reported in the presence of chitosan due to crosslinking of erythrocytes. These bind to each other via chitosan polymer chains and then repolymerize to form a cell capturing lattice, hence creating an artificial clot [27]. While in contact with chitosan, erythrocytes lose their biconcave morphology. The low-molecular-weight chitosan can directly bind with the erythrocyte's wall due to its cationic nature and is hypothesized to be the key mechanism of hemagglutination [28]. This enhanced erythrocyte agglutination in the presence of chitosan occurred due to crosslinking of erythrocytes.

In addition to these, it has also been demonstrated that chitosan films can induce platelet adhesion, aggregation, and the activation of intrinsic blood coagulation [29]. Chitosan is also known to induce intracellular reactions that elevate platelet spreading and strengthen adhesion [30]. Another

162 *Natural Polymers: Perspectives and Applications for a Green Approach*

simultaneous mechanism includes activation of the actin cytoskeleton of adhered platelets [31].

Moreover, chitosan also has a broad polydispersity index, is semicrystalline in nature, and provides oxygen permeability for faster wound healing. It has also been shown to be nontoxic and enzymatically degradable.

Erythrocytes lose their native morphology and bind to each other via chitosan polymer chains. They then repolymerize to form a cell capturing lattice, hence creating an artificial clot [32]. Therefore, compared to the existing hemostatic solutions, chitosan-based dressings are the most promising ones.

Currently, there are many commercially available chitosan-based hemostatic bandages/solutions available in the market. Some popular commercial choices for civil emergency or battlefield applications have been listed in Table 7.1 with their respective prominent mechanisms of action [33].

TABLE 7.1 Chitosan-Based Hemostatic Agents

Product Name	Material Used	Mode of Action	Manufacturer
QuickClot	Zeolite, chitosan	Water absorption, facilitation of clotting cascade	ZMedica
Celox	Chitosan granules	Swelling	Celox Medical
Hemcon	Chitosan acetate	Platelet activation	Tricol Medical
Chitoflex	Chitosan based	Works as a flexible barrier; effective for moderate to severe bleeding; known to have antibacterial nature	Tricol Biomedical
Chitoseal	Chitosan based	Contains positively charge molecules that attract negatively charged RBCs, to clot the blood	Abott Laboratories
ChitoPack C	Non-woven chitosan fibers	Regeneration of tissues	Eisai Corporation

7.3.2 CHITOSAN ALTERATIONS AND ENHANCEMENT

Dowling et al. modified chitosan by attaching a small number of hydrophobic tails to the backbone of chitosan. The hypothesized mechanism for such hydrophobically modified chitosan involved anchoring of hydrophobic portions from the polymer into the hydrophobic interiors of blood cell membranes. This results in the formation of a sample-spanning gel network as the blood cells get connected by biopolymer and hence halts blood flow. Various small and large animal injury models have verified the increased

Spray-Dried Chitosan and Alginate Microparticles 163

efficacy of hemostasis with more than 90% reduction in bleeding time with respect to rat femoral transections.

Compared to these, novel hydrophobically modified chitosan sponges are found to be even better at achieving hemostasis in swine models [34]. Sodium hydroxide (NaOH) and/or sodium tripolyphosphate ($Na_5P_3O_{10}$) treatment of chitosan is also known to accelerate blood clotting, enhance RBC adhesion, and at the same time maintain its original shape after hemostatic testing [35].

Bon Kang et al. used ultrasonication treatment to increase porosity and sorption rate. In addition, the proliferation of normal human dermal fibroblasts on the sonicated nanofiber mat was found to be higher than that on the nonsonicated material (with respect to 7 days of culture).

Despite their great potential, chitosan also has certain limitations. It is poorly soluble at physiological pH and hence projects low strength of tissue adhesion. In addition to this, it has poor stability over time. When stored at room temperature, it undergoes gradual degradation and destruction of its functional groups. This ultimately leads to irreversible loss of its physicochemical properties. Both intrinsic (degree of deacetylation, molecular weight, purity, and moisture level) and extrinsic factors (environmental storage conditions, thermal processing, sterilization, and processing involving acidic dissolution) are acknowledged as crucial parameters affecting the stability of the chitosan-based formulations [33].

However, various strategies have been reported to supplement chitosan stability. Some prominent modifications include the addition of the stabilizing agent during the preparation process, blending with a hydrophilic polymer, and the use of ionic or chemical cross-linkers.

Its chemical modifications might be via the addition of tissue adhesive materials or direct associations with hemostasis-promoting materials. While modifications can be done by the formation of the nonionic complex (polymer blends), direct chemical modification of the chitosan structure can be achieved by either physical or covalent cross-linking. Physical cross-linking makes use of inherent charge and size to establish associations. This results in the formation of PECs. Chitosan PEC possesses enhanced stability due to electrostatic interaction between the cationic chitosan and the negatively charged polymer. The presence of the negatively charged molecules is hypothesized to prevent protonation of chitosan amine groups and thus retards the rate of chitosan hydrolysis.

Moreover, compared to chemical modifications, physical modifications are mild and simple and avoid the need to end purification processes or additional catalysts/factors [33].

In our pursuit of enhancing the hemostatic action of chitosan, alginate (an anionic biopolymer) was chosen as a potential additive. The following section sheds light on the biochemistry of alginate and explains its supplementation role in the hemostatic effect of the chitosan–alginate blend.

7.3.3 ALGINATE

Alginate is a polysaccharide, a linear copolymer the consists of homopolymeric blocks of α-D-mannuronic acid and β-l-guluronic acid residues. It is widely found in the cell walls of brown algae. The material is grossly used in various industries including textile printing, pharmaceuticals, and food. A variety of alginates are extracted from various seaweeds, for example, the giant kelp *Ascophyllum nodosum* and *Macrocystic pyrifera*. They have their own commercial importance. Two bacterial genera, *Azotobacter* and *Pseudomonas*, are also found to produce Alginate. These bacterial strains have been proved very useful in the production of micro- and nanostructures for medical applications [36].

Alginate is the ionic form of alginic acid, hence found with its complementary ion. Sodium, potassium, and calcium salt of alginic acid are available in the market. The first two are naturally extracted from the cell wall of brown algae and seaweeds respectively and the calcium salt is derived from the replacement of sodium with calcium in sodium alginate.

Alginate is a very good water absorbent. It is frequently used as an additive in paper and textile industries where dehydration is needed. The food industry makes good use of it as a thickening agent for drinks, ice creams, and others. For hemostatic application also we can make good use of this quality of alginate. The thickening of blood helps to form clots.

In some articles, the wound healing properties of calcium alginate have been claimed [37]. For this reason, it is used in wound dressing applications. Calcium is primarily involved as Factor IV, in wound repair. Calcium alginate dressings are designed such that calcium is liberated early in the acute phase to promote hemostasis [37]. The wound healing properties of the material increases the usefulness of the hemorrhage control devices. Though it does not affect the clotting time or mechanism but have other benefits with respect to patients. For a patient, curing makes all the sense in the end. Clotting is just a part of it.

Calcium alginate is known to have antibacterial properties against *Escherichia coli, Staphylococcus aureus*, and *Pseudomonas aeruginosa* bacterial strains [38]. Many other variations have also been tried to enhance the antibacterial nature of alginate. Fibers synthesized using zinc and copper alginate have shown significant antibacterial properties [39]. Some composites of sodium alginate are also prepared with silver nanoparticles that are very well known for their antibacterial nature [40]. Sodium alginate, on the other hand, has been found to be working against antibiotics. The diameter of the zone of inhibition reduces in the presence of sodium alginate [41]. External wounds are more prone to the attack of bacteria that hamper the clotting efficiency. Calcium alginate-based material can better serve as a hemostatic agent than sodium alginate-based materials, in this context. Another study suggests that calcium alginate when comes in contact with blood, releases calcium ions for the exchange of sodium ions rapidly. This predominantly increases both, platelet activation and blood coagulation [42].

Alginate fibers have been found to be having good tensile strength. Preferably the dry strength of yarns, prepared using alginate, is high [42]. Although the wet strength of these fibers is poor, the alginate fibers create a moist healing environment. Moist conditions are supposed to be accelerating the wound healing process; when cross-linked with multivalent cations the strength increases dramatically [43].

Not all properties of alginate are favorable for the application of hemorrhage control. Alginate-based hemostatic wound dressing materials have been reported to be showing foreign body reactions [44, 45]. Typically it may lead to unwanted inflammation that may damage the skin cells. Though inflammation is a crucial part of wound healing, more of it burns the skin cells.

Thus, the aforementioned rationales explain our choices of the biomaterials for the hemorrhage control and wound healing application application.

The powder form (microparticles) of alginate and chitosan is used to make good interconnectivity with the wounds and helps in fast healing. We want a biocompatible material with good mechanical strength, wound healing, and faster blood-clotting properties. The combination of chitosan and alginate provides all of these features. Figure 7.2 shows some of the features of these two biomaterials. Even after spray drying the oppositely charged nature is retained and hence they are supposed to form a PEC complex in water (which is the major component (92%) in the blood). The high degradation rate of the chitosan is covered up by the mechanical strength of the alginate.

7.4 ALGINATE–CHITOSAN POLYELECTROLYTE COMPLEX

To synthesize PECs we need two oppositely charged polymers. They must be biocompatible and biodegradable, for the application in hemorrhage control. As discussed in the previous section, chitosan is a widely used biopolymer that also possesses quick-healing properties and alginate, on the other hand, provides swelling properties and strength. It is well known that chitosan is a weak polymer and also degrades relatively easily, whereas alginate shows slow degradation. Further, alginate possesses a negative charge at neutral or higher pH while chitosan is positively charged at acidic pH; thus, they can form PECs.

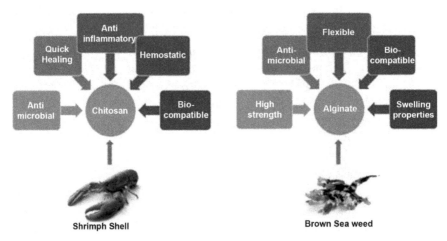

FIGURE 7.2 Chitosan and alginate distinguished choices of material for hemorrhage control.

7.4.1 SPRAY DRYING AND SYNTHESIS OF CHARGED MICROBEADS

Spray drying is a technique in which dry powder is obtained from a liquid or semiliquid substance by constant drying with hot gas. This technique is widely used in food and chemical industries and has a wide range of applications in pharmaceuticals. Thermally sensitive materials or substances are dried generally by this process. A spray dryer is a machine that follows this technique. We used it to produce microparticles of chitosan and alginate powders. The small size of the particles results in better interactions and strong PEC formation that are required for clot formation. The size of the particles that we obtained was around the micro range.

7.4.1.1 METHOD OF SYNTHESIS AND HEMOSTATIC ACTION

Chitosan is dissolved in acid and alginate is dissolved in distilled water. They are then spray dried to give positive and negative charged microbeads. These oppositely charged microbeads form PECs, when incorporated into blood (see Figure 7.3). The inlet and outlet temperatures are set to be 140 °C and 80 °C, respectively. These are the optimized values.

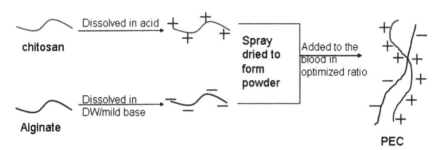

FIGURE 7.3 Synthesis and working process.

At lower temperatures, the wet solution will remain on the surface of the walls of the spray drier. Further increase in temperature may burn the samples that can adversely affect the desired properties of the materials. The other parameters related to spray drying are feed-flow rate, atomizing airflow, and aspiration rate. All these parameters may affect the size, yield, surface topography, and other properties of the microbeads. At micro- and nanolevel any modification in this parameter multiplies the effect by many folds. A detailed study of these parameters may help us getting better hemostatic efficiency. Our primary focus in this chapter is to study the PECs of the microbeads for faster blood-clotting application. Thus, we have only optimized the crucial parameters of the spray dryer and have taken the help of scholarly articles for the rest of them [46–48]. The microbeads obtained from spray drying are dissolved in blood in different ratios to form a positively charged complex that clots the blood at a faster rate. Variation in the ratio of the two components (i.e., alginate and chitosan) can affect the clotting efficiency of blood.

7.4.1.2 ANALYSIS OF THE SIZE OF SPRAY-DRIED MICROBEADS

The sizes of the spray dried chitosan and alginate particles were around 11μm and 8μm, respectively (as shown in Figure 7.4). The particle sizes significantly reduced after spray drying. The smaller size of the particles helps to form greater interconnectivity resulting in strong PEC formation. Due to the micro range of the particles they have distinct advantages as compared to other dressing devices.

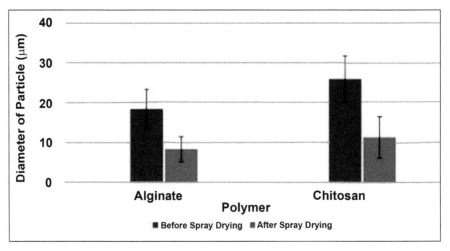

FIGURE 7.4 Particle size of chitosan and alginate before and after spray drying.

7.4.2 WHY NANO OR MICRO?

Consider an atom that is free to move anywhere in a box of size $1 \times 1 \times 1$ m^3. Now reduce the size of the box to $1 \times 1 \times 1$ cm^3. Does it make any difference to the atom? Now further reduce the size of the box to $1 \times 1 \times 1$ nm^3. What about now? Does it make any difference to the atom? Apparently yes.

Semiconductor materials start showing quantum confinement effect from 100 nm; some metal particles show surface plasmon resonance and magnetic materials start exhibiting super-paramagnetism at the order of nanometer size. What causes these changes? Micro, submicron, and nano work as a bridge between the macro- and the atomic scales and not only that, they also exhibit marvelous properties specific to this scale only. The most important thing about this bridge is its accessibility. Even in the 21st century, you find no technology that could help you to do desirable changes at the atomic level but you surely can do changes with micro- and nanoparticles.

How will a micro- or nanosize object help us to clot blood? Let me put it this way. When you are dealing with a damaged wristwatch you need small instruments as well as a magnifying glass, which means that you have to get compatible with the size. Here we are dealing with RBCs which are having a diameter of 4–6 μm. Thus we need something compatible with this scale.

Figure 7.5 shows the images of spray-dried microparticles of alginate and chitosan. Particles interacting with each other can be easily observed in the figure.

Spray-Dried Chitosan and Alginate Microparticles

With the reduction in size, the ratio of surface area to volume increases which means, now they have more area to cover. They can interact with more number of RBCs and platelets. Stabilizing the platelets and the RBCs flowing through the blood vessels is a crucial step to initialize the blood clotting. The final clot is an effect of cascades of reactions.

7.4.2.1 ELECTROSPUN NANOFIBERS FOR HEMORRHAGE CONTROL

A very common and well-known way to reach nanoscale is electrospinning. Electrospun nanofibers can provide a very high surface area to volume ratio, which can be a very important feature for a hemostatic agent. Plenty of work has been done in this field. The major problem with this method is the cost. If you wish to have very thin nanofibers you will need cutting-edge technology and very expensive needle-free electrospinning equipments. NanoSpider is one such equipment available in the market [49, 50]. Furthermore, it is quite difficult to reduce the size (diameter) of nanofibers after a limit, even by electrospinning. There are chemical methods [51] to synthesize nanofibers though it is difficult to control the homogeneity in size of the nanofibers thus created. However, ultrasonication can help to overcome the problem of heterogeneity.

A mat of fibers can be created by electrospinning and used for hemorrhage control effectively [52]. The hemostatic properties can be enhanced by using a biomaterial, with wound healing and anti-inflammatory properties, to synthesize the nanofibers. An additional drug can also be added to the system. The drug can be patient specific for example, for hemophilic patients a clotting factor VIII or IX can be added, or it could also be nonspecific for example antibacterial agent can be added to reduce the risk of contamination [53]. Some modifications in the nanofiber mat can also be done to treat specific diseased wounds such as diabetic foot ulcer (DFU). An electrospun nanofiber with huge porosity, better oxygen exchange rate, and excellent humidity absorption can effectively help to fight against DFU [54].

7.4.3 SIZE VERSUS CHARGE

"Small size" and "positive charge" are two pillars on which the whole idea of he quick blood-clotting stand. The significance of size has already been explained in the above section. RBCs are negatively charged and can be attracted by positively charged particles. Hence a positive charge micro complex can quickly clot the blood. Our findings and characteristics indicate that while the charge is important, it is less important than its size in the micron scale. You can discover the explanation in the photographs of the microparticles depicted in figure 7.5 with an environmental electron scanning microscope (ESEM). Figure 7.5a is an ESEM image of microparticulate negatively charged alginates, and Figure 7.5b shows a positive chitosan microparticles image of ESEM. The impact of dimension appears to dominate microparticle electrical repulsion. Even when repulsive electrostatic forces exist, the particles are close.

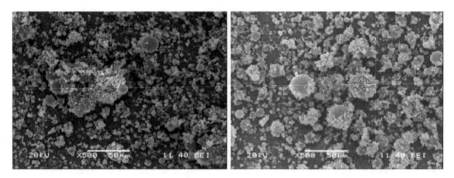

FIGURE 7.5 SEM image of spray-dried alginate (a) and chitosan (b) microparticles.

Another proof which suggests the domination of size over charge is the blood-clotting assay analysis (Figure 7.9). The results suggest that the composition with higher chitosan content increased the clotting time. Considering the effect of charge, higher chitosan content should have enhanced the clotting time. One inference that can be drawn from this result is that the arrangement of the particles matter more than the charges on their surfaces. Those with the higher chitosan content might not be forming the PEC with high efficiency.

7.5 BLOOD-CLOTTING ASSAY

While analyzing the hemostatic potential of alginate and chitosan powder and their interaction with blood, different compositions of the sample were subjected to in vitro blood assay analysis. In the following sections, we will discuss the methodology, underlying mechanism, and inference of the five blood assays that were considered in our study. Namely, blood-clotting

Spray-Dried Chitosan and Alginate Microparticles 171

assay, hemolysis assay, PT, APTT, and platelet adhesion were performed. Furthermore, a commercial chitosan-based local hemostatic product, Celox (procured from Celox Medical), was subjected to parallel analysis as control.

7.5.1 BLOOD COLLECTION

Fresh blood samples from evidently healthy adult goats (average age: 2 years) were collected from an abattoir. 3.2% trisodium citrate was used as an anticoagulant. The ratio of the blood and the anticoagulant was maintained to 9:1. 50 ml of polypropylene tube, containing 5ml of anticoagulant, was brought to the abattoir and 45ml of blood was collected into it. Blood collection was done as per the national guidelines of the Committee for the Purpose of Control and Supervision of Experiments on Animals (CPCSEA) [55].

All studies were done within 3 h of blood collection.

The following characterizations were done with the blood collected from the abattoir. These characterizations evaluated the ability of the blood to clot and the clotting time. They also revealed the hemocompatibility and efficiency of the blood clotting agents prepared by us.

7.5.2 CLOTTING TIME

This assay estimated the time that blood needs to clot with and without the samples.

We took seven sample variants-pure alginate, pure chitosan, and sample ratios (alginate:chitosan) 2:1, 1:1, 1:1.25 and 1:1.5, and Celox as a control, in equal weights (12 mg). Also, 0.25 M $CaCl_2$ was used to nullify the effect of the anticoagulant. The experiment was done for 1%, 2.5%, 5%, and 10% concentration between blood and $CaCl_2$. This means that for 1% v/v we took 100 µL of blood and 100 µL of $CaCl_2$. For 2.5% v/v, we took 100 µL of blood and 40 µL of $CaCl_2$. Similarly, sample concentration ratios of 5% and 10% were also prepared.

Pure alginate and chitosan and the different compositions of alginate and chitosan were taken in Eppendorf self-lock tubes containing blood. The addition of calcium chloride was taken as the zero time point. The blood clotting was confirmed by inverting the tubes and their clotting time was estimated for the above-mentioned compositions.

As the concentration increases from 1% to 10%, there was a gradual decrease in the clotting time of all the samples. We observed the best results at 10% where the clotting time of sample 2:1 was around 20 s. The samples showed better results compared to the control Celox at all concentrations 1%, 2.5%, 5%, and 10%.

According to the graph pure chitosan clots the blood at a faster rate than pure alginate, which seems pretty obvious because of the positively charged nature of the chitosan. The property of charge does not seem to be reflecting in the PECs of the alginate and chitosan. The sample with high alginate: chitosan ratio, that is, 2:1 (see Figure 7.6) has shown the least clotting time for 10% composition which suggests that charge is not the dominating factor here.

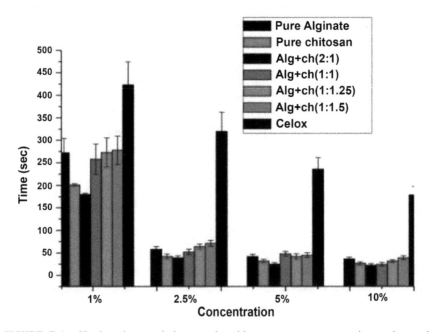

FIGURE 7.6 Clotting time variation trends with respect to concentration and sample composition.

Samples with high amount of alginate will have a higher negative charge, which should interfere with clotting by showing less interaction with negatively surfaced RBCs. This should reduce the clotting efficiency of the material, but this is not what was obtained after performing the blood-clotting assay. Instead of reduction, a considerable improvement in clotting was observed.

The expected explanation could be the distinctive properties of alginate given in section 7.3.3. Some articles have also suggested that the negative charge of material can have a positive effect on blood clotting efficiency [56].

It is a point to note that the Celox is comparatively taking more time to clot the blood than our compositions. As depicted earlier in Section 7.3, Celox is composed of granules of particles (majorly chitosan). They are not spray-dried, thus the particle's size are more than our microbeads (see Section 7.4). They also do not have any surface charge. These facts support the results of the blood clotting assay for the Celox.

7.5.3 HEMOLYSIS ASSAY

Hemolysis refers to the disruption of RBCs. Thus the hemolysis assay determines the extent of this disruption due to the presence of the test material.

To perform the hemolysis assay [55] (Figure 7.7), the stock solution was prepared by diluting anticoagulated blood by physiological saline (0.9% NaCl) in proportion 8:10. Equivalent weights (12 mg each) of Alginate-chitosan powder in different ratios were taken into polypropylene tubes and 0.5 mL of diluted blood was added to each of them. The volume inside the tubes was made up to 10 mL with saline. The negative control was prepared by mixing 0.5 mL of diluted blood and making the volume up to 10 mL utilizing saline. The positive control was prepared by adding 0.5 mL of diluted blood, 0.5 mL of 0.01 M HCl, and making up the volume to 10 mL utilizing saline. Each tube was centrifuged at 4000 rpm for 10 min. The optical density (OD) of the supernatants was taken at 540 nm.

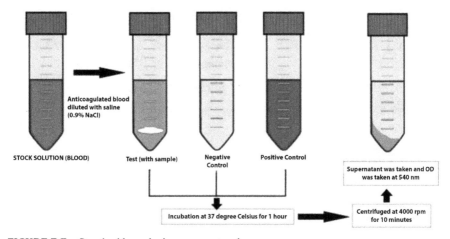

FIGURE 7.7 Standard hemolysis assay protocol.

Hemolysis rate was computed using the formula

$$\text{Hemolysis\%} = \frac{OD_{sample} - OD_{negative\ control}}{OD_{positive\ control} - OD_{negative\ control}} \times 100$$

Samples having <5% hemolysis are highly hemocompatible, 5%–20% are slightly hemolytic, and above 20% are non hemocompatible.

The hemolysis percentage of the composition 1:1.5 was observed to be 4.8 (<5%) that is considered as hemocompatible. The other compositions 2:1, 1:1, 1:1.25 were found to be 6.4%, 5.9%, and 6.32%, respectively that is close to the hemocompatible range. Figure 7.8 shows the hemolysis percentage of compositions 2:1 and 1:1. Our compositions showed better hemocompatibility than the commercially available Celox (control).

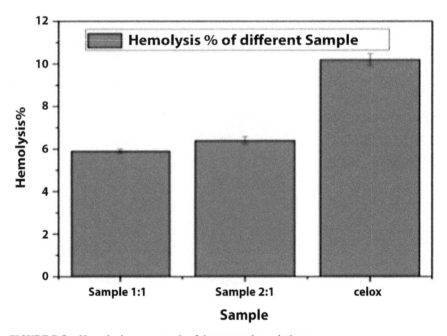

FIGURE 7.8 Hemolysis assay result of three sample variations.

7.5.4 PROTHROMBIN TIME

Whenever there is injury or bleeding in the body, the immediate response of blood clotting, called hemostasis, is initiated. A part of this process involves a series of chemical reactions called the coagulation cascade, in which various

Spray-Dried Chitosan and Alginate Microparticles 175

clotting factors are activated sequentially one after another leading to the formation of a clot.

Prothrombin is a prominent coagulation/clotting factor made by the liver. It is a blood plasma glycoprotein that is transformed into thrombin by prothrombinase (factor X). This thrombin then converts fibrinogen (another blood plasma protein) to fibrin. Fibrin and platelets work in harmony to form a clot. Therefore the prothrombin time is the time taken by the plasma to clot after the addition of tissue factor extracted from animals or autopsy patient brains or can be a recombinant tissue derivative. For our study, we have used a PT reagent prepared from human placental tissue factor combined with calcium chloride and stabilizers (Thromborel S).

In a laboratory setting, the blood in the test tube can be initiated for clotting via two pathways, namely extrinsic and intrinsic pathways. These two pathways then later merge into each other to commemorate clot formation. PT measures the function of extrinsic and the common pathways of the coagulation cascade and reflects how well they sync for clot formation [57].

The normal range for the PT is around 11–14 s. An abnormal PT can be caused by injury, liver disease, vitamin K deficiency, coagulation disorders, or treatment with blood thinners (such as warfarin and heparin). However, considerable variations are caused across different thromboplastin suppliers, storage conditions of the samples, respective incubation periods, and methods of endpoint detection. To standardize such variations, PT is expressed as an international normalized ratio (INR). A higher INR value corresponds to slower clot formation. The normal INR value is 1.1 or below.

$$INR = \frac{[\text{Patient's PT}]^{ISI}}{[\text{Control PT}]}$$

where the international sensitivity index is specific for each reagent.

For the evaluation of the PT time with respect to spray-dried chitosan–alginate PECs, fresh blood was collected from a healthy goat following the protocol given in Section 7.4.1.

Citrate in the tube binds to the calcium in the blood and prevents coagulation. This anticoagulated blood was centrifuged at 4000 rpm for 10 min and platelet-poor plasma (PPP) was extracted. The PT reagent contained calcium and thromboplastin including phospholipids and tissue factors (Figure 7.9).

The extracted PPP and PT reagent (Thromborel S, Siemens) were used in 1:2 ratio, respectively, whereas the clotting time of different samples was investigated using an analyzer (Sysmex CA-50).

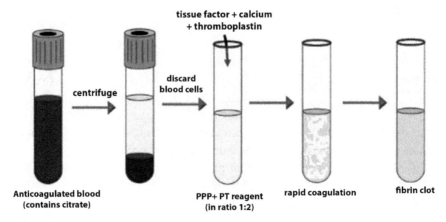

FIGURE 7.9 Schematic representation of a standard PT protocol.

Four sample test ratios of 2:1, 1:1, 1:1.25, and 1:1.5 of alginate and chitosan were considered. After performing the test in triplicates it was concluded that the PEC-based spray-dried powder effectively reduced the clotting time by accelerating the extrinsic coagulation pathway.

A comprehensive analysis of the results denotes a decrease in the PT time with increasing chitosan content.

Sample 2:1 clotting time was 14–15 s, sample 1:1, 1:1.2, 1:1.5, and Celox showed a clotting time of 15, 17, 16, and 28 s, respectively.

The decrease in chitosan content or an increase in alginate content reduces PT time. Less prothrombin time implies more clotting efficiency. So sample 2:1 showed the best results as its clotting time was 14–15 s (Figure 7.10).

Extrinsic pathway mechanism is followed in a deep wound. Thus the deep wound can be easily managed by PEC-based bandages. The clotting time depicts that the earlier onset of the extrinsic coagulation pathway than the control Celox. The time at which the onset of fibrin clot formation happens when the samples and control were tested with PPP (Platelet Poor Plasma) termed to end point. If PT exceeds more than 100 s, it is indicative of bleeding disorders. PT results are directly used for the investigation of the intrinsic and common pathways.

7.5.5 ACTIVATED PARTIAL THROMBOPLASTIN TIME

APTT characterizes blood coagulation. This includes testing activation of blood coagulation to evaluate various clotting factors of the intrinsic pathway.

This test is very similar to PT but is more sensitive to heparin and makes use of an additional activator. This implies that it will be a measure of the effectiveness of heparin (or any other anticoagulant), even more than partial thromboplastin time. Thus, APTT is an important tool for the investigation of heparin administration and/or bleeding disorders. The normal APTT value is 35s.

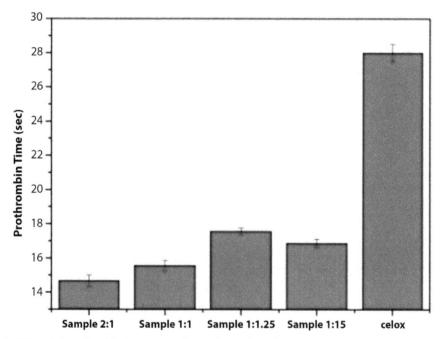

FIGURE 7.10 Graphical representation of the prothrombin time.

Prolonged APTT values are indicative of deficiencies of the coagulation factors like factors V, VIII, IX, X, XI, and XII.

Similar to the PT, here for APTT also similar steps are followed. Blood was drawn, an anticoagulant was added, and centrifuged to extract PPP. The subsequent addition of excess calcium in phospholipid suspension reverses the anticoagulant effect. Finally, to activate the intrinsic pathway, an activator is used (Figure 7.11).

The extracted PPP, calcium chloride, and PT reagent (Thromborel S, Siemens) were used in the ratio 1:1:1, respectively, whereas 200 µL of PPP was taken in Eppendorf Safe-Lock tubes. Then samples were placed in test tubes. The dressings were incubated at 37 °C for 3 min. After incubation, 100 µL of the PPP was taken into new tubes and supplemented with 100 µL of APTT reagent (Actin FSL, Siemens) and after 3 min of incubation to activate contact factors, 100 µL calcium chloride was included. The clotting time of the different samples was investigated utilizing an analyzer (Sysmex CA-50). The time was noted in "seconds" and performed in triplicates.

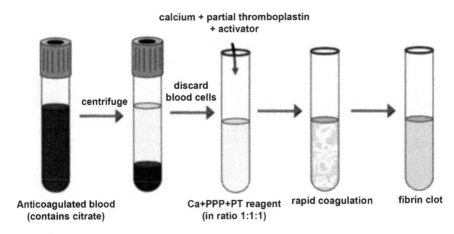

FIGURE 7.11 Schematic representation of a standard APTT protocol.

The APTT test showed that the increase in chitosan content depicts more activated partial thromboplastin time. Increased hydrophilicity inhibited protein adsorption to a certain extent and hence lowered platelet absorption and activation. This lead to the delay of APTT. Since it is sensitive to every factor of the intrinsic system along with fibrinogen and prothrombin, it is utilized to distinguish variations happening in the intrinsic pathway of blood coagulation. It is delicate to Factor VIII levels and subsequently can decide the delay in intrinsic pathway precisely.

The APTT test showed that the APTT of different samples 2:1, 1:1, 1:1.25, 1:1.5, Celox was 34, 55, 63, 69, 155 s, respectively. The observed time clearly showed that the sample was still much better than the Celox (Figure 7.12).

7.5.6 PLATELET ISOLATION AND ADHESION

For he platelet adhesion test, first, the anticoagulated blood was centrifuged at 1500 rpm for 20 min. The plasma fraction known as the supernatant was collected in a fresh tube and then subjected to centrifugation again at an rpm of 4500 for 10 min to concentrate the platelets. The platelet pellet was suspended in phosphate buffer saline (PBS) and the number of platelets per mL was calculated using a hemocytometer. Samples having 1.5×10^5 to 2×10^5 platelets/mL were poured atop the dressings. The 100 µL of calcium chloride was added per dressing to active the platelets. The dressings were fixed using 2.5% glutaraldehyde for 3 h rinsed twice in PBS to remove excess fixative and dehydrated using series of ethanol gradients (50%, 70%, 90%, 95%, and 100%). Finally, the samples were vacuum dried overnight and visualized using an ESEM. Figure 7.13 represents the same for various sample ratios of alginate and chitosan spray-dried PECs.

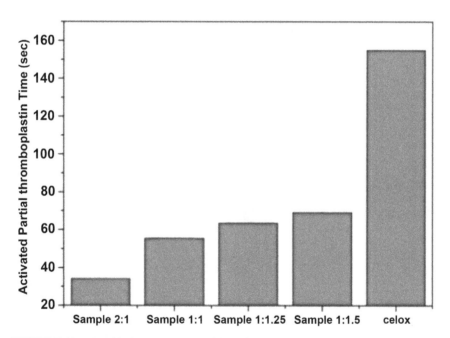

FIGURE 7.12 Graphical representation of the activated partial thromboplastin time (APTT): APTT accurately determines any changes in the intrinsic pathway of coagulation.

The standard curve was obtained at 33% (i.e., 1.5 mL plasma in 4.5 mL PBS) and the corresponding OD value was obtained at 0.186. At 0% dilution plasma is 100% and PBS is 0% and at 100% dilution the plasma is 0% and PBS is 100%. From the platelet adhesion (%) graph, we observed that from the number of platelets isolated, around 70% of platelets adhered to our samples. Composition 1:1 showed a slightly better result compared to composition 2:1 (Figure 7.14).

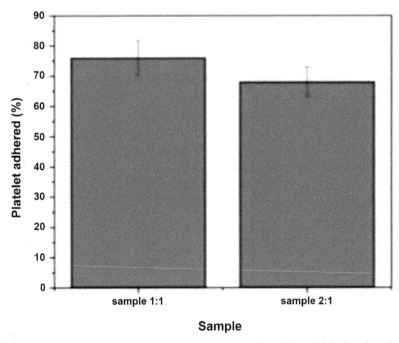

FIGURE 7.13 Image of different samples (A-2:1,B-1:1,C-1:1.5,D-1:1.25) showing platelet adhesion as obtained from ESEM.

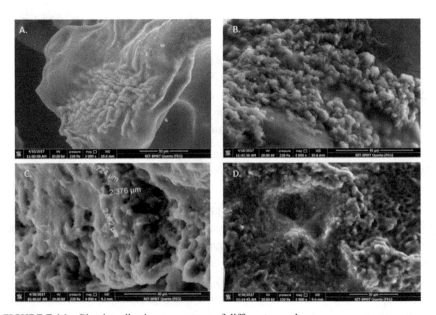

FIGURE 7.14 Platelet adhesion percentage of different samples.

Spray-Dried Chitosan and Alginate Microparticles 181

In a simultaneous study, the test showed that the platelet adhered percentage of different samples (2:1, 1:1, 1:1.25, 1:1.5) was observed as 68%, 74%, 76%, 83%. The observed results depict that more chitosan content leads to more adhesion of platelets than alginate.

7.6 SWELLING BEHAVIOR

Biomaterials may have absorption properties. When left in distilled water or PBS, they may swell and get enlarged. The void spaces available between the networks of the polymeric chains help to absorb and retain water [58]. The swelling behavior of biomaterials depend on pH, temperature, and ionic strength [59]. The high surface area to volume ratio is another crucial aspect that significantly affects the swelling ratio [60]. The high surface area allows more exchange of fluid. Thus, microparticles show greater swelling than the macro ones.

7.6.1 EFFECT OF SWELLING ON HEMORRHAGE

CELOX as mentioned previously is a blood-clotting agent available in the market. It preferably works on the swelling mechanism. Small particles absorb water in the blood that helps in the thickening of the blood and miraculously increase the frequency of encounter of RBCs and the microparticles. The fastly moving RBCs slow down which leads to a series of cascade reactions that ultimately clot the blood.

7.6.2 SWELLING ANALYSIS OF POLYELECTROLYTE COMPLEXES OF ALGINATE AND CHITOSAN

Different compositions of alginate and chitosan were taken and allowed to swell in PBS (7.4 pH) for 7 days. The weight of each of them was measured following the method of swelling with the first four readings taken at every 5 min, the second four readings for every 30 min, the third four readings for every 1 h, the fourth four readings for every 12 h, and fifth four readings for every 24 h. All of the compositions showed a similar swelling pattern with no dependency on the compositions of alginate and chitosan. The absorption rate was interestingly quite high for the first half an hour and highest for the first 5 min. This result successfully satisfies our needs. After half an hour the rate of swelling significantly decreased. A maximum of 17 times inflation was observed in the microparticles in 60–80 h after that it started degrading (Figure 7.15). Our focus is on the initial swelling behavior of the particles in which they showed significant absorbance of PBS.

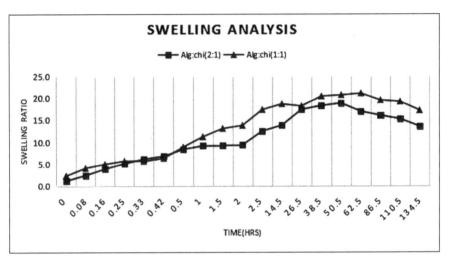

FIGURE 7.15 Swelling analysis of the 2:1 and 1:1 sample of alginate–chitosan. In the first 5 min, the rate of swelling is the highest. The microbeads were marked to enlarge 17 times at most.

7.7 USE OF PHARMACEUTICALS AND ADDITIVE STRATEGIES

The unfulfilled need for an ideal hemostatic agent as led to a plethora of biomaterials and drug targets, which can be administered directly at the injury site or can be administered intravenously. Furthermore, these agents can be biological and/or synthetic. Clinical approaches focus on the use of antifibrinolytic agents, coagulation factors, frozen/lyophilized platelets, recombinant coagulation proteins, biomimicking peptides, and others. The vast niche of pharmaceuticals ultimately manipulates fibrinolysis, coagulation pathway, and/or platelet plug formation.

When it comes to direct topical administration, hemostatic bandages can further encapsulate drugs/factors in them. Incorporation of several compounds can essentially facilitate or stabilize hemostatic pathway, although they themselves may not be prohemostatic. For external or intracavity administration of hemostat, they can be categorized into three types, based on their mechanism of action:

Spray-Dried Chitosan and Alginate Microparticles 183

1. *Factor concentrators*: These work via rapid water absorption and eventual concentration of cellular and protein components. Some commercial examples include QuikClot and TraumaDex.
2. *Mucoadhesive agents*: These agents physically block bleeding from the wound via their strong adherence to the tissues. For instance, HemCon and Celox.
3. *Procoagulant supplementers*: As the name suggests, these are factors directly delivered to the wound that will then promote hemostasis. For instance, dry fibrin sealant dressing.

In addition to the above strategies, hemostats also make use of mechanical (direct pressure), physical methods (such as cauterization), and physiological methods (such as fibrin, local thrombin, and adrenaline) [61].

Moreover, emergency or prophylactic management of internal noncompressible hemorrhage is still depended heavily on transfusion of whole blood or blood components (RBC, plasma, and platelets). These solely aim at restoration of lost blood volume and augmentation of physiological hemostasis. In addition to these transfusions, several pharmaceutical compounds can essentially facilitate or stabilize the hemostatic pathway, although they may not be pro hemostatic. For instance, desmopressin mimics natural vasopressin and hence systemically increases plasma concentrations of factor VIII and Von Willebrand factor. Also, compounds like tranexamic acid (TXA) and aprotinin are used. Aprotinin is a systemic broad-spectrum serum protease inhibitor and hence its potency to augment hemostasis is debatable, while it has been shown to benefit bleeding management [62, 63]. Moreover, TXA does not promote clotting but rather prevents plasmin-induced breakdown of the clot, while promoting clot strength and stability [64].

Although the use of whole blood or blood components (especially plasma) remains the primary intravascular hemostasis strategy, there are several significant challenges such as limited availability, high cost, and contamination risks.

7.8 FUTURISTIC ASPECTS

The reported product inherently has multiple characteristics working in harmony to achieve significant hemostasis. Its simple methodology, ease of application and effective action are only a few of their advantages. We have discussed and presented the experiment analysis of various compositions of our hemostatic agent.

184   *Natural Polymers: Perspectives and Applications for a Green Approach*

However, the scope of this product can be extended far beyond a generic daily use hemostat. The problem of cessation of blood becomes much more critical in cases of blood disorders. Many blood diseases are a direct consequence of the absence of an important cofactor or mere low levels of a certain protein. Clotting disorders occur when the body becomes unable to synthesize a sufficient amount of proteins, required for clot formation. These proteins (coagulation factors) are all synthesized in the liver. Some of these factors require vitamin K for their synthesis. Broadly speaking, a hemostatic crisis can be caused by vitamin K deficiency, severe liver disease, and antibodies (which may decrease the activity of a particular clotting factor) or can be genetic (such as hemophilia).

Exsanguination refers to the loss of blood to such an extent that it is sufficient to cause death and the patients with such coagulation disorders are at constant risk. Our product can significantly reduce this risk.

The charged microspheres associations of alginate and chitosan are potent to surface modifications, therapeutic conjugations, as well as therapeutic entrapment in the microspheres.

For instance, for a patient with a hereditary Von Willebrand disease, direct Von Willebrand factors can be incorporated within the microspheres or can be indirectly conjugated to be adsorbed on the particle surfaces. When applied, the patient's inability to form sufficient factors will be directly compensated by the product's composition itself.

Similarly, our product can potentially be incorporated with a variety of therapeutics/agents. In other words, the product not only holds the potential for customized patient care but also for topical drug delivery. Theoretical analysis indicates tailor-made topographical characteristics of the spray-dried particles.

Further enhancements can be done to increase patients' comfort. The particles can be mixed with antiseptics, analgesics, etc. Moreover, the formulation itself can be made as powder, soaked bandage, adhesive patch, and hydrogel mats.

7.9   MARKET POTENTIAL

We all are familiar with the popular phrase, "Necessity is the mother of invention." The death toll from "bleeding out" in this era of advanced medicine is outrageous. Commercialization of a viable solution, especially a medical product faces a plethora of challenges.

Spray-Dried Chitosan and Alginate Microparticles 185

A product that could successfully serve the needs of people has to pass through rigorous tests and scrutiny not only in terms of safety or efficacy but also have to face a range of social, economic, and communal circumstances.

For a hemorrhage control device, the list of customers consists of people who are prone to injury that literally include everyone. We can further section them as military forces, hospitals, and the common public.

A lot of discussion on the research and development of the product has been previously presented in the chapter. A brief knowledge of commercializing the product will enhance the utility of the text.

7.9.1 MARKET SIZE AND OPPORTUNITY

India is a developing country and so very prone to road accidents. In the year 2015 around 5 lac accidents happened in India which killed about 1.5 lac people (according to PRS Legislative Research). Broadly speaking for the statistics 90% of all accidental deaths can be prevented by early hemorrhage control [65]. Many a time patients die on the way to the hospitals due to major blood loss. The global market size of the hemorrhage control market was valued 5.3 billion USD (approximately) in 2015 and it is expected to reach 6.7 Billion USD by 2021. The compound annual growth rate of the market is 4% [66]. The market is full of potentials. There is a lot of scope in the hemorrhage control devices that are getting updated day by day from simply pressure applying mechanism to advanced nanomaterials with a tremendous scope of modifications as per the patient-specific needs of the patient as mentioned in Section 7.7.

7.9.2 COMPETITORS

A quality product is always appreciated by the customers. The product must be able to stand before the criteria decided by the customers. It must be able to satisfy their actual needs. There are many hemostatic and hemorrhage control devices available in the market. They all claim to have numerous qualities. For a new product to get a space in the market it must be better than the market available products in certain aspects. One needs to be very specific while choosing the criteria at which he/she are availing a better quality product.

A comparison between the chitosan-based commercially available hemostatic agents and our product have been given in Table 7.2.

186 *Natural Polymers: Perspectives and Applications for a Green Approach*

TABLE 7.2 Comparison of "Spray-Dried Chitosan-Alginate Powder" With Some of the Leading Hemostatic Products Available in the Market

Products ⟹ Properties ⇓	QuickClot	Celox	HemCon	Spray-Dried Chitosan-Alginate
Product type	Zeolite based	Chitosan based	Chitosan acetate based	Chitosan PEC based
Biocompatible	Yes	Yes	Yes	Yes
Clotting time(s) [68]	65–70	70–80	75–80	20–25
Mode of action (hemostasis)	Water absorption; facilitation of clotting cascade	swelling	Platelet activation	Swelling; Microsize and positively charged PEC surface which enhances RBCs adhesion; platelet adhesion; Facilitate clotting cascade
PEC	No	No	No	Yes
Cost (INR)*	500–550(3"*3")	350–400 per gram	70–80 (3"*3")	250–300 per gram (expected)

Costs of the products mentioned are in reference to official website/vendor quotations.

The comparison gives us a rough idea about our product's standing in the market. Though the price is a matter of concern, it widely depends on your choice of customers and the quality of your product. If commercialized, our product would be a high-quality military-grade product and our primary potential customers would be health care sectors and soldiers who fight on the battlefields, who are not much concerned about the price of the material but the quality. HemCon is cheap but has shown to be less effective than Celox and QuickClot in studies [68]. Further, the clotting time of our optimized composition is much better than the products available in the market. The multiple modes of action give it the instincts it needs, to survive in the market.

7.10 CONCLUSION

The alginate and chitosan powders formed by spray drying had a spherical shape, excellent water and blood absorption, high thrombin generation, and good hemocompatibility. By incorporation of alginate, the blood-clotting performance was enhanced. The dressings not only demonstrated good

Spray-Dried Chitosan and Alginate Microparticles

hemostatic performance, they can also be used in wounds with irregular depth and geometry. An analysis of the size and charge and distinctive properties of alginate helped us to solve the puzzles behind the unexpected behavior of the blood-clotting assay of various ratios of alginate and chitosan. All the characterizations suggest that there is a lot of potential in our material with a lot of scope of modifications based on the needs of the users. The product can be utilized for several blood-related diseases and disorders with some advancement. The alginate–chitosan spray-dried PEC powders are much better than the Celox (one of the most popular hemostatic agents in the market) in various aspects.

KEYWORDS

- **hemorrhage control**
- **hemostasis**
- **chitosan**
- **alginate**
- **microbeads**
- **blood clotting**
- **polyelectrolyte complex**
- **hemolysis**
- **prothrombin time**
- **nanocomplex**

REFERENCES

1. Hawk, A.J. How hemorrhage control became common sense. *J Trauma,* 2018, 85(1S), S13–S17.
2. Lemelson, J.H.; Grund, C.; Lemelson Jerome H. Tourniquet. U.S. Patent 4,321,929, March 30, 1982.
3. National Geography. This Doctor Upended Everything We Knew About the Human Heart https://www.nationalgeographic.com/archaeology-and-history/magazine/2018/01–02/history-william-harvey-medicine-heart/90 (accessed Feb 13, 2018).
4. Wiley Online Library. Discovery of the Cardiovascular System: From Galen to William Harvey. https://onlinelibrary.wiley.com/doi/full/10.1111/j.1538-7836.2011.04312.x (accessed Jul 22, 2011).

188   *Natural Polymers: Perspectives and Applications for a Green Approach*

5. Wies, C.H. The history of hemostasis. YJBM, 1929, 2(2), 167.
6. IJU. Evolution of Hemostatic Agents in Surgical Practice. http://www.indianjurol.com/article.asp?issn=0970-1591;year=2010;volume=26;issue=3;spage=374;epage=378;aulast=Sundaram (accessed Oct 1, 2010
7. Cox, E.D.; Schreiber, M.A.; McManus, J.; Wade, C.E.; Holcomb, J.B. New hemostatic agents in the combat setting. *Transfusion*, 2009, 49, 248S–255S.
8. Kheirabadi, B. Evaluation of topical hemostatic agents for combat wound treatment. *US Army Med Dep J*, 2011, 37.
9. Kheirabadi, B.S.; Acheson, E.M.; Deguzman, R.; Sondeen, J.L.; Ryan, K.L.; Delgado, A.; Dick Jr, E.J.; Holcomb, J.B. Hemostatic efficacy of two advanced dressings in an aortic hemorrhage model in Swine. *J Trauma*, 2005, 59(1), 25–35.
10. Walker, H.K.; Hall, W.D.; Hurst, J.W. Diplopia–Clinical Methods: The History, Physical, and Laboratory Examinations, 1990.
11. Park, J.; Lakes, R.S. *Biomaterials: An Introduction*. Springer Science & Business Media, 2007.
12. Suri, Y.V.; Garg, A.; Venugopalan, V.M.; Kapoor, S.; Tripathi, P.C.; Kochhar, H.K.; Mahajan, T.R. Militancy Trauma: Penetrating and nonpenetrating cardiac injury. *MJAFI*, 1997, 53(1), 30–34.
13. Jeffery, S.L. The management of combat wounds: the British military experience. *Adv Wound Care*, 2016, 5(10), 464–473.
14. Wound Source. Combat Wound Management: An Overview. https://www.woundsource.com/blog/combat-wound-management-overview (accessed Jul 11, 2018)
15. npr. For Military, Different Wars Mean Different Injurieshttps://www.npr.org/2011/06/12/137066281/for-military-different-wars-mean-different-injuries (accessed Jun 8, 2011)
16. Simak, J.; De Paoli, S. The effects of nanomaterials on blood coagulation in hemostasis and thrombosis. *Wires Nanomed Nanobi*, 2017, 9(5), 1447.
17. Daniels, R.H.; Li, E.; Rogers, E.J. Nanostructure-enhanced platelet binding and hemostatic structures. U.S. Patent 8,319,002, November 27, 2012.
18. Science Daily. Nanoparticle. https://www.sciencedaily.com/terms/nanoparticle.htm (accessed 2019).
19. Dragan, S.; Cristea, M.; Luca, C.; Simionescu, B.C. Polyelectrolyte complexes. I. Synthesis and characterization of some insoluble polyanion-polycation complexes. *J Polym Sci Pol Chem*, 1996, 34(17), 3485–3494.
20. Pogorielov, M.V.; Sikora, V.Z. Copyright© 2015 by Academic Publishing House Researcher Published in the Russian Federation. *Eur. J. Med. B*, 1, 2015, 24–33.
21. Maksym, P.V.; Vitalii, S. Chitosan as a hemostatic agent: current state. *Eur J Med Ser B*, 2015, 2(1), 24–33.
22. Benesch J.; Tengvall P. Blood protein adsorption onto chitosan. *Biomaterials*. 2002, 23, 2561–2567.
23. Smith, J.; Wood, E.; Dornish, M. Effect of chitosan on epithelial cell tight junctions. *Pharm Res*, 2004, 21(1), 43–49.
24. Harboe, M. and Mollnes, T.E. The alternative complement pathway revisited. *JCMM*, 2008, 12(4), 1074–1084.
25. Samudrala, S. Topical hemostatic agents in surgery: a surgeon's perspective. *AORN J*, 2008, 88(3).
26. Rao, S.B.; Sharma, C.P. Use of chitosan as a biomaterial: studies on its safety and hemostatic potential. *J Biomed Mater Res*, 1997, 34(1), 21–27.

Spray-Dried Chitosan and Alginate Microparticles 189

27. Arand, A.G.; Sawaya, R. Intraoperative chemical hemostasis in neurosurgery. *Neurosurgery*, 1986, 18(2), 223–233.
28. Fan, W.; Yan, W.; Xu, Z.; Ni, H. Formation mechanism of monodisperse, low molecular weight chitosan nanoparticles by ionic gelation technique. *Colloid Surf B*, 2012, 90, 21–27.
29. Wang, X.H.; Li, D.P.; Wang, W.J.; Feng, Q.L.; Cui, F.Z.; Xu, Y.X.; Song, X.H.; van der Werf, M. Crosslinked collagen/chitosan matrix for artificial livers. *Biomaterials*, 2003, 24(19), 3213–3220.
30. Mi, F.L.; Wu, Y.B.; Shyu, S.S.; Chao, A.C.; Lai, J.Y.; Su, C.C. Asymmetric chitosan membranes prepared by dry/wet phase separation: a new type of wound dressing for controlled antibacterial release. *J Membr Sci*, 2003, 212(1–2), 237–254.
31. Gu, R.; Sun, W.; Zhou, H.; Wu, Z.; Meng, Z.; Zhu, X.; Tang, Q.; Dong, J.; Dou, G. The performance of a fly-larva shell-derived chitosan sponge as an absorbable surgical hemostatic agent. *Biomaterials*, 2010, 31(6), 1270–1277.
32. Arand, A.G.; Sawaya, R., 1986. Intraoperative chemical hemostasis in neurosurgery. *Neurosurgery*, 18(2), 223–233.
33. Szymańska, E.; Winnicka, K. Stability of chitosan—a challenge for pharmaceutical and biomedical applications. *Mar Drugs*, 2015, 13(4), 1819–1846.
34. De Castro, G.P.; Dowling, M.B.; Kilbourne, M.; Keledjian, K.; Driscoll, I.R.; Raghavan, S.R.; Hess, J.R.; Scalea, T.M.; Bochicchio, G.V. Determination of efficacy of novel modified chitosan sponge dressing in a lethal arterial injury model in swine. *J Trauma*, 2012, 72(4), 899–907.
35. Kang, P.L.; Chang, S.J.; Manousakas, I.; Lee, C.W.; Yao, C.H.; Lin, F.H.; Kuo, S.M. Development and assessment of hemostasis chitosan dressings. *Carbohydr Polym*, 2011, 85(3), 565–570.
36. Rehm, B.H. Microbial production of biopolymers and polymer precursors: applications and perspectives. *HRPUB*, 2009.
37. Lansdown, A.B. Calcium: a potential central regulator in wound healing in the skin. *Wound Repair Regen*, 2002, 10(5), 271–285.
38. Ahmed, A.; Boateng, J. Calcium alginate-based antimicrobial film dressings for potential healing of infected foot ulcers. *Ther Deliv*, 2018, 9(3), 185–204.
39. Mikołajczyk, T.; Wołowska-Czapnik, D. Multifunctional alginate fibres with antibacterial properties. *Fibres Text East Eur*, 2005, 13, 35–40.
40. Sharma, S.; Sanpui, P.; Chattopadhyay, A.; Ghosh, S.S. Fabrication of antibacterial silver nanoparticle—sodium alginate–chitosan composite films. *RSC Adv*, 2012, 2(13), 5837–5843.
41. Baltimore, R.S.; Cross, A.S.; Dobek, A.S. The inhibitory effect of sodium alginate on antibiotic activity against mucoid and non-mucoid strains of *Pseudomonas aeruginosa*. *J Antimicrob Chemother*, 1987, 20(6), 815–823.
42. Rajendran, S.; Anand, S.C.; Rigby, A.J. Textiles for healthcare and medical applications. In *Handbook of Technical Textiles*. Woodhead Publishing; 2016; pp. 135–167.
43. Abhilash, M.; Thomas, D. Biopolymers for biocomposites and chemical sensor applications. In *Biopolymer Composites in Electronics*; Elsevier; 2017; pp. 405–435.
44. Odell, E.W.; Lombardi, T.; Oades, P. Symptomatic foreign body reaction to haemostatic alginate. *Br J Oral Max Surg*, 1994, 32(3), 178–179.
45. Suzuki, Y.; Nishimura, Y.; Tanihara, M.; Suzuki, K.; Nakamura, T.; Shimizu, Y.; Yamawaki, Y.; Kakimaru, Y. Evaluation of a novel alginate gel dressing: cytotoxicity to

190 *Natural Polymers: Perspectives and Applications for a Green Approach*

fibroblasts in vitro and foreign-body reaction in pig skin in vivo. *J Biomed Mater Res*, 1998, 39(2), 317–322.

46. Zhang, T.; Youan, B.B.C. Analysis of process parameters affecting spray-dried oily core nanocapsules using factorial design. *AAPS PharmSciTech*, 2010, 11(3), 1422–1431.

47. Brennan, J.G.; Herrera, J.; Jowitt, R. A study of some of the factors affecting the spray drying of concentrated orange juice, on a laboratory scale. *Int J Food Sci Technol*, 1971, 6(3), 295–307.

48. Schuck, P. Understanding the factors affecting spray-dried dairy powder properties and behavior. In *Dairy-Derived Ingredients*. Woodhead Publishing; 2009; pp. 24–50.

49. El-Newehy, M.H.; Al-Deyab, S.S.; Kenawy, E.R.; Abdel-Megeed, A. Fabrication of electrospun antimicrobial nanofibers containing metronidazole using nanospider technology. *Fiber Polym*, 2012, 13(6), 709–717.

50. Dubský, M.; Kubinová, Š.; Širc, J., Voska, L.; Zajíček, R.; Zajícová, A.; Lesný, P.; Jirkovská, A.; Michálek, J.; Munzarová, M.; Holáň, V. Nanofibers prepared by needleless electrospinning technology as scaffolds for wound healing. *J Mater Sci Mater Med*, 2012, 23(4), 931–941.

51. Jayakumar, R.; Prabaharan, M.; Kumar, P.S.; Nair, S.V.; Tamura, H. Biomaterials based on chitin and chitosan in wound dressing applications. *Biotechnol Adv*, 2011, 29(3), 322–337.

52. Unnithan, A.R.; Gnanasekaran, G.; Sathishkumar, Y.; Lee, Y.S.; Kim, C.S. Electrospun antibacterial polyurethane–cellulose acetate–zein composite mats for wound dressing. *Carbohydr Polym*, 2014, 102, 884–892.

53. Bao, J.; Yang, B.; Sun, Y.; Zu, Y.; Deng, Y. A berberine-loaded electrospun poly-(ε-caprolactone) nanofibrous membrane with hemostatic potential and antimicrobial property for wound dressing. *J Biomed Nanotechnol*, 2013, 9(7), 1173–1180.

54. Liu, M.; Duan, X.P.; Li, Y.M.;Yang, D.P.; Long, Y.Z. Electrospun nanofibers for wound healing. *Mater Sci Eng C*, 2017, 76, 1413–1423.

55. Sperling, C.; Fischer, M.; Maitz, M.F.; Werner, C. Blood coagulation on biomaterials requires the combination of distinct activation processes. *Biomaterials*, 2009, 30(27), 4447–4456.

56. Mallick, R.; Hubsch, A.; Barnes, D.G. Hemolytic adverse effects of intravenous immunoglobulin: modeling predicts risk reduction with anti-A/B immunoaffinity chromatography and to a lesser extent with anti-A donor screening. *Transfusion*, 2018, 58(12), 2752–2756.

57. Kamal, A.H.; Tefferi, A.; Pruthi, R.K., 2007, July. How to interpret and pursue an abnormal prothrombin time, activated partial thromboplastin time, and bleeding time in adults. In Mayo Clinic Proceedings; Elsevier; Vol. 82; No. 7, 864–873.

58. Bettini, R.; Colombo, P.; Massimo, G.; Catellani, P.L.; Vitali, T. Swelling and drug release in hydrogel matrices: polymer viscosity and matrix porosity effects. *Eur J Pharm Sci*, 1994, 2(3), 213–219.

59. Hench, L.; Jones, J. eds. *Biomaterials, Artificial Organs and Tissue Engineering*. Elsevier; 2005.

60. Germain, J.; Fréchet, J.M.; Svec, F. Hypercrosslinked polyanilines with nanoporous structure and high surface area: potential adsorbents for hydrogen storage. *J Mater Chem*, 2007 17(47), 4989–4997.

61. Kratz, A.; Danon, A. Controlling bleeding from superficial wounds by the use of topical alpha adrenoreceptor agonists spray: a randomized, masked, controlled study. *Injury*, 2004, 35(11), 1096–1101.

Spray-Dried Chitosan and Alginate Microparticles 191

62. Levi, M.; Cromheecke, M.E.; de Jonge, E.; Prins, M.H.; de Mol, B.J.; Briët, E.; Büller, H.R. Pharmacological strategies to decrease excessive blood loss in cardiac surgery: a meta-analysis of clinically relevant endpoints. *The Lancet*, 1999, 354(9194), 1940–1947.

63. Capdevila, X.; Calvet, Y.; Biboulet, P.; Biron, C.; Rubenovitch, J.; d'Athis, F. Aprotinin decreases blood loss and homologous transfusions in patients undergoing major orthopedic surgery. *Anesthesiol J Am Soc Anesthesiol,* 1998, 88(1), 50–57.

64. Roberts, I.; Prieto-Merino, D.; Manno, D. Mechanism of action of tranexamic acid in bleeding trauma patients: an exploratory analysis of data from the CRASH-2 trial. *Crit Care,* 2014, 18(6), 685.

65. Future Market Insight. Hemorrhage Control System Market: Global Industry Analysis and Opportunity Assessment 2016–2026. https://www.futuremarketinsights.com/reports/hemorrhage-control-system-market (accessed Aug 2019).

66. Zion Market Research. Global Hemostasis Products Market will reach USD 5.30 Billion By 2021: Zion Market Research. https://www.globenewswire.com/news-release/2016/12/12/897011/0/en/Global-Hemostasis-Products-Market-will-reach-USD-5–30-Billion-By-2021-Zion-Market-Research.html (accessed Dec 12, 2016).

67. Clay, J.G.; Grayson, J.K.; Zierold, D. Comparative testing of new hemostatic agents in a swine odel of extremity arterial and venous hemorrhage. *Mil Med*, 2010, 175(4), 280–284.

68. Kozen, B.G.; Kircher, S.J., Henao, J.; Godinez, F.S.; Johnson, A.S. An alternative hemostatic dressing: comparison of CELOX, HemCon, and QuikClot. *Acad Emerg Med*, 2008, 15(1), 74–81.

CHAPTER 8

High-Intensity Ultrasonication (HIU) Effect on Sunn Hemp Fiber-Reinforced Epoxy Composite: A Physicochemical Treatment Toward Fiber

CHINMAYEE DASH, ASIM DAS, and DILLIP KUMAR BISOYI[*]

Composite Laboratory, National Institute of Technology Rourkela, Rourkela 769008, Odisha, India

[*]Corresponding author. E-mail: dkbisoyi@nitrkl.ac.in

ABSTRACT

The present study aims to disintegrate fibrils from sunn hemp (*Crotalaria juncea*) cellulosic fiber by using a treatment method where a fusion of chemical as well as physical technique is involved. Before fabricating the composite, sunn hemp fiber was treated with a combination of alkali and high-intensity ultrasonication (HIU) that induces cavitation by means of the shock wave with high pressure and temperature. Sunn hemp fibers were subjected to 5% alkali (NaOH) solution and then were exposed under HIU vibration for 60, 90, and 180 min. The impact of treatment was researched on cellulose crystallinity, morphology, functional groups of fiber, and mechanical properties of their strengthened epoxy composites. Cellulose crystallinity was increased by increasing the exposure time of ultrasonication toward chemical impregnation. Morphology of sunn hemp fiber showed an irregular and hew surface that confirmed the removal of noncellulosic components that were well agreed by the change in functional groups. Mechanical properties of treated and untreated sunn hemp reinforced composite were likewise expanded with expanding introduction time and the best outcome was seen in the mix of 5%NaOH+180-min treated fibers. Therefore induced cavitation by the shock wave generated with HIU plays a very essential role in strengthening the mechanical properties of composites.

8.1 INTRODUCTION

While dealing with materials in day-to-day life, we must have a concern about our health and also surroundings. Human health and the surrounding environment are both complementary to each other. Grasping toxic gases, using toxic material are largely affecting our health, so the world must have to take responsibility for the protection of nature. Among different materials that have been utilized, polymeric materials give their job in various potential applications. It can bring insight into progressing product development that can be able to achieve a successful market in medical, packaging, aerospace, and consumer goods industries. The polymers may be grouped as synthetic and natural. Both have their distinct vision in different application subjects, but the world is trying to be successful in developing the material from the ecofriendly product. Synthetic polymers exhibit excellent dimensional stability with good properties like thermal, mechanical, chemical resistance, and others, but it is not ecofriendly. The reason is although it provides good strength to final material, after their life span, the exposal toward the environment is of health concern due to its inability to biodegradation. So the vision is changing to have the startup of assembling eco-accommodating material from the common item. The effort has been already started from the 19th century to build a greener world to protect the environment. We agree that the natural polymer cannot replace the workability talent of synthetic fiber fully, but the effort is still in progress to generate the product with comparable properties like synthetic and many successful reports have been submitted too [1–3]. These nature's extracted polymers from nature are often available as silk, wool, cotton, protein, cellulose and DNA. Contingent on these, the common polymer can be in the kind of animal or plant fiber or minerals fillers. Fibers are often compared to hairlike things where the aspect ratio regulates its properties. Sustainable development through the use of these natural materials is now a craze among the researchers because of their innate noble characteristics properties like a natural occurrence, cost-effective, recyclability, eco-friendly, and others. Addition of regular strands as support into the polymer network in creating polymeric composite material is now a new alternative to synthetic fibers.

High-Intensity Ultrasonication (HIU) Effect 195

8.2 ANATOMY OF LIGNOCELLULOSIC FIBER

8.2.1 *FIBER CELL STRUCTURE*

The primary concoction parts of lignocellulose fibers are holocellulose, which is the combined fraction of cellulose and hemicellulose, and apart from that there are lignin, wax, and different impurities. They are deposited in the fiber layer by layer. A schematic diagram of fiber structure has been shown in Figure 8.1. The plant cell wall consists of the primary cell wall, secondary cell wall, and tertiary cell wall. The essential primary cell is the external layer of the plant and framed first during cell development. The prime components in the primary cell wall are polysaccharides, calcium, enzymes, and others. The polysaccharides are for the most part cellulose and hemicellulose glucose polymer, however, the hemicellulose and pectin divide are prevailing in the primary cell. The prime role of the primary cell wall in the plant is to provide structural stability to plant, determine cell morphology, control cell growth, and preserve carbohydrates [4]. There is a limit to the growth of the primary wall, and this growth completion leads to the formation of the secondary cell wall. Here in the secondary wall, the composition varies from the primary one. It is appended to the primary layer and contains polysaccharides alongside lignin. But cellulose is predominant in this layer, so it grows upward rather than flattened. The secondary auxiliary cell layer is again isolated into three sublayers; external secondary layer (S1), center secondary layer (S2), and the internal secondary layer (S3), where the S2 layer directs the mechanical properties of the fiber. These sublayers contrast from each other through the microfibrillar point of helically twisted cell microfibrils that created from long-chain cellulosic polymer [5]. When fibers are treated or lignified, middle lamella, primary wall, and secondary wall are mainly affected. Lumen is the central part of the fiber, which is responsible for food and nutrients transportation throughout the plant.

8.2.2 *CHEMICAL COMPONENTS*

Cellulose $(C_6H_{10}O_5)_n$ is the prime constituent in plant fiber, and it is the building block of fiber where "n" is called the degree of polymerization. The chained network in cellulose is shaped by the D-glucose unit, which associated by β (1–4) glycosidic linkages; in this way, it is a homopolymer. It is a linear chain polymer, and the degree of polymerization is very

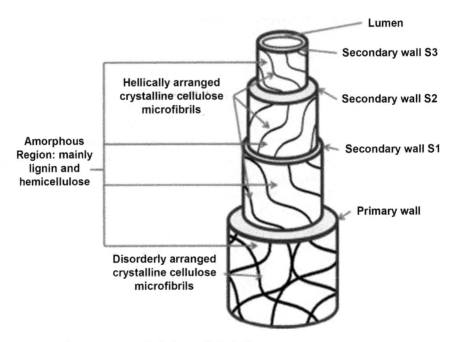

FIGURE 8.1 Structure of single lignocellulosic fiber.

high because there are almost 800–10,000 glucose units combine to form cellulose. It exhibits an unbranched structure in the fiber and possesses a high packing density. Cellulose is hydrophilic in nature because of the presence of polar hydroxyl group (OH–) from several glucose units that are formed due to intramolecular hydrogen (H) bond with moisture and intermolecular H-bond with other OH bunches inside the fiber itself. Due to this, inter- and intraconnection with the formation of a hydrogen bond lead to a large network system in the fiber and make the fiber hydrophilic [5]. Moreover, cellulose is available abundantly in the plant, and its supramolecular structure exists in two phases: crystalline and amorphous. The crystalline phase is due to the high orientation of the same glucose polymer chain, but amorphous contributions are from the presence of hydroxyl groups of other noncellulosic components. The crystalline structure of cellulose is accessible fundamentally in two structures; cellulose I and cellulose II. Cellulose I is the least stable state and has another two coexisting phases; I_α (triclinic) and I_β (monoclinic). Cellulose in algae and bacteria is enriched with I_α, where I_β is dominant in larger plants. With alkalization, cellulose I to cellulose II conversion is possible, and once the conversion happens to cellulose II, it is

High-Intensity Ultrasonication (HIU) Effect 197

then irreversible to the former state and remains stable. Again the crystalline cellulose polymorph III and IV can be shaped from cellulose on the off chance that it experiences treatment with liquid ammonia and heating [6]. A cellulose structure is shown in Figure 8.2.

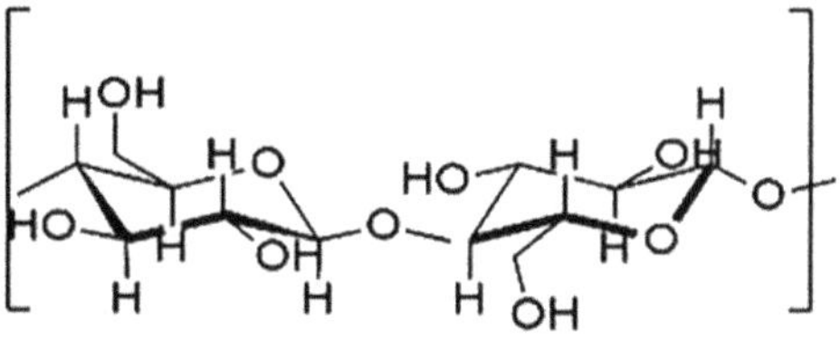

FIGURE 8.2 Cellulose structure showing the repetition of glucose units.

Hemicellulose is also a sugar-based polymer that presents with cellulose in the fiber. It appreciates various sorts of D and L sugar polymer units; D-glucopyranose, D-galactopyranose, D-manopyranose, D-glucopyranosyl-uronic acid, L-arabinosfuranose along with acetyl and methyl-substituted groups, therefore called a heteropolymer. In view of many sugar polymeric units, the hemicellulose has a lower level of polymerization (under 500) and unbranched, in contrast to cellulose. Hemicellulose is gotten from a polymer backbone of D-galactose, D-glucose and D-mannose linked by β (1–4) glycosidic linkage wherein hardwood, it contains the backbone of D-xylose connected by β (1–4) with acetyl groups at C-2 and C-4 of the xylose units [7]. Being an unbranched heteropolymer, the hemicellulose is short-tied and amorphous. It acts as a compatibilizer in fiber in order to be associated with cellulose. Hemicellulose can be extricated by alkali treatment with advanced focus on the grounds that a degree fixation may upset the holding between the chains in cellulose, which may influence the mechanical properties of the fiber. A hemicellulose structure with monomer sugars has been shown in Figure 8.3.

Lignin is an aromatic polymer of phenyl propane units, and they are highly complex, extended amorphous polymer in all cellulose-based fiber. Plant lignin is mainly composed of gymnosperm and angiosperm with the basic building blocks of guaiacyl, syringyl, and *p*-hydroxyphenyl moieties [6]. Lignin is available nearly in all the layers of fiber. It holds cellulose and hemicellulose and serves as a cementing material by deposition within the vacant places. Lignin is hydrophobic and nonpolar, so a high content of lignin in the plant has a high resistance to water retention. So coir (45–48 wt.%) [8], oil palm (17–29 wt.%) [9], and kenaf (18–27 wt.%) [10] of lignin content indicate that these fibers resist with water absorption hence use in various applications like rope making, fishnet, and others. This suggests the percentage of lignin content in some specific fiber. A structure of lignin has shown in Figure 8.4.

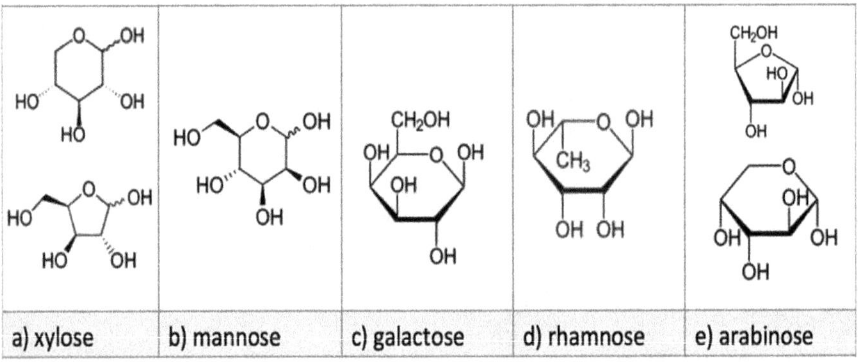

| a) xylose | b) mannose | c) galactose | d) rhamnose | e) arabinose |

FIGURE 8.3 Monomer sugars in hemicellulose.

Source: Reprinted from [39].

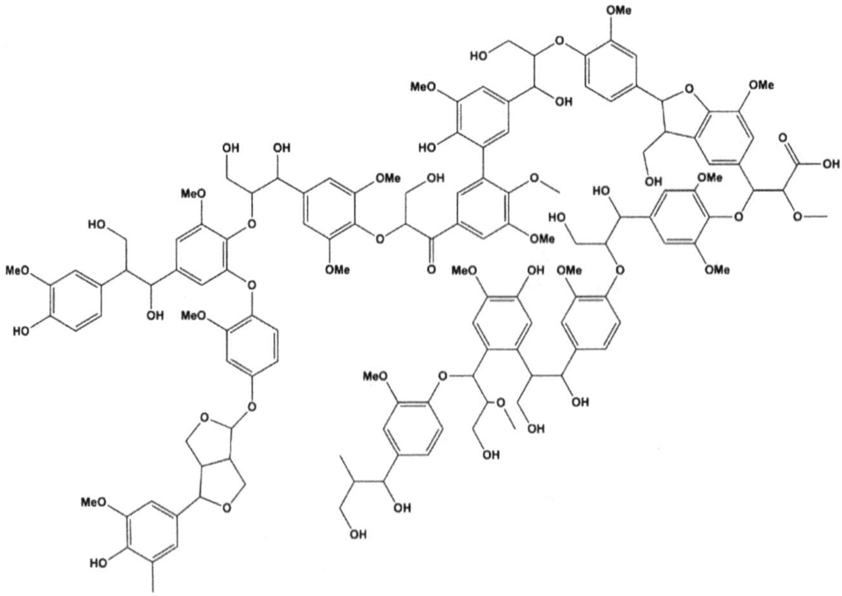

FIGURE 8.4 Structure of lignin.

High-Intensity Ultrasonication (HIU) Effect 199

Pectin is a heteropolysaccharide that dominant in the middle lamella of the plant fiber. Center lamella appends two plant cell together and normally made up of calcium and magnesium pectates. It is wholly attached to the first layer of fiber so unable to distinguish primary cells from middle lamella. Pectin is a significant component in the nonwoody type of plants and a long chain polymer having a high molecular weight of 25,000–50,000. It is soluble in water and provides flexibility to plant [6].

Wax in plant fiber acts as a covering or coating for the plant cuticle and compose of straight-chain aliphatic hydrocarbons. It reflects ultraviolet light, assists in the formation of hydrophobic surface, helps plant from disease and insects with resistance to drought. It can be separated from the plant by the mechanical process [11].

8.3 SOME WELL-KNOWN NATURAL POLYMER

All the above discussed chemical components present in almost all type of natural fiber but vary with concentrations from one to another. Some of the well-known fibers with their chemical components are described below. Depending on the substance segment rate, the properties of fiber additionally differ as it needs to be. A rundown of some chose characteristic plant fiber with their chemical parts have appeared in Table 8.1.

TABLE 8.1 Chemical Composition of Some Selected Natural Fibers

Fiber Name	Origin of Fiber	Cellulose	Hemicellulose	Lignin	Pectin
Cotton	Seed	82.7–92	2–5.7	0.5–1	5.7
Sisal	Leaf	43–88	10–13	4–12	0.8–2
Coir	Fruit	36–43	0.15–0.25	41–45	4.0
Flax	Bast	60–81	14–18.6	2–3	1.8–2.3
Bamboo	Grass	26–43	15–26	21–31	–

Source: Reproduced with permission from Ref. [52] © Science Publishing Group.

8.3.1 COTTON FIBER

Cotton (Gossypium) is considered as well known as the lord of fiber among all the plant fibers. It is built as a delicate staple fiber that developed as boll and appended to the seed of the plant. Cotton is the only plant that is highly enriched with cellulose (more than 90%) with a very small content of lignin. So it has high absorbency toward water and moisture. Successful cultivation of cotton plant needs warm weather, moderate rainfall, and heavy fairy soils.

So the cotton farming is most preferable in tropics and subtropics of the Northern and Southern hemispheres [12]. China and India are the largest producers of cotton, and the annual production is about 34 million and 33.4 million bales, respectively. Cotton fiber has its own regular fondness to water, substantial scraped spot obstruction, long cellulose microfibrils that gives quality and solidness to the end-use things, high level of degree of polymerization (9000–15,000), weak resistance to acids, insects, sunlight (prolonged exposure may weaken fibers) but excellent resistance to alkali and organic solvents. In the material industry, cotton is richly utilized in the creation of garments, weaving, and reaches out to paper industry, angling nets, espresso channels, and others [13]. In the composite industry, the cotton fiber as reinforcement is mostly not favorable because of its dominance in the textile industry, high absorbency of water, and cost. A close view of cotton fiber with its structure is shown in Figure 8.5.

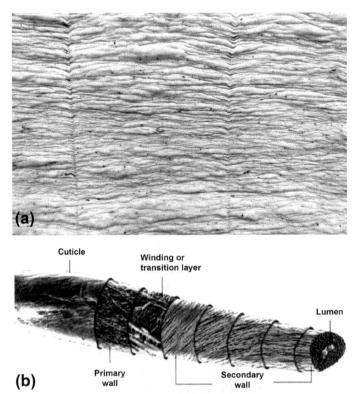

FIGURE 8.5 (a) Close view of cotton fibers (b) structure of cotton fiber.

Source: Reprinted from permission from a) https://www.nourluxury.com/blogs/news/non-organic-cotton-bedding-toxic b) Richardson, David and Sue. Asian Textile Studies: Cotton, http://www.asiantextilestudies.com/cotton.html#g

8.3.2 SISAL FIBER

Sisal (*Agave sisalana*) belongs to the family of Agave, which produce rosette of fleshy leaves of 1–1.5 m long and about 10 cm wide. Each fiber of sisal contains an average of 1000 fibers and has a life span of 7–10 years. Globally, Brazil is the largest country in the production of Sisal and produces about 150.6 tons sisal annually. Sisal fibers are generally coarse and robust, durable, and able to stretch, provide suitable insulation medium, excellent resistance to salt water, therefore preferred for making ropes and twines. Aside from these, sisal shows its flexibility in the method for paper creation, geotextiles, carpets, handicrafts, and others [14]. Being an ecofriendly fiber and 100% biodegradable, the sisal reinforced composites are used in the automobile industry by replacing the glass fiber and asbestos. Because of the unbending nature and quality in fiber, sisal fiber offers a great mechanical base to the composite material. A close view of sisal fiber with a microscopy image of its cross-section is shown in Figure 8.6.

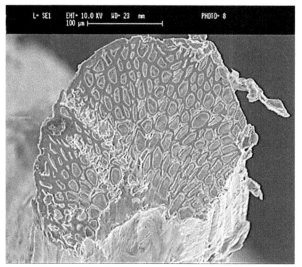

FIGURE 8.6 Cross-sectional view of sisal fiber
Source: Reprinted with permission from Ref. [53]. © Springer Nature.

8.3.3 COIR FIBER

Coir (*Cocos nucifera*) fiber is mainly found in both inner and outer shells of the coconut fruit belongs to the family of Plamae. The fibers are typically

4–12 inches long and 0.1–0.5 mm in diameter. The color transition of fiber from pale to yellow is the indication of lignin deposition on their walls. The surface of the coconut shell is on the whole rough with cross markings, while the fibers are coated with a waxy layer [15]. The strands are possible in darker and white shading, where the darker hued filaments are increasingly developed and contain more lignin than cellulose while white color fibers are weak and harvested before ripening often spun to make yarn in mats and rope. Sri Lanka and India collect about 90% of the coir fiber every year. The main features that lie within this fiber are it is durable and relatively waterproof due to the high accommodate of lignin and good resistance to salt water. Coir is a prevalent fiber since the early ages and has multifaceted uses in the packaging industry, bedding, flooring, carpets. The composite from coir fibers are used as making of ecofriendly vehicles, door panels, and others. A close view of coconut shells with a cross-sectional and longitudinal view of coir fiber have shown in Figure 8.7.

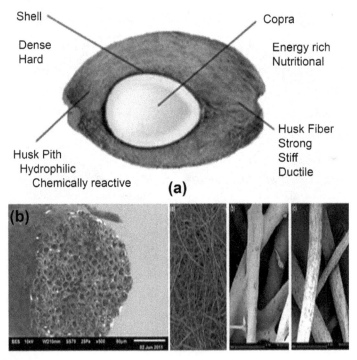

FIGURE 8.7 (A) Close view of coconut shell (b) cross-sectional and longitudinal view of coir fiber.

Source: https://textilelearner.net/coconut-coir-fiber-properties-manufacturing/

8.3.4 FLAX FIBER

Flax (*Linum usitatissimum*) is an individual from the Lenum family, and the strands are started from stems of the plant. Bast fibers possess a high content of cellulose after cotton seed fibers. Within 8 weeks of sowing, the plant can reach a height of 10–15 cm and grows to several centimeters. So the gathered fibers appear with approximate the normal length of 70–80 cm and 12–16 µm width [16]. It requires few fertilizers and pesticides to grow. The fiber is mainly utilized in dress clothing, produces high-quality paper due to high cellulosic content, reinforcing with polymer yield composite material used in paneling, automobiles, insulation, and others. Russia is the biggest maker of flax fiber drove by Canada and creates about 2.65 million tons of fiber annually. A close perspective on flax fiber with microscopy picture has appeared in Figure 8.8.

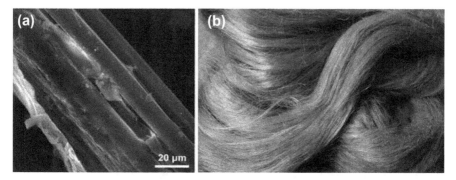

FIGURE 8.8 (a) FESEM image of flax fiber and (b) close view of flax fiber. a) Reprinted with permission Ref. 43. © 2017 Springer Nature.

8.3.5 BAMBOO FIBER

Bamboo (*Bambusoideae*) belongs to the grass family of Poaceaea, and about 90% of the total weight of the bamboo tree contains its fiber only while another part contains a limited amount of ash, fat, protein, pectin and tannins. In Asia, the large production of Bamboo is mainly from China, India, Indonesia, Philippines, Myanmar, and Vietnam. The Bamboo tree has the potential to grow faster attains a height of about 90 cm and stops around 6–7 months after the emerging of shoots. The shoot grows vertically and from a culm that resembles a hollow cylinder with an appearance of a ring like a form from outside, and this ring divides the culm from the inner side called

Diaphragm. The space between the two rings is called internodes. The fibers are aligned longitudinally to the length of the culm and make its strength when it attains maturity [17]. The essential element of bamboo filaments incorporates higher explicit compressive quality and elasticity, its elasticity increases with age, good to moisture absorption. As a reinforcing fiber, they are used to make composite material because of the biodegradability, renewable, and ecofriendly. These fibers have gained popularity in the construction industry, scaffolding, musical instruments, furniture, and panels. A close view of bamboo fiber with surface morphology is shown in Figure 8.9.

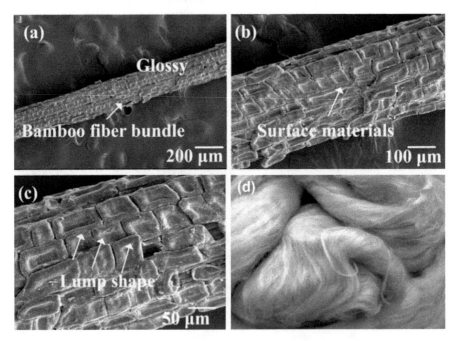

FIGURE 8.9 (a)–(c) FESEM images of untreated bamboo fiber (d) Close view of Bamboo fiber.

Source: Reprinted from Ref. 44. © 2018 by the authors. Licensee MDPI, Basel, Switzerland. (http://creativecommons.org/licenses/by/4.0/).

8.4 CONSTRAINTS INVOLVED IN NATURAL FIBERS

For the last numerous years, including late years, regular fibers have earned a remarkable position among scientists and researchers. Regular filaments as the strengthening material that utilization in the creation of the polymer composites, which offer the yield of lightweight, eco-accommodating, solid, and biodegradable by-products. Numerous common fortifying composites have demonstrated to give great mechanical and physical properties because of their constituting desired product properties (lightweight, high strength to weight ratio, greater fatigue, and high toughness). Developing countries take enough opportunity to make low-cost, high strength, and ecofriendly composite products due to their economic concern. These raw materials have some inherent difficulties that are procured by nature only. Prior to fabricating the composites, one should concern about the selection of fibers and matrix as their individual properties regulate different properties of the end-use products. The main difficulties in fiber include chemical composition, harvesting time, soil type, extraction method, treatment, storage procedure, fiber variety, aspect ratio, voids contents, high water, and moisture absorption that lead to poor wetting between fiber and matrix, fiber orientation relative to the fiber axis and fiber dispersion within matrix. In the case of the matrix, the selection should be accordingly the thermosetting or thermoplastic polymer matrices or petroleum-based bioderived matrices types [18]. Thermoplastic polymer matrices do have a good impact as well as damage tolerance where thermosetting polymers are acceptable saturator and a superior cement to fortifying fibers, great thermal steadiness with the prudent economy than thermoplastic matrices. Fibers are hydrophilic while the available, human-made matrix is hydrophobic. This conflict between hydrophilic and hydrophobic phases deplore the adhesions at the interface, which affects the final products. So in order to get compatibility and enhance the adhesion, various pretreatment techniques are performed on the fibers. Such remedies generally attempt to take out noncellulosic segments with polluting influences from fibers, excite the reactive groups from fiber structure, clean the surface, and increase the fiber roughness to bind with the matrix.

8.5 PRETREATMENT METHODS

There are lots of pretreatment methods reported and the influence of the treatment on the fiber has been investigated as well. Some of the pretreatment techniques are discussed below.

8.5.1 ALKALIZATION

In alkalization method, it uses various alkaline base chemicals like sodium hydroxide (NaOH), potassium hydroxide (KOH), calcium hydroxide (CaOH), magnesium hydroxide (MgOH), and others to facilitate the modification in cellulose macromolecular structure. Alkali sensitive OH-groups react with the intercom chemicals and change the periodicity in the crystalline domain of the cellulose chain from the structure. The response inside the fiber during the alkalization is demonstrated as follows:

$$\text{Fiber-OH} + \text{NaOH} \longrightarrow \text{Fiber-O}^-\text{Na}^+ + \text{H}_2\text{O}$$

The alkali concentration beyond an optimization value may hinder H-bond and form the amorphous cellulosic stage at t he cost of crystalline cellulose. Native cellulose I transform to thermodynamic stable cellulose II upon the alkali treatment [19]. A high centralization of soluble base alkali influences the various physical properties of composite adherently. Cai et al. treated Abaca fiber with 5, 10, and 15 wt.% NaOH alkali for 2 h and studied their effect on the mechanical properties of the abaca-epoxy composite. They found treatment with a mellow concentration (5 wt.%) of alkali to abaca fiber upgraded cellulose crystallinity of fiber and tensile strength of composite that is benefitted for making abaca/epoxy polymer composite [20]. Fiore utilized 6 wt.% NaOH for 48 and 144 h over kenaf fiber and contemplated elasticity, which resulted that the treatment with alkali for 48 h was able to wipe out the surface and made the headway of the mechanical qualities of the composite. An adjustment in surface morphology of untreated and alkali kenaf fiber have appeared in Figure 8.10.

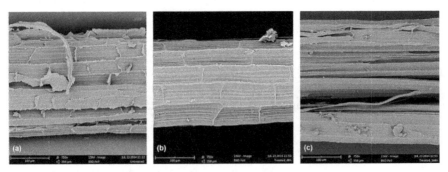

FIGURE 8.10 SEM image of (a) untreated fibers, (b) 48 h treated NaOH fiber, (c) 144 h treated fiber.

Source: Reprinted from Ref. [45]. © 2015 Elsevier.

8.5.2 SILANIZATION

The silanization method uses multifunction silane molecules as a coupling agent to modify fiber surface. Silane chemicals attempt to associate with the fiber through the Siloxane connect and were seen as adequate coupling operator among the various coupling agents. During treatment, it undergoes various activities: hydrolysis, condensation, and bond formation. The hydrocarbon chains formed by the application of silane retain the swelling of the fiber by crosslinking. Study shows that silane-treated fiber composite provides better mechanical results than alkali-treated fiber composite [21, 22]. The chemical reaction takes place during silanization, as shown below. Principal change occurs in *Eulaliopsis binata* fiber upon silane treatment is demonstrated in Figure 8.11. Silane reaction in cellulosic fiber is shown below [23].

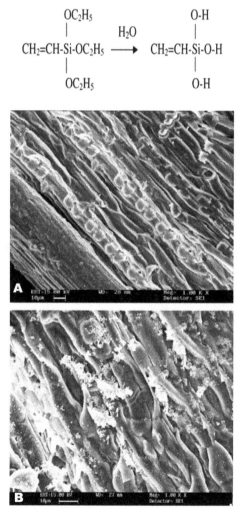

FIGURE 8.11 SEM image of (a) untreated (b) silane treated *Eulaliopsis binata* fiber.
Source: Reprinted with permission from Ref. [46]. © 2014 Elsevier.

8.5.3 PERMANGANATE TREATMENT

Permanganate treatment to various natural fibers is often completed by potassium permanganate ($KMnO_4$) in an acetone arrangement. This reaction forms highly reactive permanganate ions with cellulose OH groups and initiates cellulose-manganate network. It likewise responds with other noncellulosic segments inside the fiber and removes them out, thus reduces the hydrophilic nature of the fiber. Mohammed researched the impact of permanganate treatment with different 0.033, 0.066, and 0.125 wt.% in 6 wt.% NaOH on sugar palm fiber and fiber/polyurethane composite and found that 0.125% treated fiber offers great outcome over mechanical strength in composite [24]. The reaction takes place in the fiber during permanganate treatment, as shown below, and surface morphology of sisal fiber upon permanganate treatment is shown in Figure 8.12 [25].

$$\text{Fiber–OH} + KMnO_4 \longrightarrow \text{Fiber–O–H–O–Mn–OK}^+ \overset{O}{\underset{O}{\parallel}}$$

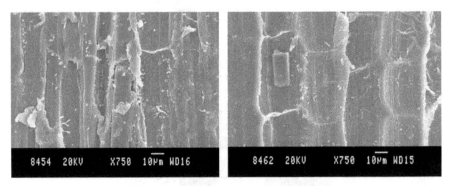

FIGURE 8.12 Changes in surface morphology of sisal fiber (a) untreated (b) silane treated.
Source: Reprinted with permission from Ref. [47]. © 2009 Elsevier.

8.5.4 PLASMA TREATMENT

The above-talked about medications utilize different synthesized chemicals to change the fiber structure. There are many other kinds of chemical treatment techniques that have been reported, such as acetylation, peroxification, benzoylation, isocyanization, acrylation, and many others. Apart from these methods, there is some physical method also used to isolate the fibrils, which include corona and plasma treatment, laser and UV bombardment, heat treatment, and others.

Plasma treatment is an effective physical method that has been used to alter the earth benevolent characteristic fibers. Plasma, which is described as the fourth state of matter, consists of negative and positive charged particles, neutral atoms, and molecules. To design the plasma process, some terms have been used such as low-pressure plasma, cold plasma, nonequilibrium plasma, and glow discharge. The salient features of plasma treatment are that it is a dry modification process so more reliable and safe, the volumetric properties of treated material remain unaffected understudy, and the processing time is short [26]. Bozaci et al. performed argon and atmospheric plasma treatment with power variation (100, 200, 300 W) on flax fiber and conducted fiber pull out and roughness tests. He found that 300 W-treated flax fiber has a better result in comparison to untreated and 100, 200 W-treated fiber. A gradual change of surface on the flax fiber with plasma treatment has shown in Figure 8.13.

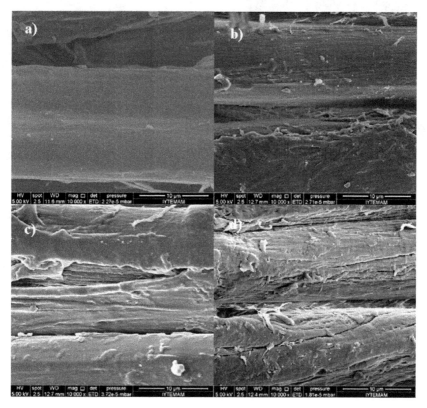

FIGURE 8.13 SEM image of (a) untreated, (b) 100 W, (c) 200 W, (d) 300 W argon treated flax fibers.

Source: Reprinted with permission from Ref. [48]. © 2013 Elsevier.

8.5.5 ENZYMATIC TREATMENT

The developing temperance of achievable enzymatic treatment innovation improves the general materialness of natural fiber in polymer composite application. This process is mainly characterized by their high selectivity, specificity, and mild process condition where no chemicals are used. Some industrial enzymes are cellulases, hemicellulases, pectinases, which are active in cellulosic plants for bio scouring, which can eliminate the unwanted surface impurities from fiber and offer other benefits like cost reduction, low energy, and water management, improved product quality [27]. The surface images of enzymatic treatment over hemp fiber are shown in Figure 8.14.

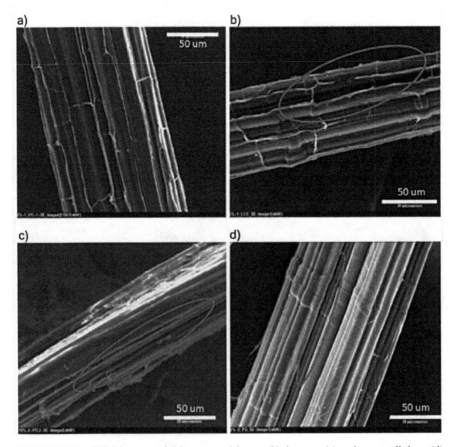

FIGURE 8.14 SEM images of (a) untreated hemp (b) laccase (c) xylaane+cellulase (d) polygalacturronase. Circle mark symbolizes fiber separation after treatment.

Source: Reprinted with permission from Ref. [49]. © 2018 Elsevier.

8.6 SUNN HEMP FIBER AND ITS POLYMER COMPOSITE

8.6.1 SUNN HEMP FIBER

Sunn hemp (*Crotalaria juncea*), also known as Indian hemp, or Brown hemp, belongs to the family of Fabaceae, is a multipurpose purpose plant that holds an emerging area as a fiber crop as well as green manure crops. The fibers are very much finer than jute, sisal, and coir. Those strands are predominantly removed from the stem part of the plant and their need prevails in cordage, fishing nets, ropes, and canvas applications. It has a high resistance to root-knot nematodes hence improve soil crop via nitrogen fixation. India is the largest producer of sunn hemp fiber, and also the cultivation is predominant in Brazil, Bangladesh, Taiwan, and Africa. This fiber has enormous possibilities to be used in various diversified potential areas such as furniture, building materials, load-bearing elements, and insulating materials.

8.6.2 STRUCTURE AND FIBER PROPERTIES

Sunn Hemp plant is the fastest growing plant and attains a height of about 3 cm in 4–5 months after plantation; therefore there is the minimal issue of the event while it comes to crude material assortments. These Sunn hemp strands are separated from the stem segment of the plant. There are both primary and secondary fibers in the growing part of the stem, but the primary fiber has a significant contribution. The auxiliary fiber has a low aspect proportion, hence less substantial contributions in the regulation of fiber properties. Each single fiber cell has a central lumen is attached, and the region at the interface of two cells is middle lamella. The length of the essential cell shifts from 3.9 to 4.3 mm and width 26 to 40 μm, where the optional or auxillary fiber cell has a length of 1.55 mm and a distance across of 30 μm. Sunn hemp fiber has been reported to be more ecofriendly due to its instinct high fiber length per diameter proportion (500–600), low density (1.4–1.6 g/cm^3), high cellulose content (70–78 wt.%), abundant availability, durability than synthetic fibers that serves a remarkable criterion in the use of composite fabrication [28]. A group of sunn hemp fiber with its plant is shown in Figure 8.15.

8.6.3 TREATMENT METHOD

All naturally emerged fibers are inherently hydrophilic and other problems that already been discussed in the previous section (Section 8.4). These significant issues bring about poor bond with the hydrophobic networked matrix that prompts poor crosslinking [29]. Sunn Hemp is additionally not an issue free fiber, despite the fact that it delivers the best threads. Being a cellulosic fiber and low lignified, it has high absorbency towards air and moisture, which at last influences the composite items by diminishing its quality. So it is necessary to treat the fiber in order to reduce the hydrophilicity prior to developing composites. Numerous pretreatment strategies have been utilized to improve the Sunn hemp fiber structure. Krishnan et al. studied the influence of alkali treatment on the fiber orientation polyester composite and discovered a higher estimation of elastic and flexural properties if there should arise an occurrence of 0 degrees situated fiber while 90 degrees arranged filaments yield a superior effect quality on the composite [30]. Turkmane et al. found alkali-treated Sunn hemp fiber increased tensile, flexural, and impact properties of the epoxy composite as well [31]. The present paper focuses on the assessment of a physicochemical treatment where a high-intensity ultrasonicator (HIU) processor has used for physical treatment and alkali used for chemical treatment.

FIGURE 8.15 (a) Close view of sunn hemp fiber (b) sunn hemp plant.

8.7 HIGH-INTENSITY ULTRASONIFICATION

8.7.1 WORKING PRINCIPLE

The effect of ultrasonication on different products has been an emerging trend and gaining popularity among the researchers. A schematic diagram of the ultrasonic machining process to remove the material from its surface in the presence of abrasive particles has shown in Figure 8.16a. Figure 8.16b shows our laboratory ultrasonication machining (OSCAR Ultrasonic Processor Sonapros, Model PR-1000) during ultrasonic vibration.

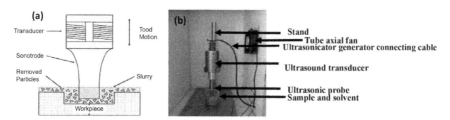

FIGURE 8.16 (a) Schematic diagram of ultrasonic machining process (b) laboratory ultrasonic.

High frequency (18–40 kHz) and low amplitude electrical energy generated by the HIU processor are converted to mechanical vibration by a piezoelectric transducer that connects to a probe tip. The vibration from the tip causes cavitation due to sudden formation and collapse of bubbles within the fluid where high temperature, pressure, and shock wave are involved. Due to millions of bubbles formation, there is an etching effect that partner with fiber surface, which helps with removing the undefined segment from the fiber strands. The cavitation makes fiber surface rough in order to have better contact with foreign materials (polymer matrix).

This unique technique is useful in many applications like particle homogenization, cell disintegration, liquid degassing, particle dispersion, and others. Ultrasonication technique is used to reduce small particles in the fluid and accomplish consistency just as strength, consequently valuable in molecule homogenization. This technique is useful in disintegrating the cell or tissue into small debris to expose the intracellular material, therefore increase the surface area for foreign material penetrations. It is also valuable in the way of degassing and particle dispersions. In the degassing process, the HIU technique able to remove small air suspended bubbles from the liquid while the cavitation generates shear force and breaks the agglomerate particles into dispersed one [32]. Some authors reported the fruitful outcome of the physicochemical treatment, which proves as a suitable technique to enhance the mechanical properties of the composite. Krishnaiah et al. studied tensile properties of alkali-treated and combined treatment of alkali and HIU toward sisal fiber reinforced polypropylene composite and found the combined treatment of alkalization and Ultrasonication improved tensile modulus and tensile strength of a composite by 50% and 10% respectively [33]. Alkali treatment and combined alkali-sonication treatment were performed over oil

214 *Natural Polymers: Perspectives and Applications for a Green Approach*

palm fiber by Beg et al., and the flexural test was completed over henequen/polypropylene biocomposite. They found similar results like Krishnaih [34]. Alam has done a structural, thermal, and mechanical investigation of simultaneous ultrasound and alkali-treated oil palm empty fruit branch fiber-reinforced polylactic composites. Significant enhancement of tensile strength, modulus, and impact strength were observed from jointly ultrasound and alkali-treated composite with agreeable structural and thermal evidence [35]. So motivated from the literature work, the present work focus on the join treatment of HIU-NaOH and their impact on fiber supra-fine structures just as on the composites was investigated.

8.8 EXPERIMENTAL INVESTIGATIONS

8.8.1 *MATERIALS*

Sunn hemp fiber was obtained from CSIR-Central Research Institute for Jute and Allied Fibers (CSIR-CRIJAF), Kolkata, India. General-purpose epoxy resin (L-12) and Hardener (K-6) were acquired from Himedia Laboratories Private Limited, India. For the treatment of sunn hemp fiber, sodium hydroxide (NaOH) was collected from Merck Life Science Pvt. Ltd, Mumbai, India.

8.8.2 *METHODS*

At first, the sunn hemp fiber was cleaned with deionized (DI) water to remove the surface dust and air-dried for 24 h to extract the moisture. This dried fiber was named as Raw Sunn Hemp (RSH) fiber. The fiber treatment procedure has shown in Figure 8.17.

For treatment, the RSH fibers were soaked in the 5% NaOH solution for 1 h to excite the reactive sites of cellulose hydroxyl groups. Aftermath, the fibrous solution exposed under HIU vibration for 60, 90, and 180 min. Finally, the fibers were collected, washed, and vacuum dried. The treated fibers were named NRSH, NHIU1RSH, NHIU2RSH, and NHIU3RSH, where NHIU denote NaOH HIU treated RSH and the number indicates the expose period. X-ray diffractogram of untreated and treated RSH fiber was recorded by CuK_a radiation X-ray diffractometer (Ultima IV- Rigaku, Japan). Data were taken at 2θ range from 5 to 50 degrees and scanning rate of 3 degrees/min with step size 0.05. Functional groups were investigated by Fourier Transmission Infrared (FTIR) spectroscopy (PerkinElmer Spectrum) from 4000 to 400 cm^{-1} wavenumbers, and the surface morphology was carried out by using (Nova Nanosem 450, Japan) Field Emission Scanning Electron Microscopy (FESEM) at 10 kV accelerating voltage.

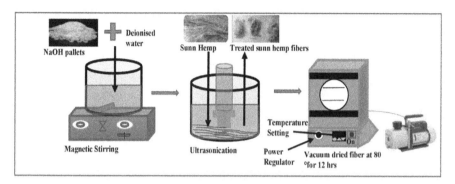

FIGURE 8.17 Treatment of sunn hemp fiber by HIU processor.

8.8.3 COMPOSITE FABRICATION

The composites were set up by Hand-layout technique, and the procedure has shown in Figure 8.18. There are many techniques used to fabricate composite. Spray Layup, Wet Layup, Vacuum Bagging, Filament winding, Pultrusion, and Resin transfer are some of them. Composite properties may shift starting with one manufacture strategy then onto the next relying on the instruments used, the difference in stress from the matrix framework to fiber, wetting, fitting mechanical interlocking between fiber/matrix network at interface, and so on. Here, the hand layout method technique is adopted in order to avoid the complicated equipment, cost-effective, simple to use, and any combination of fiber matrix can be incorporated. Initially, a deliberate measure of sunn hemp fiber was blended with epoxy resin, which was degassed in the vacuum in order to lower the viscosity level. The mixture was mechanically stirred and sonicated at least for 1 h for homogenizing. In the next step, the mixture was completed with its catalyst hardener prior to making composite. The epoxy and hardener ratio should be maintained at 10:1. The final composite was restored at atmospheric temperature for 24 h.

Two molds of dimensions (51×40×3) mm^3 and (200×60×3) mm^3 were prepared according to the ASTM-D3039 and ASTM-D7264 standard for flexural and tensile measurement, respectively. Mechanical tests were carried out using a Universal Testing Machine (Instron-5967, United Kingdom) with Environmental Chamber. Specimens were cut with dimension (48×13×3) mm³ and (170×19×3) and tested at a loading rate of 5mm/min. The ultimate flexural strength of the composite was determined using the following equation.

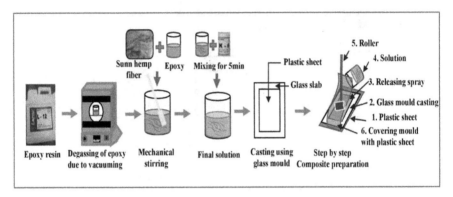

FIGURE 8.18 Composite fabrication process.

$$\text{Flexural Strength} = \frac{3FL}{2bd^2}$$

where F = force at the fracture point, L = length of the support scan, b and d = width, and diameter of specimen.

8.9 RESULTS AND DISCUSSIONS

8.9.1 STRUCTURAL ANALYSIS

Two pronounced peaks have appeared at 2θ angle of 22.6, and 15.6° that are attributed to crystalline cellulose I region district through the minimum intensity of diffraction at 2θ of 18.6° is ascribed to the presence of noncellulosic components in fiber. As a semicrystalline material, it is necessary to know the relative amount of cellulose present in the fiber. So cellulose crystallinity percentage was calculated using Segal's peak height method, where he considered the highest peak [36] and the size of cellulose crystallites is calculated using Scherrer's formula The calculated crystallinity index (CI) and crystallite size (L) for RSH fiber was found to be 72% and 31.2 Å, respectively. But after the treatment, the CI as well L were observed to be increased with increment of HIU vibration exposing period and showed the following trend RSH (72%, 31.2Å) < NRSH (80.6%, 42.6 Å) < NHIU1RSH (80.7%, 42.7 Å) < NHIU2RSH (81.4%, 43.2 Å) < NHIU3RSH (82.2%, 44.5 Å). A very minimal increment of CI and L were seen for NHIU1RSH fiber than NRSH fiber and a gradual increase with increasing exposure time. Highest CI and L values were observed in NHIU3RSH fiber. Alkali treatment was able to swell the fiber by removing the hydrogen bond from the structure where the etching effect caused by HIU vibration disintegrate cellulose within the fiber that eliminates noncellulosic amorphous water absorptive components from the fiber. Due to treatment, the cellulose structure got modified by rearrangement of the amorphous polymer chain in the direction of crystalline cellulose. So the combined treatment of HIU and alkali is effective in improving the crystallinity of cellulose polymer chain that can be an indication to enhance the strength of the fiber. Increment in crystallinity lessens the chance of crystal defects that lead to an increase in the crystallite size (Figure 8.19).

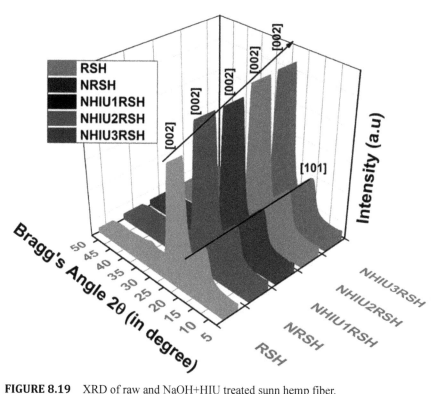

FIGURE 8.19 XRD of raw and NaOH+HIU treated sunn hemp fiber.

FTIR spectra for both untreated and treated RSH fibers have been shown in Figure 8.20. The broadband around 3000–3500 cm^{-1} is referred to OH-stretching due to the presence of hydrophilic substances in fiber. The peaks at 2924 and 2852 cm^{-1} are due to C–H stretching, which explains the presence of aromatic rings of lignin and wax, respectively. But upon the treatment, the lignin peak shifted for NRSH, NHIU1RSH, NHIU2RSH, NHIU3RSH, and wax peak disappeared, which indicates the partial removal of lignin and a complete removal of wax from the fiber. Furthermore, height at 1732 cm^{-1} in RSH fiber is due to C=O bond stretching, which suggests the presence of hemicellulose but vanishes latter with treatment. Peaks at 1633, 1061, and 892 cm^{-1} are due to O–H stretching due to absorbed water in cells, in-plane C–O stretching due to carbohydrates, and β—the glycosidic linkage of holocellulose, respectively. About 1633 cm^{-1} peak shifted toward higher wavenumber sides was observed in combined treated fiber due to the lessening of hydroxyl groups. So, this joined treatment over RSH fiber was found to have great impact on elimination of amorphous compounds from fiber structure.

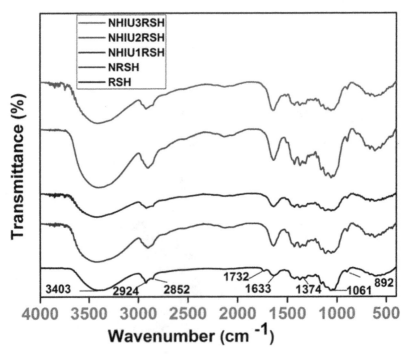

FIGURE 8.20 FTIR spectra of raw and NaOH+HIU treated sunn hemp fiber.

FESEM images for both untreated and treated RSH fibers have shown in Figure 8.21. Figure 8.21a shows a flat fiber surface with a firm arrangement of fibers due to the presence of all chemical components in the fiber. On treatment with alkali, the fiber surface seems to be rough, and fibrils start to rupture, as shown in Figure 8.21b, which explains the removal of the very least amount of noncellulosic compounds. In Figure 8. 21c, the roughness of the NHIU1RSH fiber surface increases. The fiber ruptures increase in the case of NHIU2RSH due to the removal of more amount of inorganic materials, as shown in Figure 8.21d. Finally, a prominent fiber splitting with fewer impurities was observed in NHIU3RSH, as shown in Figure 8.21e. This is due to the reduction in fiber diameter by etching that led to the elimination of the maximum amount of cementing materials that increased the active surface area between fibrils. Therefore, the combined treatment of HIU and NaOH with high exposure time strengthening the interfacial adherence between fiber and matrix in the composite.

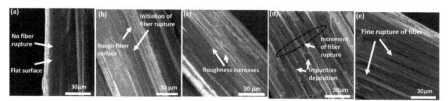

FIGURE 8.21 FESEM image of (a) RSH (b) NRSH (b) NHIU1RSH (c) NHIU2RSH (d) NHIU3RSH fiber.

8.9.2 MECHANICAL PROPERTIES ANALYSIS

Flexural strength quality (FS) and tensile strength quality (TS) for both untreated and treated RSH strands composites have been shown in Figure 8.22. It was observed that the combined treatment of HIU and NaOH shows a better FS and TS in comparison to RSH composite. The reason can be attributed to substantial interfacial holding among fiber and matrix due to the formation of cavities and cell disintegration, which causes the removal of cementing materials from fibers. Furthermore, with an augmentation of HIU vibration uncovering time, the FS and TS esteems are step by step expanding. This is due to the elimination of the maximum amount of fiber impurities and increased surface area between the fibers. Along with this, the treatment increases the fiber aspect ratio by decreasing the fiber diameter, which allowed the smooth passage of the matrix. An excellent chemical, as well as physical bonding due to combined treatment of HIU and NaOH, leads to a better FS and TS in the composite.

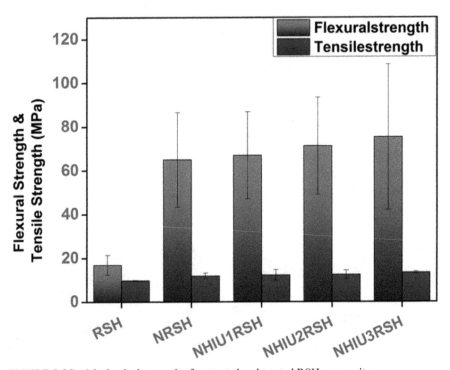

FIGURE 8.22 Mechanical strength of untreated and treated RSH composite.

8.10 CONCLUSION

Sunn hemp fiber, a characteristic polymer, has increased enough consideration in view of its natural expanded physical and chemical properties and its miscellaneous application in the area of automobiles, housing, panels, and insulations as composite. For better adhesion with polymer matrix, the effort has been taken in the reorganization sunn hemp fiber structure employing combined treatment of HIU and NaOH, where the time of HIU vibration was varied. XRD and FTIR study confirmed the increment of cellulose crystallinity and removal of non-cellulosic amorphous components like hemicellulose, lignin, and pectin from fiber, respectively. FESEM study of treated fiber showed a rough and ruptured fiber surface that expanded the fiber surface region bringing about better interfacial grip. The flexural and

High-Intensity Ultrasonication (HIU) Effect   221

tensile strength of combined treated fiber composite was found to be higher and more effective for NHIU3RSH composite due to better adherence and increased cellulose crystallinity. Therefore this modification technique is sufficient to enhance the various properties of composite in order to furnish them with useful potential applications.

ACKNOWLEDGMENT

The author would like to acknowledge Director, National Institute of Technology Rourkela, Rourkela-769008, India, for the financial support and allowing to conduct research.

KEYWORDS

- **ultrasonication treatment**
- **composite**
- **adhesion**
- **interface**

REFERENCES

1. Mohanta, N.; Acharya, S.K. Effect of alkali treatment on the flexural properties of a Luffa cylindrica- reinforced epoxy composite. *Sci. Eng. Compos. Mater.* 2018,25, 85–93.
2. Hossain, S.I.; M.N. Hasan, M.N.; Hassan, A. Effect of chemical treatment on physical, mechanical and thermal properties of ladies finger natural fiber. *Adv. Mater. Sci. Eng.* 2013, 2013, 1–6.
3. Oushabi, A.; Sair, A.; Hassani, F.O.; Abboud, Y.; Tanane, O.El.; Bouari, A. The effect of alkali treatment on mechanical, morphological and thermal properties of date palm fibers (DPFs): study of the interface of DPF-polyurethane composite. *South Afr. J. Chem. Eng.* 2017, 23, 116–123.
4. Li, S.; Bashline, L.; Lei, L. Cellulose synthesis and its regulation. *Am. Soc. Plant Biol.* 2014, 1–21.
5. Chen, H. Chemical Composition and Structure of Natural Lignocellulose. Chemical Industry Press, Beijing, 2014, 25–72.
6. Park, S.; Baker, J.O.; Himmel, M.E.; Parilla, P.A.; Johnson, D.K. Cellulose crystallinity index: measurement techniques and their impact on interpreting cellulase performance. *Biotechnol. Biofuels.* 2010, 3, 1–10.

7. Han, J.S.; Rowell, J.S. Paper and composites from agro-based resources. http://www.ncbi.nlm.nih.gov/entrez/query.fcgi?cmd=Retrieve&db=PubMed&dopt=Citation&list_uids=18171670 (accessed 2008).

8. Satyanarayana, K. G.; Arizaga Gregorio, G.C. Biodegradable composites based on lignocellulosic fibers—an overview. *Prog. Polym. Sci.* 2009, 34 982–1021.

9. Khalil, H.P.S.; Abdul, Hanida S.; kang, C.W.; Fuad N.A., Nik. Agro-hybrid composite: the effects on mechanical and physical properties of oil palm. *J. Reinf. Plast. Compos.* 2007, 26.

10. Leão, A.; Sartor, S.M.; Cláudio, J. Natural fibers based composites—technical and social issues. *Mol. Cryst. Liq. Cryst.* 2006, 448, 37–41.

11. Komuraiah, A.; Kumar, N.; Prasad, B. Chemical composition of natural fibers and its influence on their mechanical properties. *Mech. Compos. Mater.* 2014, 50, 359–376.

12. Cotton Fiber Properties An Extensive Technical Guide. *Barnhardt Purified Cott.* https://www.barnhardtcotton.net/technology/cotton-properties/ (accessed 2019).

13. Non organic Cotton. In Nour Luxury. Retrived August 31 st , 2018 2019, from https://www.nourluxury.com/blogs/news/non-organic-cotton-bedding-toxic.

14. Joseph, K.; Filho, R.D.T.; James, B.; Thomas, S. A review on sisal fiber reinforced polymer composites. *Rev. Bras. Eng. Agric. E Ambient.* 1999, 3, 367–378.

15. Satyanarayana, K.G.; Kulkarni, A.G.; Rohatgi, P.K. Structure and properties of coir fibres. *Proc. Indian Acad. Sci.* 1981, 4, 419–436.

16. Yan, L.; Chouw, N.; Jayaraman, K. Flax fibre and its composites. *Compos. B J.* 2014, 56, 296–317.

17. Cotton in Asian Textile Studies. Retrieved 24 th January, 2016, from http://www.asiantextilestudies.com/cotton.html.

18. Pickering, K.L.; Efendy, M.G.A.; Le, T.M. A review of recent developments in natural fibre composites and their mechanical performance. *Compos. A.* 2016, 83, 98–112.

19. John, M.J;, Anandjiwala, R.D. Recent developments in chemical modification and characterization of natural fiber-reinforced composites. *Polym. Compos.* 2008, 29, 187–207.

20. Cai, M.; Takagi, H.; Nakagaito, A.N; Li, Y.; Waterhouse, G.I.N. Effect of alkali treatment on interfacial bonding in abaca fiber. *Compos. A.* 2016, 90, 589–597.

21. Asim, M.; Jawaid, M.; Abdan, K.; Ishak, M.R. Effect of Alkali and silane treatments on mechanical and fibre-matrix bond strength of kenaf and pineapple leaf fibres. *J. Bionic Eng.* 2016, 13, 426–435.

22. Sepe, R; Bollino, F.; Boccarusso, L.; Caputo, F. Influence of chemical treatments on mechanical properties of hemp. *Compos. B J.* 2018, 133, 210–217.

23. Kalia, S.; Kaith, B.S; Kaur, I. Pretreatments of natural fibers and their application as reinforcing material in polymer composites—a review. *Polym. Eng. Sci.* 2009, 49, 21–25.

24. Mohammed, A.A.; Bachtiar, D.; Rejab, M.R.M.; Hasany, S.F. Effect of potassium permanganate on tensile properties of sugar palm fibre reinforced thermoplastic polyurethane. *Indian J. Sci. Technol.* 2017, 10, 1–5.

25. Kabir, M.M. Chemical treatments on plant-based natural fibre reinforced polymer composites: an overview. *Compos. B Eng.* 2012, 43, 2883–2892.

26. Sun, D. Surface modification of natural fibers using plasma treatment. *Biodegradable Green Composites.* John Wiley & Sons Inc, 2018: pp. 18–38.

27. Bledzki, A.K.; Mamun, A.A.; Jaszkiewicz, A.; Erdmann, K. Polypropylene composites with enzyme modified abaca fibre. *Compos. Sci. Technol.* 2010, 70, 854–860.

28. Liu, M.; Thygesen, A.; Summerscales, J.; Meyer, A.S. Targeted pre-treatment of hemp bast fibres for optimal performance in biocomposite materials: a review. *Ind. Crops Prod.* 2017, 108, 660–683.

High-Intensity Ultrasonication (HIU) Effect 223

29. Routa, J.; Misra, M.; Tripathy, S.S.; Nayak, S.K.; Mohanty, A.K. The influence of fibre treatment on the performance of coir polyester composites. *Compos. Sci. Technol.* 2001, 61, 1303–1310.

30. Krishnan, T.; Jayabal, S.; Krishna, V.N. Tensile flexural impact and hardness properties of alkaline-treated Sunnhemp fiber reinforced polyester composites. *J. Nat. Fibers.* 2018, 1–11.

31. Turukmane, R.N.; Nadiger, V.G.; Bhongade, A.L.; Borkar, S.P. Studies on treated sunn hemp and treated jute fibre reinforced epoxy composites. *Int. J. Adv. Eng. Res. Sci.* 2016, 3, 2349–6495.

32. Ultrasonics : Applications and Processes. Heelscher-Ultrasound Technol. https://www.hielscher.com/technolo.htm (accessed 2019).

33. Krishnaiah, P.; Thevy, C.; Manickam, S. Enhancements in crystallinity thermal stability tensile modulus and strength of sisal fibres and their PP composites induced by the synergistic effects of alkali and high intensity ultrasound (HIU) treatments. *Ultrason. Sonochem.* 2017, 34, 729–742.

34. Lee, H.S.; Cho, D. Effect of Natural Fiber Surface treatments on the interfacial and mechanical properties of henequen/polypropylene biocomposites. *Macromol. Res.* 2008, 16, 411–417.

35. Alam, A.K.M.M.; Beg, M.D.H.; Prasad, D.M.R.; Khan, M.R.; Mina, M.F. Structures and performances of simultaneous ultrasound and alkali treated oil palm empty fruit bunch fiber reinforced poly (lactic acid) composites. *Compos. A.* 2012, 43, 1921–1928.

36. Segal, L.; Creely, J.J.; Martin, A.E.; Conrad, C.M. An empirical method for estimating the degree of crystallinity of native cellulose using the X-ray diffractometer. *Text. Res. J.* 1959, 29 786–794.

37. Pereira, P.H.F.; Rosa M., de F.; Cioffi, M.O.H.; Benini, K.C.C de C.; Milanese, A.C.; Voorwald, H.J.C.; Mulinari, D.R. Vegetal fibers in polymeric composites: a review. *Polímeros.* 2015, 25, 9–22.

38. Cellulose formula. https://www.softschools.com/formulas/chemistry/cellulose_formula/464/(accessed 2019).

39. Dutton, J.A. Hemicellulose, Altern. Fuels from Biomass Sources. https://www.e-education.psu.edu/egee439/node/664 (accessed 2019).

40. Irmer, J. Lignin—a natural resource with huge potential, Bioeconomy BW. https://www.biooekonomie-bw.de/en/articles/dossiers/lignin-a-natural-resource-with-huge-potential/ (accessed 2019).

41. Laing R. M.; Selected mechanical properties of sisal aggregates (Agava sisalana). J. Mater. Sci. 2006 41, 511–515.

42. Wazir, H. An Overview of Coconut Coir Fiber. https://textilelearner.net/coconut-coir-fiber-properties-manufacturing/ (accessed 2019).

43. Lazic, B.D.; Pejic, B.M.; Kramar, A.D.; Vukcevic, M.M.; Mihajiovski, K.R.; Rusmirovic, J.D.; Kostic, M.M.; Influence of hemicelluloses and lignin content on structure and sorption properties of flax fiber. *Cellulose.* 2018, 25, 697–708.

44. Zhang, K.; Wang, F.; Liang, W.; Wang, Z.; Duan, Z.; Yang, B. Thermal and mechanical properties of bamboo fiber reinforced epoxy composites. *Polym. Artic.* 2018, 10, 1–18.

45. Fiore, V.; Bella G., Di.; Valenza, A. The effect of alkaline treatment on mechanical properties of kenaf fibers and their epoxy composites. *Compos. B.* 2015, 68, 14–21.

46. Thakur, M.K.; Gupta, R.K.; Thakur, V.K. Surface modification of cellulose using silane coupling agent. *Carbohydr. Polym.* 2014, 111, 849–855.

47. Sreekumar, P.A.; Thomas, S.P.; Saiter J., marc.; Joseph, K.; Unnikrishnan, G.; Thomase, S. Effect of fiber surface modification on the mechanical and water absorption

characteristics of sisal/polyester composite fabricated by resin transfer molding. *Compos. A J.* 2009, 40, 1777–1784.

48. Bozaci, E.; Sever, K.; Sarikanat, M.; Seki, Y.; Demir, A.; Ozdogan, E. Effects of the atmospheric plasma treatments on surface and mechanical properties of flax fiber and adhesion between fiber–matrix for composite materials. *Compos. B.* 2013, 45, 565–572.

49. Prez J., De.; Willem, A.; Vuure, V.; Ivens, J.; Aerts, G.; Voorde I. van, De. Enzymatic treatment of flax for use in composites. *Biotechnol. Rep.* 2018, 20, 1–18.

50. R. Division *et al.*, "Sunnhampa," *Wikipedia*, 2014. [Online]. Available: https://sv.wikipedia.org/wiki/Sunnhampa.

51. P. Mechanics, T. Rotary, A. Advantages, D. References, C. Material, and A. Zirconia, "Ultrasonic machining," *Wikipedia*, 1955. [Online]. Available: https://en.wikipedia.org/wiki/Ultrasonic_machining.

52. Ekundayo, G.; Adejuyigbe, S. Reviewing the development of natural fiber polymer composite: a case study of sisal and jute. *Am. J. Mech. Mater. Eng.* 2019, 3, 1–10.

53. 53. Carr, D. & Cruthers, N. & Laing, Raechel & Niven, B.. (2006). Selected mechanical properties of sisal aggregates (Agava sisalana). Journal of Materials Science - J MATER SCI. 41. 511-515. 10.1007/s10853-005-2189-z.

CHAPTER 9

Tribological Behavior of Coconut Shell-Fly Ash-Epoxy Hybrid Composites: An Investigation

BIBHU PRASAD GANTHIA[1*], MAITRI MALLICK [2], SUSHREE SASMITA[3], KAUSHIK UTKAL[4], and IPSITA MOHANTY[3]

[1]*Electrical Engineering, IGIT, Sarang, Dhenkanal, Odisha, India*

[2]*Civil Engineering, GIET, Bhubaneswar, Odisha, India*

[3]*School of Civil Engineering, KIIT, Bhubaneswar, Odisha, India*

[4]*Electrical Engineering, GCE, Keonjhar, Odisha, India*

Corresponding author. E-mail: jb.bibhu@gmail.com

ABSTRACT

Natural fiber (NF), which mainly reinforces polymer composites that hold up to the most recent study, offered a potentially ecological and alternatively low cost to synthetic fiber-reinforced compounds. Over the past decade, major companies, industries, and firms, such as construction, automotive, and construction firms, have described significant strides in the development and of new and innovative natural fiber-reinforced composite (NFRC) materials. Hence, the production and availability of NFs in the present world now encourage researchers to study and develop their practical potential and utility as a disciplinary source, which refers to the specifications, parameters, and criteria required in the tribological applications. The key points of this investigation as well as practical analysis here are to demonstrate effective tribal behavior in obtaining their use and application of NFRC in various areas, where the trilogy plays a distinctive and dominant role. In this process composites are prepared by making appropriate dye at home. For this process, fly ash content is kept constant at 5 wt.% weight and the rate of coconut shell

226 *Natural Polymers: Perspectives and Applications for a Green Approach*

dust content varies from 10 wp, 15 wp, and 20 wp. Here it is observed after wearing a short run and test in a pin on the disc machine and it is estimated that composite 10 with shell dust is perfect for short-term fabrication and abrasive applications.

9.1 INTRODUCTION

In recent times polymers have become important materials for various use and applications due to their smooth properties like lightweight, smooth processing, and less cost-effectiveness. Therefore significant efforts are taken to use polymers in various industrial usability applications, utilizing a variety of matrix reinforcements, including fibers incorporated into different polymers to enhance their physical, tensional, and mechanical parametric properties. Fiber-reinforced polymer matrix compounds are widely used due to their light mass weight, violability, high strength durability, maximum hardness, good air corrosion, and low mass friction coefficient in many effective applications [1, 2]. On the other hand, matrix composites are having environmental benefits and extend beyond the production stage. Since fibers are lightweight, they reduce carbon dioxide emissions in addition to fuel consumption. Generally, the two main advantages of natural fiber composites (NFCs) are the product's recyclability, economical and biodegradability [3, 4], after a useful life.

Natural fiber (NF) reinforced polymer composites have emerged as a potentially ecological and economical and less expensive than synthetic compounds. Therefore over the past decade, several large industries, such as the construction, automotive, and packaging industries have shown major considerable interest in the development of new natural fiber-reinforced composites (NFRCs) [3]. The availability of NFs and ease of manufacturing prompted researchers to study the feasibility of the application as a reinforcement and to the extent that the specifications in the tribal applications could be satisfied. However, information on the performance of NFRC materials is lacking in the research literature [7, 12]. Therefore, the purpose of this investigation is to analyze and to demonstrate the nature of NF-reinforced compounds. They determine their usability, effectiveness for various domestic and industrial applications where tribology plays a very important role [9]. In this context the attempt is to develop composites of coconut dust shells and ash of fly particles reinforced in epoxy matrix that can have significant tribological applications [11]. The composite of coconut dust of shells and ash reinforced (fly ash particles) in epoxy have been manufactured by developing the suitable die in-house [12]. The fly ash content is kept constant at 5 wt.% and the coconut shell dust content is different from 10%, 15%, and 20% of the weight in the epoxy matrix. After the initial short-run wear test in a pin-on-disc machine, it finds that the composite sample with 10% shell dust is found to be the best for short-run frictional applications. The main contents of this chapter consist of a brief overview of composites, NF-reinforced polymer composites, the actual experimental method used to determine the composite mix with the best tribological performance [1, 13].

9.2 HISTORICAL OVERVIEW

9.2.1 PAST

After being able to control the machinery and control the fire, continuous spinning was the most important and demanding development of mankind, allowing it to survive, acquire outside the zone of tropical climates, and spread the earth surface. Flexible fabrics (FF) made from locally farmed and cotton-like fibers; hemp and jute were a major step in comparison to animal skins [4]. Highly natural resources were used to see progress in the context of natural resource use, output in the first composite-reinforced straw walls, and adhesive layers of wood and bones. Made of bows and chariots made of more durable and strong materials like fiber and soon replaced metals like ancient compounds [2, 15, 22].

9.2.2 PRESENT

Beginning in the early agricultural societies and forgotten almost centuries later, a real recovery in the lightweight structure used to solve many technical problems during the second half of the 20th century [2, 33]. After being fully used for their electrical properties (insulator and radar dome), the use of composites to improve the structural performance of spacecraft and military aircraft became popular in the last two decades of the last century [17]. Development is the use of composites to protect humans from fire and impact, and the recent trend of environmentally friendly design has given rise to a new generation of NFs in comprehensive technology [1, 32].

228 *Natural Polymers: Perspectives and Applications for a Green Approach*

9.2.3 FUTURE

In the future, it is envisaged that the composite design will be developed in a more integrated design, resulting in more construction in terms of parameters such as mass, definite shape, specific strength, hardness, long durability, and cost. So that the impact of design changes is immediately visible to users on each of these parameters.

9.3 CONCEPT OF COMPOSITE

The fibers or particles that form another matrix in the matrix formulation are an excellent example of modern material. They are mostly regular and structural. Fragments are a composite material where different layers of material give them a special role to perform in a particular working composite material. Fabrics have no matrix to fall back on, but fibers of different compounds play a different role together. The reinforcing materials usually withstand the maximum load and complement the required properties. In matrix-based structural composites (SC), technically the matrix serves two important purposes, namely, restriction of space and instability by binding the enrichment phase—to suppress the constituent regulatory content under an applicable force [2, 18]. There are many matrix materials (MMs) related to the industrial application. They can reduce temperature variations according to inputs, can be insulated or resistant to electricity, have moisture sensitivity, and so forth. Additionally, they can offer weight gains, ease of handling, and other features that depend on the purpose for which the matrix is chosen. There are solid potential metric materials that adjust the tension by adding other circles to provide strong bonds for the reinforcement phase [1, 2]. With the remarkable success of some nonessential materials, fibers, metals, and polymers, their applications have been found as MMs in the design of SC. These materials are flexible as long as stress and stress failure are minimized. Compounds of composites cannot be formed from circles with different geometry extension properties because it can adversely affect the interface. The interface is a source of communication between reinforcement and MMs [11]. In some cases, this region is a separate step. Whenever there is an extra gap, there should be two communications between each side of the interface and the adjacent circle. Some surfaces provide a cohesive interface when surfacing conflicting surfaces. The choice of the deformed method depends on the matrix properties and the effect of the matrix on the reinforcement properties. One of the important

Tribological Behavior of Coconut Shell-Fly Ash-Epoxy 229

things about composite selection and fabrication is that the component must be inactive [1, 12, 22].

9.3.1 COMPOSITES

A composite material consisting of two or more ingredients, which contains significantly different physical or chemical properties, when combined, produces substances with different properties than individual components. Individual components are isolated and isolated in the structure. New materials may be preferred for a number of reasons, for example, materials that are stronger, lighter, or lower priced than traditional materials. Recently, researchers have begun to actively incorporate sensing, acting, computing, and communication into compounds, known as robotic materials [21]. Composite materials are commonly used for buildings, bridges, and structures. They are used as boats, swimming pool panels, race car bodies, shower stalls, bathtubs, storage tanks, imitation granite, and decent marble sinks and counter. The most modern example performs routinely in spacecraft and aircraft demand environments [1, 11].

9.3.2 CLASSIFICATION OF COMPOSITES

Composite materials are normally grouped by two particular levels:

1. The major composite classes include first is metal matrix composites (MMCs). Second ceramic matrix composites (CMCs) and third organic matrix composites (OMCs). The term organic matrix alloys is generally thought to consist of two classes of compounds, namely, carbon matrix compounds commonly called carbon–carbon compounds and polymer matrix composites [1, 23].
2. The second level of classification is reinforcement form—laminar composite (LC) and particulate form composite PC and fiber-reinforcement composite [2, 24].

Fiber resurfaced composites consist of fibers embedded in MM. Such a proportional ratio of fiber or short fiber is considered comprehensive if its properties vary with the length of the fiber. On the other hand, when the fiber length is such that the length is not increased further, the elastic modulus of the composite is considered to stabilize the fiber [3, 8, 19]. The fibers are small in diameter and are easily twisted when pushed axially, though their

230 *Natural Polymers: Perspectives and Applications for a Green Approach*

nerve properties are very good. These fibers should be helped to prevent individual fibers from turning and laughing. LCs include layers of material through a matrix. Sandwich structures fall into this category. Particulate composites are particles that are subdivided or embedded in the body of the matrix. The particles can be in the form of flakes or powders. Examples of this category are concrete and wooden particleboards [1, 23, 24].

9.3.2.1 ORGANIC MATRIX COMPOSITES

Polymers produce ideal materials because they are easy to process, lightweight, and have the desired mechanical properties. The two types of polymers are thermostats and thermoplastics [2, 26]. Thermostats have features such as well strongly bonded three-dimensional molecular structures after fitness. They swell instead of melting when hardened. Simply changing the basic ingredients of the resin is sufficient to adapt it to therapeutic conditions, and it also determines its other properties. Thermostats can be long-lasting even in partially healed conditions, making them very flexible. As such, they are most suitable as matrix bases for the advanced state of the fiber blanket mixture. Thermostats find a wide range of applications in shredded fiber compost form, especially when starting with a premixed or molding compound with specific quality and aspect ratio fibers. As in the case of epoxy, polymer, and phenolic polyamide resins [4, 12, 21].

Thermoset materials are unique in the fiber-composite matrix, without which research work and development area in the field of structural engineering can be disconnected. Aerospace components, automobile parts, defense systems, and many more that make great use of this type of fiber composite. Epoxy MM is used in printed circuit boards and similar areas. Direct condensation polymerization after rearrangement reactions for the formation of heterocyclic bodies is the method most commonly used to generate thermoset resins [25]. In both ways, the production of water, reaction, hinders the production of false-free composites. These voids have negatively affected the composite properties in terms of strength and dielectric properties. There are two main classes of polyester phenolic and epoxy thermoset resin. Epoxy resin is widely used in filament-wound composites and is suitable for molding prepress [1, 3, 9]. They are reasonably strong for chemical attacks and are well-followed, which gradually shrinks during treatment and does not emit volatile gases. However, these benefits

Tribological Behavior of Coconut Shell-Fly Ash-Epoxy

make the use of epoxies expensive. In addition, they cannot be expected to be above 140 °C. The use of these materials is not allowed in high technology areas where the service temperature is high. Polyester resin, on the other hand, is more easily accessible, cheap, and widely used in the field. Liquid polyesters are stored at room temperature for months, sometimes for many years and only the addition of a catalyst can fix matrix content in a short time. They are used in automobiles and structural applications. The treated polyester is usually rigid or flexible as the case may be transparent and transparent. Polyesters are stable against environmental changes and chemicals. Depending on the need for resin formation or application service, they may be used up to 75 °C or higher. Other benefits of polyester include easy compatibility with some glass fibers and can be used with plastic [24] certification. Fragrant polyamides are most sought after candidates because advanced modern fiber composites for structural applications demand permanent exposure to 200–250 °C [1, 2].

9.3.2.2 METAL MATRIX COMPOSITES

MMCs, although presently generating considerable high interest in the research community, are not widely used by their plastic counterparts. The high strength, fracture hardness, and hardness are represented by the matrix of their polymer counterparts. They can withstand higher temperatures in a corrosive environment than polymer alloys. Most metals and alloys can be used as matrix and require reinforcement materials that need to be stable in terms of temperature and reactivity [5, 12, 15]. However, the guiding principle for selection depends on the matrix of the matrix. Generally, light metals form a matrix for temperature use, and in addition to the above factors, reinforcement is also a feature of higher modules. Most metals and alloys make a good matrix. However, in practice, the choice of low-temperature applications is not very high. Only lighter metals are responsible, whose lower density is beneficial. Titanium, aluminum, and magnesium are currently popular matrix metals, especially useful for aircraft use. If the material of the metal matrix has to offer high strength, they need high modulus reinforcement. As a result, the weight-to-weight ratio of composites can be higher than that of most alloys. The melting, physical, and mechanical properties of the mixture at different temperatures determine the service temperature of the mixture. Most metals, ceramics, and alloy mixtures can be used with the melting point blend matrix [2, 12, 14, 27].

232 Natural Polymers: Perspectives and Applications for a Green Approach

9.3.2.3 CERAMIC MATRIX COMPOSITES

Ceramics can be depicted as strong solid and these solid materials exhibit highly efficient and very strong ion bonding in general and in some cases in harmony. High melting point, good corrosion resistance, high-temperature stability, and high compressive strength ceramic-based MM applications require a material that does not give way to temperatures above 1500 °C. Naturally, CMC is the obvious choice for high-temperature applications. However, the high modulus of elasticity and high tension stress, which is mostly ceramic, has led to the failure of reinforcement efforts to achieve strength improvement [12, 17, 22]. This is because at the stress level where the ceramics explode, there is insufficient matrix expansion that prevents the load from moving into the blanket. And unless the percentage of fiber is high enough, the mixture can fail. The material acts as a reinforcer to utilize the high tensile strength of the fiber to enhance the matrix's load-bearing capacity. However, the addition of high-strength fibers to weak ceramics is not always successful and often results in a weak alloy [1, 28].

9.3.2.4 NATURAL FIBER-REINFORCED COMPOSITES

Natural fiber polymer composite (NFPC) is a composite material consisting of a polymer matrix adorned with high strength NFs such as jute, oil palm, sisal, kenaf, and sun. Generally, polymers can be classified into two types, thermoplastics and thermostats. Thermoplastic MM is composed of one- or two-dimensional molecules, so the tendency of these polymers to become more and more diluted with increasing heat and return their properties during cooling. On the other hand, thermostats polymers can be interpreted as highly cross-linked polymers that are cured using only heat, or heat and pressure, and/or light irradiation [12, 29]. This structure gives thermoset polymer good properties such as high flexibility, large strength, and modules to improve the desired final properties. Thermoplastics widely used for biofibers are polythene, polypropylene, and polyvinyl chloride. Most of the thermostating matrix used here is as phenol, polyester, and epoxy resin. Various factors can affect NFPC characteristics and performance. The hydrophilic nature of NFs and fiber loading also affects composite properties. In general, high fiber loading is required to achieve good NFPC properties. Generally, the increase in fiber content improves the tensile properties of the composites. Another important factor that greatly influences the properties and surface properties of the mixture is the process by which the parameters are developed. For this reason, appropriate process techniques and parameters must be strictly chosen to achieve the best qualities of comprehensive preparation. The chemical composition of NFs also has a great impact on the properties of composites, representing the percentage of cellulose, hemicellulose, lignin, and wax [2, 22, 29].

Like fiberglass, there is a class of material that is a permanent filament, or as long as a piece of thread, which is separated into long pieces. They can be tied into wires, thread, or rope. They can be used as components of a composite material. They can also be mat mats to make paper or felt products. Figure 9.1 shows the classification of NFs. NFs include plant, animal, and mineral sources [1, 31, 50]. NFs can be classified according to their origin:

1. Animal fibers: They include wool, goat hair, horsehair, silk, and avian fiber.
2. Mineral fibers: These are naturally found or slightly modified fibers obtained from minerals. They can be further classified as ceramics, asbestos, and metallic fiber.
3. Plant fibers: These are usually composed of cellulose. This fiber can be further classified as follows:
   a. Fiber of seeds—cotton
   b. Fibers of leaves—agave
   c. Fibers skin—the fibers collected from the skin or bark around the trunk of their plant like flax.
   d. Fiber from fruits—coconut fiber
   e. Fiber stalk—wheat

Over the past two decades, NF has received considerable attention due to the following qualities:

1. Low density leads weight minimization.
2. Good thermal insulation and mechanical properties.
3. Tool wear minimization.
4. High availability, price is less, easy disposal.
5. Calorific value is good.
6. Low specific gravity, high strength to weight ratio.
7. Nonabrasive, nontoxic, and biodegradable.
8. Renewable and recyclable.

Under the above circumstances, it is easy to investigate the tribological behavior of NFRCs mainly made out of waste materials like coconut shell/fly ash and others.

9.4 THEORETICAL INVESTIGATIONS

The mixture is usually prepared based on the calculation of the weight fraction or volume fraction of the matrix or filler material [32, 50]. Since the random distribution of filler materials has been taken into consideration the various properties of the composite, as mentioned below are found after the principle of mixing.

Weight Fraction: Reinforcement
$$w._r = W._r \, (W._r + W._m + W._f) \times 100$$
Weight Fraction: Mat:
$$w._m = W._m / (W._r + W._m + W._f) \times 100$$
where

$W._r$ = weight reinforcement
$W._m$ = weight matrix
$W._f$ = weight filler
$W._c$ = weight composite
$W._c = W._r + W._m + W._f$

Further as per rule of mixtures,
Composite Density Calculation:

$$\rho._c = \rho._m v._m + \rho._r v._r + \rho._f v._f \tag{9.1}$$

where

$\rho._c$ = density composite
$\rho._m$ = density matrix
$\rho._r$ = density reinforcement
$\rho._f$ = density filler
$v._m$ = volume fraction matrix
$v._r$ = volume fraction reinforcement
$v._f$ = volume fraction filler.

Here $v._m = V._m \, (V._m + V._r + V._f + V._v) \times 100$,
$$v._r = V._r \, (V._m + V._r + V._f + V._v) \times 100$$
Composite Volume Calculation:
$$V._c = V._m + V._r + V._f + V._v$$
where

$V._m$ = volume matrix
$V._r$ = volume reinforcement
$V._f$ = volume filler
$V._v$ = volume void

Assuming modulus reinforcing efficiency as unity and as per the rule of mixtures:

Composite Modulus of Elasticity Calculation:

$$E_c = E_m v_m + E_r v_r + E_f v_f \qquad (9.2)$$

where

E_r = modulus of elasticity reinforcement
E_m = modulus of elasticity matrix
E_f = modulus of elasticity filler

Composite Strength Calculation:

$$\sigma_c = \sigma_m v_m + \sigma_r v_r + \sigma_f v_f \qquad (9.3)$$

where

σ_r = strength reinforcement
σ_m = strength matrix
σ_f = strength filler.

The properties of epoxy, coconut shell dust, and fly ash are shown in Tables 9.1, 9.2, and 9.3, respectively. The properties of the composite have been found out using the above relations of the rule of mixtures as indicated in Table 9.4.

TABLE 9.1   Properties of Epoxy

Properties	Value
Density (g/cc)	1.2
Elastic modulus (GPa)	20
Tensile strength (MPa)	75

TABLE 9.2   Properties of Coconut Shell Dust

Properties	Value
Density (g/cc)	0.6
Elastic modulus (GPa)	521
Tensile strength (MPa)	18.03

TABLE 9.3   Properties of Fly Ash

Bulk density (g/cc)	0.9–1.33
Specific gravity	1.6–2.6
Plasticity	Lower or nonplastic
Compression index (C_C)	0.05–0.4
Compressive strength	43 MPa

236 Natural Polymers: Perspectives and Applications for a Green Approach

TABLE 9.4 Properties of Composites

Composites	Density (g/cc)	Elastic Modulus (GPa)	Tensile Strength (MPa)
Specimen-A	1.125	69.1	63.403
Specimen-B	1.095	94.15	60.5545
Specimen-C	1.065	119.2	57.706

9.4.1 EXPERIMENTAL WORK

9.4.1.1 FRICTION, WEAR, AND LUBRICATION

The friction, which is an elastic criterion that has the same characteristics as the friction, is due to the complex and multiple plexus sets of microscopic interactions between surfaces that are mechanically oriented and slide against each other. These interactions are the result of the material, the geological and geographical properties of the surfaces, and the overall conditions under which the surfaces are designed to slide against each other, such as loading, temperature, environment, and type of contact. All the mechanical, surface contact and physical, chemical and geometric aspects of the surrounding environment affect the surface interactions and thus the topological properties of the system [2, 9, 12]. For most surfaces in relative sliding or rolling contact, the actual contact area is much smaller than the nominal contact area. This burden is triggered by a number of small spatial activities that make up the real contact area, and the friction and wearable behavior that result from the interaction between these spatial contact properties. In these local contact areas, conditions are characterized by high pressure and shear stress; material production pressure, high local (short) short-term temperature, and even high shear deformation and high shear rate. In such cases, the local mechanical properties of the material may be very different from what is found, for example, usually in tensile testing. The importance of oxide layers, small amounts of pollution, local phase changes, and so forth is even more important than the mechanical testing on a large scale [3, 33, 34].

9.4.1.2 LABORATORY WEAR TESTING

An apparatus for testing wear property is referred to as a wear tester, a tribotester. "Tribo" refers to prepping, rubbing, and lubricating. Various testing methods and procedures are used in laboratories around the world and are described in the technical literature [34, 48]. One management

Tribological Behavior of Coconut Shell-Fly Ash-Epoxy

difference compared to the other, one clothing tester will always include two components and move relative to each other. This movement can be driven by a motor or by an electromagnetic device. For convenience purposes, the material or component being investigated is generally referred to as the sample. The other is called the counterface [35, 36].

9.4.1.3 A PIN-ON-DISC WEAR TESTER

In a tester wearing a pin on type of disc, a pin is mounted against a flat rotating disk specimen that describes the way the machisne wears circular wear. Schematic of a pin-on-disc wear test and the arrangement of samples were shown in Figure 9.1. The machine can be used to evaluate the wear and abrasive properties of the material in pure sliding conditions. Can act as either a disc or a pin pattern, while the other as a masked face. The pin can be used with different geometries in this method [36]. An easy way is to use the hair of commercially available materials such as bearing steel, tungsten carbide or alumina (Al_2O_3) as a counter face [49].

The equation for wear rate calculation is [Archard's law (1953)]

$$V_{.i} = k_i F s. \tag{9.4}$$

where F mentioned as the normal load, s is the sliding distance, V_i. provide the wear of volume and k_i is the specific of wear rate coefficient. The index i identifies the surface.

The k-value is given in m^3/Nm or m^2/N, sometimes in mm^3/Nm. From design view the wear displacement h is more convenient than V. With $h_i = V_i/A$, the contact pressure $p = F/A$, where A is the area subjected to wear then

$$h_i = k_i p s \tag{9.5}$$

The sliding distance s can be replaced by $s = vt$ work for calculation. v is the mean value calculation for the slide rate and t provides the running time. The reason is k-value depends on various parameters and this factor calculation is to be determined experimentally [38, 39, 48].

9.4.1.4 WEAR MEASUREMENT/QUANTIFICATION

Wear in measurements are used to determine the amount of material to be removed after a wear test. Exposure to what is worn is also intended for precise tests of weight (mass), volume reduction, or linear dimension

(LD) change, depending on the type of clothing, size of geometry and test pattern, and sometimes on availability. Measuring facility [36, 37] common measurement techniques include the use of precision balancing to measure weight (mass) loss, profiling surfaces, or depth of wear, or part of a matrix. Use a microscope to measure the cross-sectional area so that the volume of the matrix is reduced or linearly determined [39, 46].

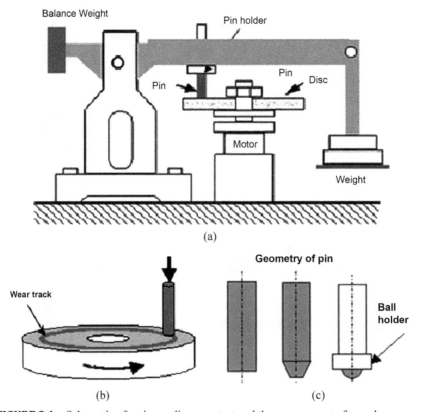

FIGURE 9.1 Schematic of a pin-on-disc wear test and the arrangement of samples.

9.4.1.5 MASS LOSS

Weight loss through precision balancing is an easy way to measure wear, especially if the worn surface is irregular. The measuring sample is carefully cleaned, and the weight is measured before and after the test. The weight difference before and after the test represents the weight loss caused by clothing [7, 8]. The unit can be grams (g) or milligrams (mg).

Tribological Behavior of Coconut Shell-Fly Ash-Epoxy 239

9.4.1.6 VOLUME LOSS

Wear of volume is generally used to measure the depth, length, width, and or geometry of the wearer track/stain from the wear profile. Level profilter, for example, a stylus type, or sometimes a scale microscope, is used to measure. Weight loss reporting unit is mm^3. Reduced wear compares better texture of clothing between different density materials. However, it is not easy to measure volume loss (VL) when the way of wearing is wrong. In this case, mass loss (ML) can be estimated first, and VL will be calculated if the material is the same and its density is known.

9.4.1.7 LINEAR DIMENSION LOSS

Measurement of wear through transverse dimensions is very useful in many engineering situations, where dimensions such as dimensions, length, or diameter are more important for routine work [6]. Level profilter, for example, the type of stylus, micrometer, or microscope can be used. Linear damage unit can be millimeter (mm) [8, 10].

9.4.1.8 WEAR RATE

Wear rate is calculated, resulting in ML, VL, or LD change that is the normal force applied to the unit and or the sliding distance of the unit [30].

9.4.1.9 WEAR RESISTANCE

Wearing resistance is a term often used to describe the characteristics of an antiwear material. However, the scientific meaning of clothing resistance is ambiguous, and there is no specific unit to explain wear resistance. However, widespread loss or VL reversal is sometimes used as (relative) wear resistance. The value of wear reduction for an experimental reference material can also be used as relatively wear resistance [2, 5] compared to the material examined under the same testing conditions. In another case, if a numerical value of wear resistance is calculated, this means the weight ratio must be clearly stated. In addition to the aforementioned methodology, surface analysis by measuring topography, debris analysis, and microstructural changes in the material as well as the proportion of wear is corrected. However, gravity analysis is the simplest form of measurement of wear [2, 7, 43].

9.4.2 MATERIALS AND METHODS

The following materials were collected from different sources for the composite lattice with different coconut shell materials (10%, 15%, and 20%) and as a percentage of fly ash (5%).

9.4.2.1 FLY ASH

Sample fly ash (Figure 9.2) was collected from National Thermal Power Corporation, Angul and tested in an R&D laboratory.

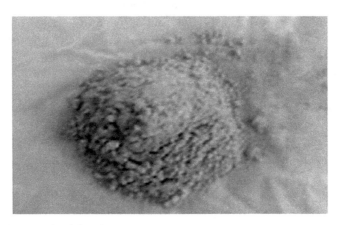

FIGURE 9.2 Sample of fly ash.

The components of the fly ash according to its testing parameters and test value percentage are given in Table 9.5.

9.4.2.2 COCONUT SHELL DUST

Coconut shells (Figure 9.3) were obtained from the local market in Talcher, Odisha. It was crushed to the desired grain size in Ploverser at IMMT, BBSR. Shell dust (Figure 9.4) was used at random orientation in the epoxy matrix with grain sizes between 45 and 75 μm.

TABLE 9.5 Constituents of Fly Ash

Sl. No.	Testing Parameter	Test Values (%)
1	Loss of ignition	1.2
2	Silica	56.5
3	Ferris oxide	11.0
4	Alumina (Al_2O_3)	17.7
5	Magnesia	5.4
6	Calcium oxide	3.2

FIGURE 9.3 Coconut shells.

FIGURE 9.4 Coconut shell dust.

242   *Natural Polymers: Perspectives and Applications for a Green Approach*

The properties of coconut shell powder are calculated experimentally in percentages as shown in Table 9.6.

TABLE 9.6   Properties of Coconut Shell Powder

Lignin	29.4%
Pentosans	27.7 %
Cellulose	26.6%
Moisture	8%
Solvent extractives	4.2%
Uronic anhydrides	3.5%
Ash	0.6%

9.4.2.3   COMPOSITE PREPARATION

The dielectric casting dimension, 30 mm height and 8 mm diameter, is shown in Figure 9.5. The samples are prepared with different weight percentages of coconut shell dust, that is, 10%, 15%, and 20% [39] [40]. The size of the coconut dust was measured by a shaker and found 70 m. For the fly ash portion of the fly ash (5%) the resin and hardener were thoroughly stirred to minimize air penetration. Epoxy and hardener were mixed in a 10:1 ratio. The coconut dust was then added to the slurry for various weight ratios and mixed thoroughly. The sludge was then put into a mold box to achieve the desired thickness. For smooth removal and easy movement of composite sheets, silicon spray liquid was applied to the inner zone surface of the mold. Precautions were taken to reduce the formation of air bubbles. Hence, the patient was allowed to hospitalize for the cure for 24 h at room temperature. After 24 h the samples of trace were removed from the die, according to the standard cut to different sizes [36–39].

9.4.2.4   WEAR TESTING

The wear test was carried out at Metallurgy and Materials Engineering Department, IGIT, Sarang using Pin-on-disc wear and friction testing machine (Figure 9.6) having specification as follows:

Make: MAGNUM ENGINEERS
SL.NO./YEAR OF MFG.:120/2013
Model no.-TE-100
POWER INPUT :230V, 15AMPS, 50HZ, Disc size–65 dia.

Tribological Behavior of Coconut Shell-Fly Ash-Epoxy 243

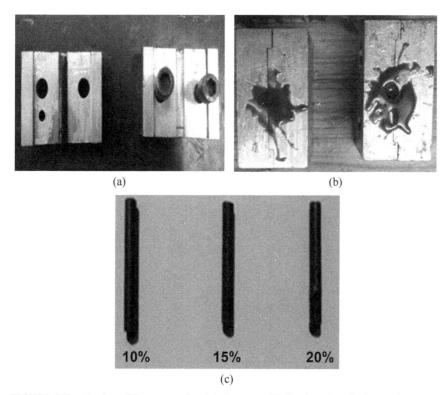

FIGURE 9.5 Casting of the composites (a) die open, (b) die closed, and (c) cast pins.

FIGURE 9.6 Pin-on-disc wear and friction testing.

During the experiment, the track radius, speed, and time were kept constant to assess the friction coefficient at 40 mm, 400 rpm, and 5 min, respectively. This is shown in Figure 9.7. The load varied from 0.5 to 2 kg in different samples.

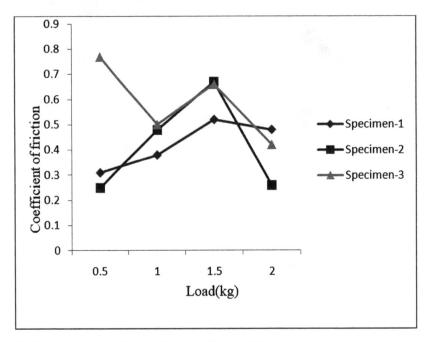

FIGURE 9.7 Variation of load with the coefficient of friction.

The wear in micron was measured keeping all the above parameters constant as described earlier. The variation of wear with load is shown in Figure 9.8. Similarly, the variation of wear with sliding distance is indicated in Figure 9.9.

9.4.3 RESULTS AND DISCUSSION

It is observed from the results obtained from load against coefficient of friction that the sample composite with 10% shell dust exhibits a steady rate of increase in friction coefficient and then decreases. Though the specimen with 15% shell dust showed lowest value initially, sudden increase and decrease in friction coefficient are rather inconclusive. However, highest friction coefficient is observed in 20% shell dust composite. Furthermore, the wear

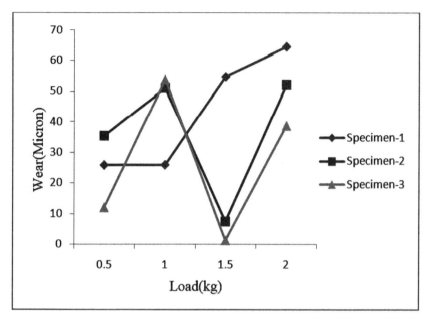

FIGURE 9.8 Variation of wear w.r.t load.

behavior of 10% shell dust composite (Figure 9.8) is found to be steadily increasing and probably stabilizes at a higher load condition. Wear rate of the other two composite samples is random. The same behavior is also observed in wear versus sliding distance graph (Figure 9.9).

9.5 CONCLUSIONS

After conducting the wear tests and the results obtained from the short-run tests the following conclusions may be derived. From the study of friction and wear coefficient, it is found that the lowest values of composite with 10% shell dust are found and it is expected that the wear will become stable after a reasonable amount of time. The random variability of the friction coefficient and the wear of other composites indicate that some defects may have occurred during the casting of the composites or that the steel disc surface has changed due to repeated tests on the same track. So with 10% coconut shell dust and 5% fly ash, it is expected that they will work satisfactorily in short-term friction. It is expected that long-wearing tests may produce better results so more specific conclusions can be drawn.

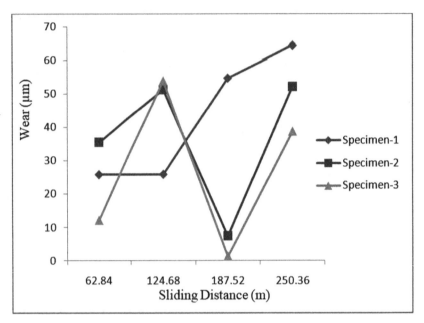

FIGURE 9.9 Wear versus sliding distance.

KEYWORDS

- tribological
- epoxy matrix
- fly ash
- biodegradability
- natural fiber (NF)

REFERENCES

1. Ward, I.M. and Hine, P.J., "*The Science and Technology of Hot Compaction,*" Polymer, Vol. 45, 2004, pp. 1413–1427.
2. Vijayaram T.R., "*Synthesis and Mechanical Characterization of Processed Coconut Shell Particulate Reinforced Epoxy Matrix Composite,*" Metal World, 2013, pp. 31–34.
3. Sapuan, S.M., Harimi, M., Maleque, M.A., "*Mechanical Properties of Epoxy/Coconut Shell Filler Particle Composites,*" Arb. J. Sci. Eng., Vol. 28, 2003, pp. 171–181.

4. Singla, M. and Chawla, V, *"Mechanical Property of Epoxy Resin Fly Ash Composite,"* JMMCE, Vol. 9, 2010, pp. 199–210.
5. Husseinsyah, S. and Mostapha, M., *"The Effect of Filler Content on Properties of Coconut Shell Filled Polyester Composites,"* J. Mal. Polym, Vol. 6, 2011, pp. 87–97.
6. Bhaskar, J. and Singh. V.K, *"Physical and Mechanical Properties of Coconut Shell Particle Reinforced-epoxy Composite,"* J. Mater. Environ. Sci., Vol. 4, 2011, pp. 227–232.
7. Mulinari, D.R., Baptista, C.A.R.P., Souza, J. V. C. and Voorwald, H.J.C. *"Mechanical Properties of Coconut Fibers Reinforced Polyester Composites,"* Proc. Eng., Vol. 10, 2011, pp. 2074–2079.
8. Kumar, S and Kumar, B.K, *"Study of Mechanical Properties of Coconut Shell Particle and Coir Fibre Reinforced Epoxy Composite,"* Int. J. Adv. Eng. Res, Vol. 4, 2012, pp. 39–62.
9. Singh, A, Singh, S, and Kumar, A, *"Study of Mechanical Properties and Absorption Behavior of Cocconut Shell Powder-Epoxy Composites,"* Int. J. Mater. Sci. Appl., Vol. 2, 2012, pp. 157–161.
10. Dan-Asabe. B, Madakson. P.B, Manji. J, *"Material Selection and Production of aCold-worked Composite Brake Pad,"* World J. Eng. Pure Appl. Sci., Vol. 2, 2012, pp. 92–97.
11. Girisha, C., Sanjeeba, M., Gunti, R., *"Tensile Properties of Natural Fibre-reinforced Epoxy-hybrid Composites,"* Int. J. Mod. Eng. Res., Vol. 2, Issue 2, 2012, pp. 471–474.
12. Olumuyiwa, J. Talabi, Agunsoye, Issac, S., Samuel, Sanni., O., *"Study of Mechanical Behaviour of Coconut Shell Reinforced Polymer Matrix Composite,"* J. Minerals Mater. Charact. Eng., Vol. 11, 2012, pp. 774–779.
13. Gummadi, J., Kumar, G.V, Gunthi, R, *"Evaluation of Flexural Properties of Fly ash Filled Polypropylene Composites,"* Int. J. Mod. Eng. Res., Vol. 2, Issue 4, 2012, pp. 2584–2590.
14. Kartikeyan, A., and Balamurgan, K. *"Effect of Alkali Treatment and Fibre length on Impact Behaviour of Coir Fibre Reinforced Composite,"* J. Sci. Ind. Res., Vol. 71, 2012, pp. 627–631.
15. Subham, P and Tiwari, S.K, *"Effect of Fly ash Concentration and its Surface Modification on Fibre Reinforced Epoxy Composite's Mechanical Properties,"* Int. J. Sci. Eng. Res., Vol. 4, 2013, pp. 1173–1180.
16. Singh, H.P, *"Strength and Stiffness Response of Itanagar Fly ash Reinforced With Coir Fiber,"* Int. J. Res. Sci., Eng. Innov. Technol., Vol. 2, 2013, pp. 4500–4509.
17. Raja, R.S, Manisekhar, K, Manikandan, V, *"Effect of Fly ash Filler Size on Mechanical Properties of Polymer Matrix Composites,"* Int. J. Minerals Met. Mech. Eng., Vol. 1, 2013, pp. 34–38.
18. Bhaskar, J and Singh, V.K, *"Water Absorption and Compressive Properties of Coconut Shell Particle Reinforced-Epoxy Composite,"* J. Mater. Environ. Sci., Vol. 4, 2013, pp. 113–118.
19. Salmah, H., Marliza, M., The, P.L., *"Treated Coconut Shell Reinforced Unsaturated Polyster Composites,"* Int. J. Res. Sci., Eng. Innov. Technol, Vol. 13, 2013, pp. 94–103.
20. Thaker, N, Srinivasulu, B, Subash, C.S, *"A study on Characterization and Comparison of Alkali Treated and Untreated Coconut shell Powder Reinforced Polyster Composites,"* Int. J. Sci. Eng. Res., Vol. 2, 2013, pp. 469–473.
21. Srivastava, N, Singh, V.K, Bhaskar, J, *"Composite Behavior of Walnut (Juglans L.) Shell Particle Reinforce Composites,"* Usak Uni. J. Mater. Sci., Vol. 1, 2013, pp. 23–30.

22. Ramesh, M., Palanikumar, K., and Reddy, K.H. "*Comparative Evaluation on Properties of Hybrid Glass Fiber—Sisal/Jute Reinforced Epoxy Composites,*" Proc. Eng., Vol. 51, 2013, pp. 745–750.
23. Ozsoy, N., Ozsoy, M., Mimaroglu, A., "*Comparison of Mechanical Characteristics of Chopped Bamboo and Chopped Coconut Shell Reinforced Epoxy Matrix Composite materials,*" Eur. Int. J. Sci. Technol., Vol. 3, 2014, pp. 15–20.
24. Sudharsan, B., Reddy, S.S.K. and Kumar, M.L, "*The Mechanical Behavior of Eggshell and Coconut Coir Reinforced Composites,*" Int. J. Eng. Trends Technol., Vol. 18, 2014, pp. 9–13.
25. Muthukumar, S. and Lingaduari, K. "*Investigating the Mechanical Behavior of Cocconut Shell and Groundnut Shell Reinforced Polymer Composite,*" Glob. J. Eng. Sci. Res., Vol. 1, 2014, pp. 19–23.
26. Rudramurthy, Chandrashekara. K, Ravishankar, R, Abhinandan. S, "*Evaluation of the Properties of Eco-friendly Brake Pad Using Coconut Shell Powder as Filler Materials,*" Int. J. Res. Mech. Eng. Technol., Vol. 4, 2014, pp. 98–106.
27. Vignesh, K, Natarajan, U, Vijayasekar, A, "*Wear Behaviour of Coconut Shell Powder and Coir Fibre Reinforced Polyester Composites,*" J. Mech. Civ. Eng., 2014, pp. 53–57.
28. Gopinath, P and Suresh, P., "*Mechanical Behaviour of Fly ash Filled, Woven Banana Fibre Reinforced Hybrid Composites as Wood Substitute,*" Int. J. Mech. Prod., Vol. 4, 2014, pp. 113–115.
29. Nagarajan, V.K, Devi, S.A, Manohari, S.P, Santha, M.M, "*Experimental Study on Partial Replacement of Cement With Coconut Shell Ash in Concrete,*" Int. J. Sci. Res., Vol. 3, 2014, pp. 631–661.
30. Durowaye, S.I, Lawal, G.I, Akande, M.A, Durowaye, V.O, "*Mechanical Properties of Particulate Coconut Shell and Palm Fruit Polyester Composites,*" Int. J. Mater. Eng., Vol. 4, 2014, pp. 141–147.
31. Akindapo, J.O., Harrison, A, Sanusi, O.N, "*Evaluation of Mechanical Properties of Coconut Shell Fibres as Reinforcement Material in Epoxy Matrix,*" Int. J. Eng. Res. Technol., Vol. 3, 2014, pp. 2337–2347.
32. Deshpande, S and Rangaswamy, T, "*Effect of Fillers on E-glass/jute Fibre Reinforced Epoxy Composites,*" Int. J. Eng. Res. Appl., Vol. 4, 2014, pp. 118–123.
33. Choudhry, A.K., Gope, P.C, Singh, V.K, Verma, A, Suman, A.R, "*Thermal Analysis of Epoxy Based Coconut Fiber-Almond Shell Particle Reinforced Biocomposites,*" Adv. Manuf. Sci. Technol., Vol. 38, 2014, pp. 38–50.
34. Gelfuso, M.V, Thomazini, D, Jesuza, J.C.S, Lima junior, J.J, "*Vibration Analysis of Coconut Fibre-pp Composites,*" Jour. Mater. Res., Vol. 17, 2014, pp. 267–273.
35. Divya, G.S., Khakandaki, A, Suresha, B., "*Wear Behavior of Coir Reinforced Treated and Untreated Hybrid Composite,*" Int. J. Innov. Res. Dev., Vol. 3, 2014, pp. 632–639.
36. Bongarde, U.S., Shinde, V.D., "*Review on Natural Fiber Reinforced Polymer Composites,* Int. J. Eng. Sci. Innov. Technol.", Vol. 3, 2014, pp. 431–436.
37. Dinesh, K.R., Jagadish, S.P., Thimmanagouda, A., *Characterization and Analysis of Wear Study on Sisal Fibre Reinforced Epoxy Composite,*" Int. J. Adv. Eng. Technol., Vol. 6, 2014, pp. 2745–2757.
38. Panyasart, K., Chaiyut, N., Amornsakchai, T., Santawitee, O., "*Effect of Surface Treatment on the Properties of Pineapple Leaf Fibers Reinforced Polyamide 6 Composites,*" Energy Proc., Vol. 56, 2014, pp. 406–413.

39. Ramesh, M., Atreya, T.S. A., Aswin, U. S., *Eashwar, H. and Deepa, C. "Processing and Mechanical Property Evaluation of Banana Fiber Reinforced Polymer Composites,"* Proc. Eng., Vol. 97, 2014, pp. 563–572.
40. Saleh, A.N, Al-Maamori, M.H, Al Jabory, M.B, *"Evaluating the Mechanical Properties of Epoxy Resin With Fly Ash and Silica Fume as Fillers,"* Adv. Phys. Theor. Appl., Vol. 30, 2015, pp. 1–7.
41. Vignesh, K, Sivakumar, K, Prakash, M, Palanive, A., and Sriram, A., *"Experimental Analysis of Mechanical Properties of Alkaline Treated Coconut Shell Powder Polymer Matrix Composites,"* Int. J. Adv. Eng., Vol. 4, 2015, pp. 448–453.
42. Venkatesh B, *"Fabrication and Testing of Coconut Shell Powder Reinforced Epoxy Composites,"* Int. J. Adv. Eng. Res. Dev., Vol. 2, 2015, pp. 89–95.
43. Pattanaik, A, Mohanty, M.K, Sathpathy, M.P, Mishra, S.C, *"Effect of Mixing Time on Mechanical Properties of Epoxy-Fly Ash Composite,"* J. Mater. Metall. Eng., Vol. 2, 2015. pp. 11–17.
44. Srivastava, A and Maurya, M. *"Preparation and Mechanical Characterization of Epoxy Based Composite Developed by Bio Waste Material,"* Int. J. Res. Eng. Technol., Vol. 4, 2015, pp. 397–400.
45. Kumar, M.L, Krishnaiah, D, Killari, N. *"Investigation on Mechanical Properties of CET Composite Materials,* "Int. J. Emerg. Trends Eng. Res., Vol. 3, 2015, pp. 516–519.
46. Maheswaran, Hemanth M., Velmurugan, M., Vijaybabu, K., Prabhu, S., Palaniswamy, E. *"Characterization of Natural Fiber Reinforced Polymer Composite,"* Int. J. Eng. Sci. Res. Technol., Vol. 4, 2015, pp. 362–369.
47. Kumar, B.S, Kumar, D.K, Shankara Babu, CH.S, Kumar, P.V. *"Effect of Mechanical Properties on Polyester Typha Fibre in Composition of Wood Powder and Coconut Shell Ash,"* Int. J. Eng. Technol. Sci. Res., Vol. 2, 2015, pp. 91–99.
48. Biswas, S., Shahinur, S., Hasan, M. and Ahsan, Q. *"Physical, Mechanical and Thermal Properties of Jute and Bamboo Fiber Reinforced Unidirectional Epoxy Composites,"* Proc. Eng., Vol. 105, 2015, pp. 933–939.
49. Ramachandran M., Bansal, S. and Raichurkar, P. *"Experimental Study of Bamboo Using Banana and Linen Fibre Reinforced Polymeric Composites,"* Perspect. Sci., Vol. 8, 2016, pp. 313—316.
50. Korniejenko, K., Fraczek, E., Pytlak, E. and Adamski, M. *"Mechanical Properties of Geopolymer Composites Reinforced With Natural Fibers,"* Proc. Eng., Vol. 151, 2016, pp. 388–393.
51. Gopal, P., Raja, V. K. B., Chandrasekaran, M. and Dhanasekaran, C. *"Wear Study on Hybrid Natural Fiber Epoxy Composite Materials Used as Automotive Body Shell,"* ARPN J. Eng. Appl. Sci., Vol. 12, 2017, pp. 2485–2490.
52. Nordin, N. A., Yussof, F. M., Kasolang, S., Salleh, Z. and. Ahmad, M. A. *"Wear Rate of Natural Fibre: Long Kenaf Composite,"* Proc. Eng., Vol. 68, 2017, pp. 145–151.

CHAPTER 10

Cellulose Acetate-TiO$_2$-Based Nanocomposite Flexible Films for Photochromic Applications

T. RADHIKA

Centre for Materials for Electronics Technology [C-MET],
Ministry of Electronics and Information Technology (MeitY),
Athani, Shoranur Road, Thrissur, Kerala 680581, India.
E-mail: rads12@gmail.com

ABSTRACT

Flexible films of nanocomposites of cellulose acetate-metal oxide ceramics and cellulose acetate-modified metal oxide ceramics containing an intelligent ink were prepared through the method of solution casting. The photochromic applications of films were tested under various irradiation conditions using UV–vis spectrophotometry. Upon irradiation, these films exhibited rapid photochromic response. Films containing more ceramics exhibit enhanced decoloration/recoloration under all the studied irradiation conditions, with a high rate of decoloration/recoloration under simulated solar irradiation. The information confirms that the amount of ceramics, the concentration of intelligent ink, and the solvent greatly influence the photochromic properties of the prepared films. The cellulose acetate-ceramics nanocomposite films that are flexible and transparent with enhanced decoloration/recoloration properties especially under solar irradiation are promising materials for use in photo reversible printed electronics applications.

252 *Natural Polymers: Perspectives and Applications for a Green Approach*

10.1 INTRODUCTION

10.1.1 BIOPOLYMER—AN OVERVIEW

Synthetic polymers produced from petrochemicals are not biodegradable. Persistent polymers are causing significant environmental pollution, which is harmful for wildlife when these are disposed into nature. For example, the sea life is adversely affected by the disposal of nondegradable plastic bags. In the products with small life span, such as for engineering applications, catering, packaging, surgery, and hygiene, the use of long-lasting polymers is not adequate. Toxic emissions from incineration of plastic waste present environmental issues. Since plastic represents a huge part of waste collection at the local, regional, and national levels, significant savings of compostable or biodegradable materials can be generated. The replacement of the conventional plastic by degradable polymers for short-term applications (packaging, agriculture, etc.) is the major interest to the society now. Since sustainable development policies tend to expand with the diminishing reserve of fossil fuel, recently biodegradable and bio-based products have gained great interest. Bio-based polymers may contribute to the sustainable development with minor environmental impact. The market of these environmentally friendly materials is expanding rapidly, 10%–20% per year [1].

Biopolymers are long-chain compounds. These are formed by the polymerization of long-chain molecular subunits. A biopolymer is an organic polymer and is having various applications in different fields.

In the future, the biodegradable polymers could be recognized to overcome the limitation of the petrochemical resources. Starch, proteins, cellulose, and plant oils are derived from agricultural products and are natural polymers. These are the major resource of renewable and biodegradable polymer materials in many industrial applications to replace petrochemicals due to increased environmental concern. However, biopolymers show inferior performances and in terms of functional and structural properties as compared to synthetic polymers, when processed with traditional technologies. The blends of biopolymers can be an alternative for currently used synthetic polymers. Starch, chitosan, cellulose, gelatin, polylactic acid (PLA), PHAs, and others are the most common biopolymers. Biopolymer is an organic polymer that is produced naturally by living organisms. Depending on its origin and production, the biobased polymers are divided into three categories: (1) directly extracted from biomass, (2) produced from monomers, and (3) synthesized by microorganisms [14]. One of the major advantages of biopolymers is that these are capable of biodegradation at accelerated rates, breaking down into simple molecules

(CO_2, H_2O, or CH_4) found in the environment under the enzymatic action of microorganisms, in a defined period of time. Renewable resources derived polymeric materials are classified according to the method of production or its source: polymers directly extracted or removed from biomass such as polysaccharides and proteins; polymers produced through classical chemical synthesis initially from renewable bio-based monomers, for example, PLA, and polymers synthesized through microorganisms or genetically modified bacteria such as polyhydroxyalkanoates and bacterial cellulose (Figure 10.1).

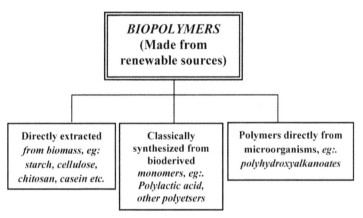

FIGURE 10.1 Types of biopolymers.

10.1.2 CELLULOSE

Cellulose is a biopolymer and the most common organic compound. A wide variety of plant materials have been studied for the extraction of cellulose cotton, potato, tubers, sugar beet pulp, soyabean stock, and banana rachis. Cellulose constitutes about 33% of plant matters. The cotton has cellulose content of 90%, for wood it is 50%. Cellulose has no taste, no odor, and is hydrophilic with the contact angle of 20°–30°, is insoluble in water and most organic solvents. Cellulose is a linear polymer of anhydroglucose and one of the most abundantly occurring natural polymers on earth. It is one of the complex carbohydrates consisting of 3000 or more repeating glucose units. Cellulose is the basic structural component of plant cell walls, nondigestible by humans [14]. It is chiralin nature and is biodegradable. The structure of cellulose is shown Figure 10.2.

Because of its regular structure and array of hydroxyl groups, cellulose is infusible and is also crystalline. In addition, it tends to form hydrogen-bonded

254   *Natural Polymers: Perspectives and Applications for a Green Approach*

crystalline microfibrils and fibers and is used in the packaging context in the form of paper or cardboard. Cellulose can be dissolved and swollen to films coatings or filaments using a solvent. Cellulose derivatives can be prepared by reacting the three hydroxyl groups on the anhydrous glucose unit partially or totally with different reagents. Earlier works on cellulose derivatives have been done in 1993.

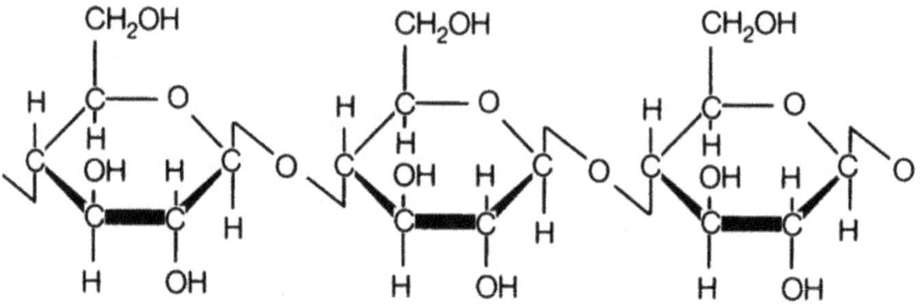

FIGURE 10.2   Structure of cellulose.

Celluloses derivatives are more resistant to microbial attack and enzymatic cleavages than native cellulose forms [14]. Cellulose and derivatives are the most important biopolymers. Due to its biocompatibility, biodegradability, and renewability it can be used in a large spectrum of applications such as manufacturing of paper, textiles, membranes for water purification, and/or biomedical applications such as magneto-responsive composites, bio-imaging materials, or support for catalysts [10, 18, 19, 30].

10.1.3   CELLULOSE ACETATE

One of the most important derivatives of cellulose is cellulose acetate (CA). Several applications of CA have been found including modern coating, controlled release, optical film, and membranes as well as the traditional textile field in the forms of fiber, film, and plastic [11, 12]. CA along with cellulose diacetate and cellulose triacetate is used widely in food packaging as a rigid wrapping film. The tensile strength of commercial cellulose acetate is 41–87 MPa and the elastic modulus is found to be 1.9–3.8 GPa [14]. Figure 10.3 represents the structure of CA.

CA can be prepared through the esterification of cellulose, which is abundant in agricultural waste like straw and biomass residues. CA is biodegradable, nontoxic, and biocompatible [13]. Moreover, CA is having certain mechanical

Cellulose Acetate-TiO$_2$-Based Nanocomposite   255

strength that facilitates its processing into films, membranes, and fibers from either melts or solutions [9, 15].

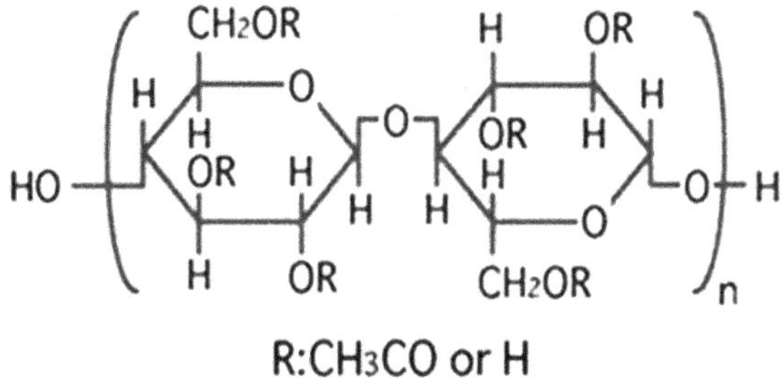

$$R:CH_3CO \text{ or } H$$

FIGURE 10.3   Structure of cellulose acetate.

10.1.4   *METAL OXIDE CERAMICS*

Metal oxide ceramics like TiO$_2$ are nontoxic, low cost, and efficient photocatalyst. They have been widely used in many other fields, such as electrochemistry, high-performance hydrogen sensors [13, 31], hydrophilic coating surfaces [32], detoxification processes, dye-sensitized solar cells, and so on [28, 29]. Metal oxide nanoparticles have played and will continue to play an important role in photocatalytic applications due to their high specific surface area, which is beneficial for promoting the diffusion of reactants and products, as well as for enhancing photocatalytic activity by facilitating access to reactive sites on the surface of the photocatalyst [30]. The nanosized TiO$_2$ was introduced in the cellulose biopolymer matrix for the modification of materials properties [8]. TiO$_2$ can be considered as a good semiconductor photocatalyst [33]. Modified metal oxide ceramic TiO$_2$ shows absorption in the visible region [34–37]. This is found to have applications in photochromic activities [33]. A redox dye incorporated biopolymer with modified metal oxide ceramics (MMOC) nanocomposite was used for easy fabrication of photochromic films [20–23].

10.1.5   *CHROMISM*

Chromism is a process in which a natural, organic, or inorganic compound undergoes reversible color change. Chromism arises due to a change in the electron states of molecules, especially in the Π or d-electron state. This phenomenon is induced by various external stimuli such as chemical or physical that causes the change of color or the restriction of light transmission. Moreover, chemicals that undergo reversible color changes with the application of an external stimulus have been increased. These found uses in a wide variety of low and high technology areas.

The major classes of chromism based on various stimuli are as follows:

1. Thermochromism is the most common of all chromism. It is induced by heat and exhibits a change of temperature.
2. Photochromism is happening by the light irradiation. This is happening due to the isomerization of two different structured molecules, formation of color centers in materials through light-induced, precipitation of metal particles in a glass, or through other mechanisms.
3. Electrochromism is happening due to the gain and loss of electrons. It occurs in compounds with oxidation–reduction active centres, such as metal ions or organic radicals.
4. Solvatochromism is happening due to the polarity of the solvent. Most of these are metal complexes.

Photochromism can be defined as a chemical process in which a material undergoes a reversible change between two states having different absorption spectra, that is, various colors. The change in one direction occurs in the influence of electromagnetic radiation, generally UV light, and in the other direction by altering/removing the light source and or also by thermal means. Usually the change in color in the forward direction is to longer wavelength, that is, bathochromic, and reversibility of this change is a key to the various applications of photochromism.

Photochromism can be used to describe the process in which materials undergo reversible photochemical reaction where the wavelength of an absorption band in the visible part of the electromagnetic spectrum changes drastically. In many cases an absorbance band is present in only one form. Some of the most common processes involved in photochromism are pericyclic reactions, cis-trans isomerizations, intramolecular hydrogen transfer, intramolecular group transfers, dissociation processes, and electron transfers (oxidation-reduction) [7]. Two states of the molecule should be thermally stable under ambient conditions for a particular time. By definition photochromism is reversible. Photochromic materials have been widely used as building materials due to their novel properties as heat insulation, radiation protection, energy saving, and decoration [24]. A rewritable paper that can be used multiple times without any additional inks for printing is an attractive alternative and can bring enormous economic and environmental merits to modern society.

Cellulose Acetate-TiO$_2$-Based Nanocomposite 257

Materials showing photochromism are called photochromic materials. These are mainly classified as natural, organic, and inorganic. Naturally occurring photochromic materials are photoreceptors, minerals, and others. Rhodopsin, phytochrome, and others are class of photoreceptors showing photochromism and sodalite, tectosilicate, and others are class of minerals showing photochromism. Sodalite ($Na_4Al_3Si_3O_{12}C_1$) is an aluminosilicate showing photochromism [4]. Various organic photochromic materials are azobenzene, spiropyrans, spirooxazines, stilbenes, nitrones, naphthopyrans, etc. These materials undergo *cis–trans* isomerization or photocyclization or heterolytic fission during light exposure.

There has been much interest in inorganic photochromic materials due to their promising application in the display, imaging, and solar energy conversion. Transition metal oxide ceramics such as WO_3, TiO_2 [25–27], and MoO_3 show color changes upon exposure to either sunlight or UV irradiation [5, 38].

As evidenced by its tripled global consumption over the past three decades, paper plays a very important role in communication as well as information storage. According to recent international surveys, 90% of all information in businesses is currently retained on paper; however, most of the prints are disposed after only one-time reading, which significantly increases business operating cost on both paper and ink cartridges. It also creates huge environmental problems such as deforestation, solid wastes, and pollution to air, water, and land [7]. Hence, rewritable paper that can be used multiple times and does not require additional inks for printing is an attractive alternative with enormous economic and environmental merits to modern society [3]. The organic dyes undergoing reversible color switching based on photoisomerization were proposed for potential use as the imaging layer in rewritable printing media. However, because of major challenges only limited progress have been made in this area (1) when dyes are present in solid media instead of solution, color switching is much slower, since molecular mobility is restricted; (2) under ambient conditions, many switchable dyes retain color for only several hours, which is too short for reading; (3) the toxicity of switchable dyes is an issue for daily use and (4) most switchable dyes are expensive since these involve complex synthesis. Developing rewritable papers based on new color-switching mechanism is of high interest now. Redox dyes change color reversibly on redox reactions. If redox reactions can be controlled, the switchable dyes may serve as promising imaging media for the development of rewritable paper [2, 7].

A dye can be switched between colors in an oxidizing environment and colorless in a reducing environment. A photocatalytically active material can be used to enable the decoloration of a dye under UV irradiation [10]. Additional reducing agents like hydroxyl ethyl cellulose are usually used as a sacrificial electron donor, which is able to scavenge the holes generated from the excitation of photoactive material under UV irradiation [16]. There is no report says using catalyst and dye as an imaging layer for fabrication of photoswitchable rewritable paper. This is because, due to the presence of excessive reducing agents, the recoloration process could not be initiated. In an attempt, catalyst with appropriate ligands was used to promote decoloration of a dye solution from color to colorless under UV radiation and it can reach to its original color on visible light irradiation [17]. Compared to the photo isomerizable chromophores, the system of catalyst/dye/water can rapidly change color with high reversibility and excellent repeatability. It is having the merits like low toxicity and low cost. But due to the spontaneous recoloration process under visible light, this system is incompatible for use as an imaging layer. Moreover, the simple deposition of catalyst solution and dye on a solid substrate may led to a flaky film. It can remain colorless for < 6 h after decoloration under UV, due to the quick oxidation of the leuco form by ambient oxygen [6]. Therefore a new mechanism is required to effectively stabilize the leuco form of dye and to maintain the colorless state for significantly longer periods.

Methylene blue (MB) can be switched between blue color (oxidizing environment) and colorless (reducing environment). TiO_2 could be used to enable the decoloration of MB under light irradiation [40]. Spontaneous recoloration process makes this system incompatible with its potential use in rewritable paper [7, 40].

The experimental part includes (1) hydrothermal synthesis of nanosize metal oxide ceramics and MMOC (2) preparation of cellulose acetate, CA-metal oxide ceramics and based nanocomposite films, (3) structural characterization of the nanosize metal oxide ceramics and cellulose acetate-metal oxide ceramics (CA-MOC) and cellulose acetate modified metal oxide ceramics (CA-MMOC) composite films and (4) evaluation of properties of CA-MBMOC and CA-ink immersed MMOC (CA-MBMMOC) photochromic films.

10.2 EXPERIMENTAL

10.2.1 PREPARATION OF CELLULOSE ACETATE FILMS

Flexible CA films were prepared through solution casting method. About 0.5 g of CA was dissolved in acetic acid with stirring until it becomes homogeneous. It was poured onto a Petri dish and dried at 333 K for 12 h. The dried CA films thus formed were removed from the Petri dish.

10.2.2 PREPARATION CELLULOSE ACETATE-METAL OXIDE CERAMICS AND CELLULOSE ACETATE MODIFIED METAL OXIDE CERAMICS FILMS

Metal oxide ceramics were added into acetic acid along with CA (0.5 g) and stirred until it reaches a homogeneous solution. Concentration of metal oxide ceramics was varied from 1 to 5 (wt.%). This was poured onto a Petri dish and dried at 333 K for 12 h. The dried films of CA-MOC were removed from the Petri dish. The same procedure was repeated using 1–5 (wt.%) of MMOC also in order to get CA-MMOC films.

10.2.3 PREPARATION CA-MBMOC AND CA-MBMMOC PHOTOCHROMIC FILMS

About 0.5 g of CA was dissolved in acetic acid with constant stirring. Metal oxide/MMOC and an organic ink that consists of a dye, solvent, and an electron donor were added into this solution and continued the stirring until it becomes homogeneous. The homogeneous solution was poured onto a Petri dish and dried at 333 K for 12 h. The photochromic films of CA-MBMOC and CA-MBMMOC formed were removed from the Petri dish.

10.3 MATERIAL CHARACTERIZATION

Characterization of the materials is of prime importance in fundamental research and is a lively and highly discipline in material science. The characterization techniques reveal the surface parameters, structure, composition, and reactive efficiency of the materials. Several approaches can be adopted to investigate the fundamental relation between the state of a material and its properties. Physical methods like thermogravimetry-differential scanning colourimetry (TGA-DSC), DR UV–vis spectroscopy, UV–vis spectroscopy, FT-IR spectroscopy, Raman spectroscopy, and scanning electron microscopy (SEM) analysis were employed to characterize the prepared materials.

260   *Natural Polymers: Perspectives and Applications for a Green Approach*

10.3.1   EVALUATION OF PHOTOCHROMIC PROPERTIES

The evaluation of properties of photochromic films was carried out using UV–vis and solar simulator irradiation. The recoloration/decoloration rate of the films was monitored visually as well as using UV–vis spectroscopy.

10.4   RESULTS AND DISCUSSION

10.4.1   CHARACTERIZATION OF MATERIALS

10.4.1.1   THERMAL ANALYSIS (TGA-DSC)

TGA-DSC analysis (not shown) of the MMOC exhibits weight loss around 400 °C due to the evaporation of the surfactant used during the hydrothermal preparation. There is not much weight loss observed after 600 °C that suggests that the ceramics is stable in this temperature range [9].

10.4.1.2   DR UV–VIS SPECTROSCOPY

DR UV–vis spectroscopy was used to probe the band structure or molecular energy levels in materials since the UV light excitation creates the photogenerated electrons and holes. Thus information about the properties of metal oxide can be obtained from DR UV–vis spectroscopy. These studies also help to understand the bandgaps of materials at pre- and postevolution conditions. The spectra obtained for the prepared materials are shown in Figure 10.4.

The DR UV–vis spectrum of MMOC shows red shift toward visible region (corresponding to bandgap energy of 2.3 eV). The absorption is due to the charge transfer from the VB formed by 2p orbitals of the oxide anions (O^{2-}) in the metal oxide to the CB formed by 3d t2g orbitals of the metal cations. The observed prominent change in the band edge that shifts from UV to visible region (2.3 eV) extends the wavelength response range to the visible region. In addition, it increases the number of electrons and holes (generated through photoreaction) to participate in the photocatalytic reaction.

10.4.1.3   RAMAN SPECTROSCOPY

Raman spectroscopy can be used to clearly characterize the metal oxide structure. Figure 10.5 shows the Raman spectrum of the MMOC.

Cellulose Acetate-TiO$_2$-Based Nanocomposite

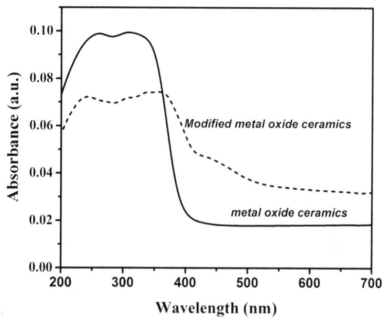

FIGURE 10.4 DR UV–visible spectra of ceramics.

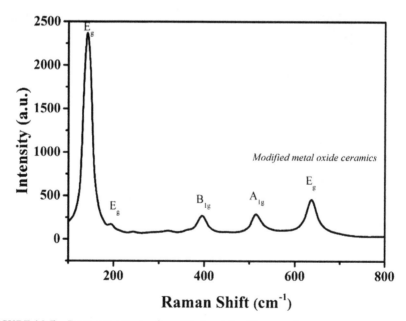

FIGURE 10.5 Raman spectrum of modified metal oxide ceramics.

The prepared MMOC exhibits six Raman active modes (A1g + 2B1g+ 3 Eg). The peaks are around ~ 144, 191, 393, 513, and 636 cm^{-1}. The band at 144 cm^{-1} is the strongest of all observed bands. All the modes observed correspond to single phase in MMOC. Thus Raman spectrum confirms the results evidenced by XRD.

10.4.1.4 SE MICROGRAPHS

The SE micrographs of the modified metal oxide are shown in Figure 10.6. As the micrographs reveal, the MMOC possess uniform morphology.

FIGURE 10.6 SE micrographs of modified metal oxide ceramics.

10.4.2 CHARACTERIZATION OF FILMS

The films of CA-MMOC and photochromic CA-MBMMOC obtained are shown in Figure 10.7.

10.4.2.1 FT-IR SPECTROSCOPY

Figure 10.8 exhibits FT-IR spectrum of CA-MMOC film. The spectrum shows a number of absorptions below 2000 cm^{-1}. The major absorption features appeared can be attributed as: 1732 cm^{-1} (C=O), 1430 cm^{-1} (O=C–OR), 1367 cm^{-1} (–CH2), 1218 cm^{-1} (C–O), 1031cm^{-1} (C–O–C) and 902 cm^{-1} (–CH). The peaks appeared at 2927 cm^{-1} (–CH$_2$) and 3441 cm^{-1} (–OH)

Cellulose Acetate-TiO$_2$-Based Nanocomposite 263

are characteristics bands of H2O. The presence of small peaks in the region 400–500 cm^{-1} can be attributed to the MMOC. The results demonstrated that only physical blending and no chemical reaction occurred between the metal oxide and CA film.

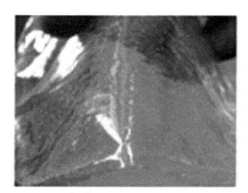
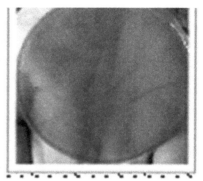

FIGURE 10.7 Films of CA-MMOC and CA-MBMMOC.

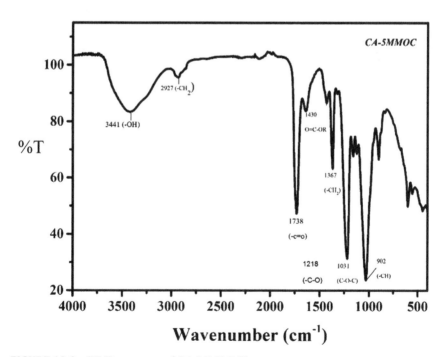

FIGURE 10.8 FT-IR spectrum of CA-MMOC film.

10.4.2.2 UV–VIS SPECTRA OF FILMS

The transparency of films was investigated from the UV–vis transmission spectra. The transmission spectra of CA-MOC films of different metal oxide concentration (1–5 wt.%) are shown in Figure 10.9 and the transparency percent calculated are presented in Table 10.1. From the UV–vis transmission spectra, it is clear that CA film performs high transparency (100%) than the CA-MOC films. The metal oxide content was varied from 1 to 5 (wt.%). The transparency rate keeps on decreasing with increasing the amount of metal oxide addition.

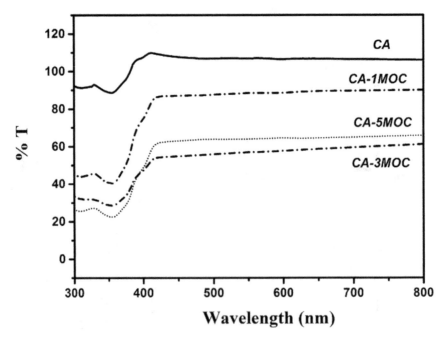

FIGURE 10.9 UV–visible transmittance spectra of CA-MOC films.

TABLE 10.1 Transparency of CA-MOC Films

Films	Transparency (%)
CA	100
CA-1MOC	89
CA-3MOC	60
CA-5MOC	65

The transmission spectra of CA-modified metal oxide ceramic (CA-MMOC) films are shown in Figure 10.10 and the transparency percent calculated are presented in Table 10.2. The same trend was observed in CA-MMOC films also. With increase in the addition of MMOC, the percent transparency keeps on decreasing. It is also found that CA-MOC films are more transparent than CA-MMOC.

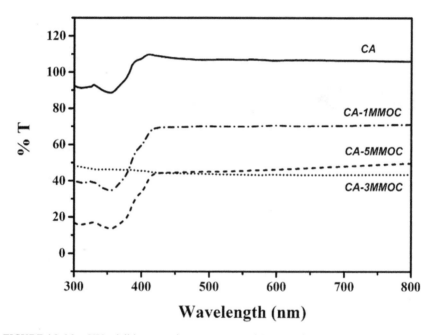

FIGURE 10.10 UV–visible transmittance spectra of CA-MMOC films.

TABLE 10.2 Transparency of CA-MMOC Films

Films	Transparency (%)
CA	100
CA-1MMOC	71
CA-3MMOC	43
CA-5MMOC	49

10.4.2.3 SE MICROGRAPHS

The SE micrographs of CA-1MOC film is shown in Figure 10.11. The SE micrograph of CA-1MOC film shows uniform distribution of ceramics on the CA-MOC nanocomposite film. Presence of metal oxide was observed on the surface of the film.

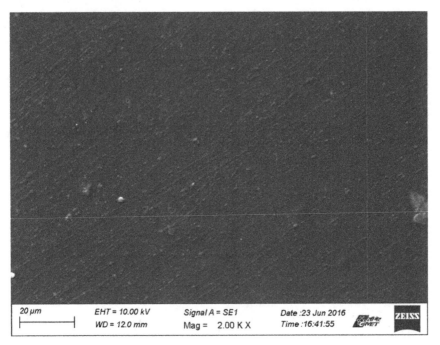

FIGURE 10.11   SE micrograph of CA-MOC film.

10.5   EVALUATION OF PHOTOCHROMIC PROPERTIES

The transmission spectra of photochromic ink immersed (cellulose acetate-modified metal oxide ceramics, CA-MBMOC) films are shown in Figure 10.12 and the calculated transparency is presented in Table 10.3.

CA film performs high transparency (100%) than the CA-MBMOC films. In addition, the transparency of the film decreases with increasing the concentration of the metal oxide.

The transmission spectra of CA-ink immersed MMOC (CA-MBMMOC) films are shown in Figure 10.13 and the transparency are presented in Table 10.4.

Films of CA-MBMMOC shows more transparency compared to CA-MBMOC films. For example, CA-1MBMOC shows percentage transparency of 84 whereas the CA-1MBMMOC shows transparency of 90% for the same concentration.

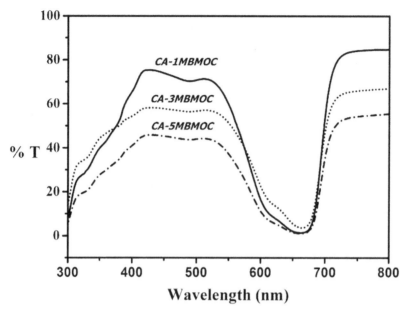

FIGURE 10.12 UV– visible transmission spectra of CA-MBMOC films.

TABLE 10.3 Transparency of CA-MBMOC Films

Films	Transparency (%)
CA	100
CA-1MBMOC	84
CA-3MBMOC	66
CA-5MBMOC	55

10.5.1 PHOTOCHROMIC PROPERTY UNDER UV IRRADIATION

The films were kept under UV irradiation for specific time and observed the rate of the decoloration/recoloration. The results obtained for studies conducted for composite films are shown in Figure 10.14. Under UV irradiation, CA-1MBMOC film decolorized within 20 min and recolorized within 24 h.

In the case of CA-5MBMOC, the same trend is observed. The obtained results are shown in Figure 10.15. It is observed that the decoloration process takes place in 25 min whereas the recoloration occurred within 1 h. The fast decoloration process occurred due to the presence of increased metal oxide ceramic content.

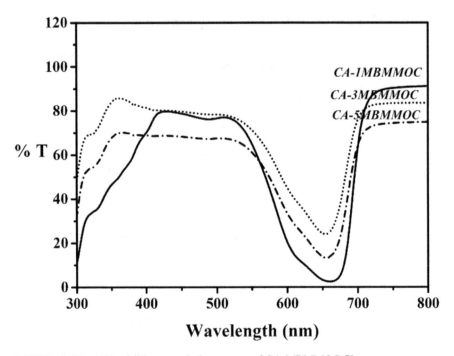

FIGURE 10.13 UV– visible transmission spectra of CA-MBMMOC films.

TABLE 10.4 Transparency of CA-MBMMOC Films

Films	Transparency (%)
CA	100
CA-1MBMMOC	90
CA-3MBMMOC	83
CA-5MBMMOC	74

The photochromic evaluation of CA-MBMMOC films were also carried out following the same procedure. The observations are shown in Figure 10.16. Under UV irradiation, CA-1MBMMOC film took 60 min to decolorize and recolorized within 12 min. For these films the time required for recoloration process is much lower compared to CA-1MBMOC films.

Photochromic evaluation of CA-5MBMMOC film was also conducted under the same conditions. As shown in the image (Figure 10.17), the film was decolorized within 25 min and recolorized in 60 min under UV irradiation.

Cellulose Acetate-TiO$_2$-Based Nanocomposite 269

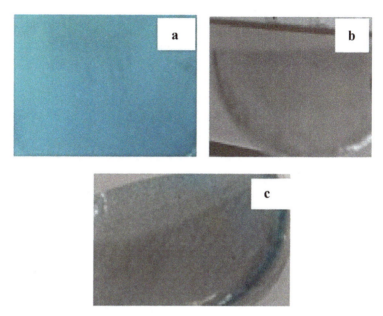

FIGURE 10.14 Photochromic evaluation of CA-1MBMOC film under UV irradiation. (A) Original, (B) decoloration at 20 min, and (C) recoloration at 24 h.

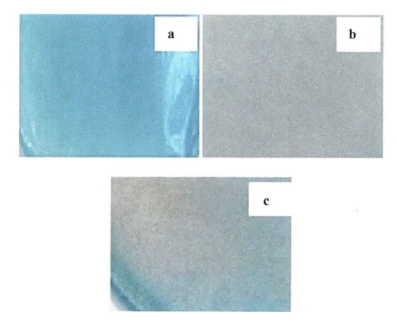

FIGURE 10.15 Photochromic evaluation of CA-5MBMOC film under UV irradiation. (a) Original, (b) decoloration at 25 min, and (c) recoloration at 1 h

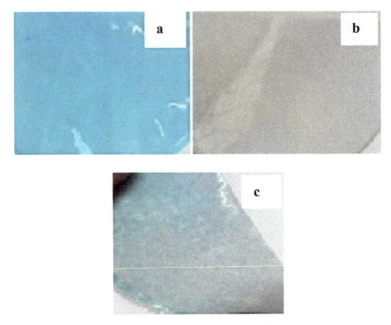

FIGURE 10.16 Photochromic evaluation of CA-1MBMMOC film under UV irradiation. (a) Original, (b) decoloration at 60 min, and (c) recoloration at 12 min.

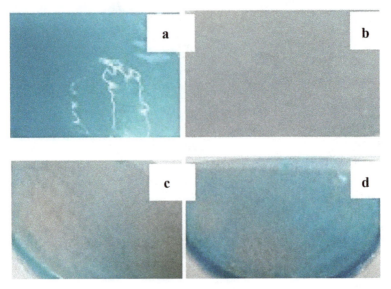

FIGURE 10.17 Photochromic evaluation of CA-5MBMMOC film under UV irradiation. (a) Original, (b) decoloration at 25 min, (c) recoloration at 10 min, and (d) 1 h

In the case of CA-MBMOC films under UV irradiation, the complete decoloration occurs in 25 minutes. However, the recoloration process was not spontaneous. The recoloration process took long time. Moreover, in the case of Ca-MBMMOC films the decoloration as well as recoloration occur very fast and the rate of the process increased with increased amount of metal oxide ceramics in the film. The parameters that can affect the decoloration/recoloration process are (1) intensity of the lamp, (2) dispersion of metal oxide on the CA film, (3) mode of casting, and (4) the nature of the organic ink.

10.5.2 VISIBLE IRRADIATION

To test the photochromic properties of the films, the decoloration/recoloration rate for a specific period of time was monitored under visible light irradiation. The visible observation obtained for studies conducted with CA-MBMOC films are shown in Figure 10.18. Under visible light irradiation, CA-1MBMOC film was decolorized within 10 min and almost completely recolorized in 1 h.

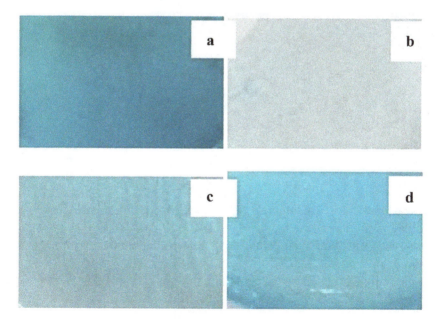

FIGURE 10.18 Photochromic evaluation of CA-1MBMOC film under visible irradiation. (a) Original, (b) decoloration at 10 min (c) recoloration at 5 min, and (d) 1 h.

The evaluation of photochromic properties conducted using CA-5MBMOC films are shown in Figure 10.19. It is observed that the decoloration and recoloration are very fast. The recoloration process is completely reversible in this film.

FIGURE 10.19 Photochromic evaluation of CA-5MBMOC film under visible irradiation. (a) Original, (b) decoloration at 10 min, (c) recoloration at 3 min, and (d) 30 min.

The photochromic evaluation of CA-1MBMMOC films carried out under visible irradiation is depicted in Figure 10.20. Under visible light irradiation, the CA-1MBMMOC nanocomposite film was decolorized within 20 min and recolorized in 30 min.

The visible observations of photochromic evaluation of CA-5MBMMOC film are shown in Figure 10.21. Under visible light irradiation, the CA-5 MBMMOC film was decolorized within 10 min. and fully recolorized within 10 min.

Cellulose Acetate-TiO$_2$-Based Nanocomposite 273

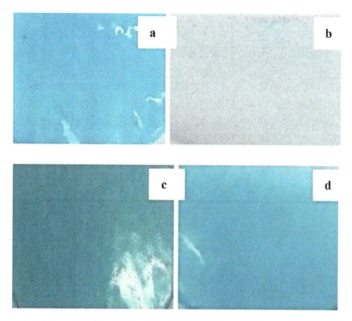

FIGURE 10.20 Photochromic evaluation of CA-1MBMMOC film under visible irradiation. (a) Original, (b) decoloration at 20 min, (c) recoloration at 4 min, and (d) 30 min.

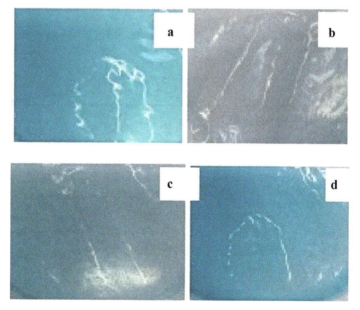

FIGURE 10.21 Photochromic evaluation of CA-5MBMMOC film under visible irradiation. (a) Original, (b) decoloration at 10 min, (c) recoloration at 8 min, and (d) 10 min.

10.5.3 DISCUSSION

10.5.3.1 UNDER VISIBLE IRRADIATION

The decoloration/recoloration characteristics of the films were studied by UV–visible spectroscopy. The spectra obtained when the sample is kept in visible light are presented in Figure 10.22. Under visible irradiation, as observed for the CA-1MBMOC film, absorbance spectrum also confirms decoloration within 10 min.

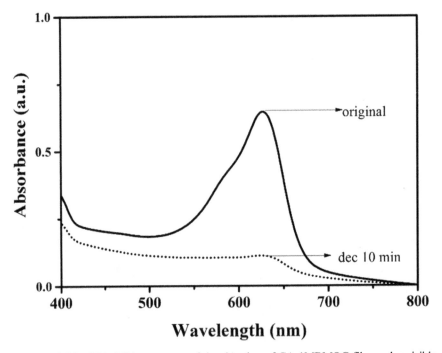

FIGURE 10.22 UV-visible spectrum of decoloration of CA-1MBMOC film under visible irradiation.

It is found that the recoloration process is much more precise in modified ceramics. Under visible irradiation, the MMOC film was decolorized within 10 min and recolorized within 90 min (Figure 10.23). The UV–vis spectral analysis confirms the decoloration and recoloration of the films under visible irradiation.

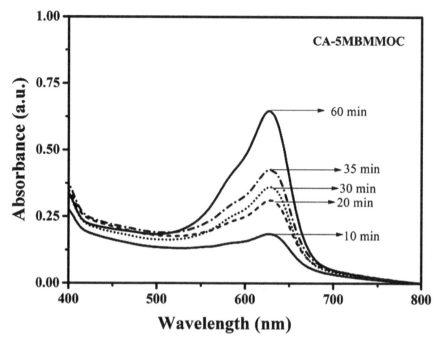

FIGURE 10.23 UV–visible spectrum of recoloration of CA-5MBMMOC film under visible irradiation.

10.5.3.2 UNDER SOLAR SIMULATOR

To evaluate the photochromic property of the prepared films, the films were kept under solar simulator and analyzed the UV–vis spectra using UV–vis spectrophotometer. The decoloration/recoloration results observed for representative films for different concentrations of metal oxide are presented in Figure 10.24. The decoloration of modified ceramics was happened within 1 min and that of metal oxide ceramics took 5 min (Figure 10.25).

Under solar simulator (Figure 10.26) the decoloration was so fast (~1 min) for CA-5MBMMOC films. Recoloration was also fast compared to UV and visible irradiations.

The photochromic printing under sunlight is shown in Figure 10.27. The photochromic film with a printed pattern on the top was exposed under sunlight for 5 min. The part other than darkened with printed pattern became colorless and darkened pattern got printed on the film.

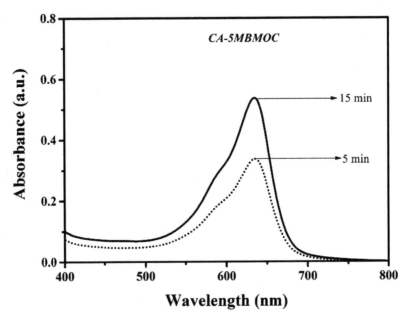

FIGURE 10.24 UV–visible spectrum of decoloration of CA-5MBMOC film under solar simulator.

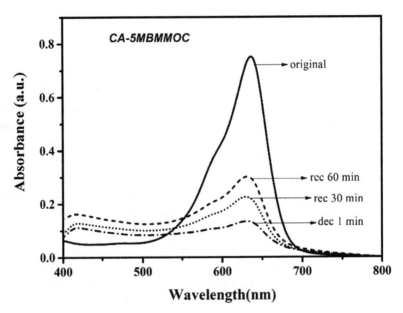

FIGURE 10.25 UV–visible spectrum of decoloration/recoloration of CA-1MBMMOC film under solar simulator.

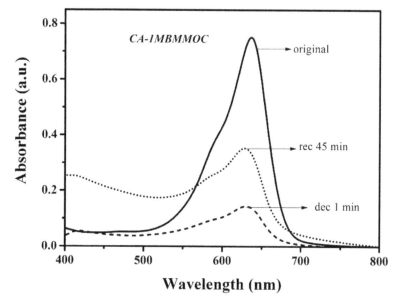

FIGURE 10.26 UV–visible spectrum of decoloration/recoloration of CA-5MBMMOC under solar simulator.

FIGURE 10.27 Photochromic printing under sunlight.

10.6 CONCLUSION

Present works concentrate on the preparation and characterization of biopolymer-ceramics nanocomposite films for versatile applications. Nanosized metal oxide ceramics and MMOC were successfully prepared by hydrothermal method, employing metal oxide precursor. Systematic investigation of the physico-chemical characterization of the materials were carried out using powder XRD, DR UV–vis spectroscopy, FT-IR spectroscopy, Raman spectroscopy, TGA-DSC analysis, SEM, and others. The photochromic property of the prepared films was studied under UV, visible light, and solar simulator irradiation.

From TGA-DSC analysis, the calcination temperature of metal oxide ceramics was selected as ~773 K. The extended absorbance into visible range suggests it could expand the response to visible region, and these films can be more effective for solar light applications. The FT-IR spectra confirm the complete dissolution of CA and absence of any other peak in the CA-metal oxide ceramics film shows that there is only physical blending of metal oxide occurs with CA. Raman analysis supports the formation of single phase metal oxide ceramics. The SEM images show uniform morphology for the prepared CA-metal oxide ceramics film and confirm the presence of ceramics in the composite.

The prepared films were tested for its photochromic properties using UV, visible, and solar simulator. In the case of UV irradiation, the complete decoloration took around 20 min. However, the recoloration process was slow. Under visible light irradiation, fast decoloration and recoloration were occurred. Under solar simulator both the decoloration and recoloration were faster than visible light (~1 min) irradiation. The work provides emphasis on various aspects of biopolymer-metal oxide ceramics nanocomposite film preparation, characterization, and evaluation of photochromic properties.

KEYWORDS

- photochromic
- cellulose acetate
- flexible film
- nanocomposite
- TiO_2

REFERENCES

1. Luc Averous and Eric Pollet, Environmental Silicate Nano-Biocomposites; 2012, Springer Science & Business Media, Chapter 2, 13–33.

Cellulose Acetate-TiO2-Based Nanocomposite   279

2. Andrew, M.; David, H. A solvent based intelligence ink for oxygen, Analyst *2008*, 133, 213–218.

3. Chengfeng Jin, Runshi Yan and Jianguo Huang, Cellulose substance with reversible photo- responsive wettability by surface modification, J. Mater. Chem. *2011*, 21, 17519.

4. Carvalho, J. M.; Norrbo, I.; Ando, R. A.; Brito, H. F.; Fantini, M. C. A.; Lastusaari, M. Fast, low-cost preparation of hackmanite minerals with reversible photochromic behavior using a microwave-assisted structure-conversion method, Chem. Commun. *2018*, 54 (53), 7326–7329.

5. Suzuko, Y.; Hiroki, I.; Dai, S.; Kenta, A. Photochromic properties of tungsten oxide/ methylcellulose composite film containing dispersing agents, Appl. Mater. Interfaces *2015*, 7, 26326–26332.

6. Kazuya, N.; Hiroaki, K.; Munetoshi, S.; Tsuyoshi, O.; Hideki, S.; Taketoshi, M.; Masahiko, A.; Akira F. UV/thermally driven rewritable wettability patterns on TiO_2-PDMS composite films, Appl. Mater. Interfaces *2010*, 2, 9, 2485–2488.

7. Wenshou, W.; Ning, X.; Le, H.; Yadong, Y. Photocatalytic colour switching of redox dyes for ink-free light-printable rewritable paper, Nature Commun. *2014*, 1–7.

8. Wenshou, W.; Miaomiao, Y.; Le, H.; Yadong, Y. Nanocrystalline TiO_2-catalyzed photoreversible colour switching, Nano Lett. *2014*, 14, 1681–1686.

9. Shu-Dong, W.; Qian, M.; Hua, L.; Ke, W.; Ke, Q. Z. Robust electrospinning cellulose acetate @ TiO_2 ultrafine fibers for dyeing water treatment by photocatalytic reactions RSC Adv. *2013*, 1–3.

10. Alexandra, W.; Hanna, T.; Steffen, K.; Mathias, U. Routes towards catalytically active TiO_2 doped porous cellulose, RSC Adv. *2015*, 5, 25866–35873.

11. Jinghua, Y.; Doug, D.; Nancy, M. C.; Ray, J. N. Jr. Characterization of cellulose acetate films, Pharm. Technol. *2009*, 33, 3, 88–100.

12. Yanxia, Z.; Tiina, N.; Carlos, S.; Julio, A.; Ingrid, C. H.; Orlando, J. R. Cellulose nanofibrils: from strong materials to bioactive surfaces J. Renew. Mater. *2013*, 1, 3, 195–210.

13. Xiujuan, J.; Jing, X.; Xianfu, W.; Zhong, X.; Zhe, L.; Bo, L.; Di, C.; Guozhen, S. Flexible TiO_2/ cellulose acetate hybrid film as a recyclable photocatalyst, RSC Adv. *2014*, 4, 12640–12648.

14. Bhanu, M.; Anu, K.; Harsha, K. Natural polymer based cling films for food packaging. Int. J. Pharm. Pharm. Sci. *2015*, 7, 10–18.

15. Alexandra, W.; Dimitri, V.; Mathias, U.; Two step and one step preparation of porous nanocomposite cellulose membranes doped withTiO_2 RSC Adv. *2015*, 5, 88070–88078.

16. Yulia, G.; Sheng-Hao, H.; Wei-Fang, S. Monitoring time and temperature by methylene blue containing polyacrylate film, Sens. Actuators B Chem. *2010*, 144, 49–55.

17. Ho, S. L.; Joong, T. H.; Donghoon, K.; Meihua, J.; Kilwon, C. Photoreversibly switchable superhydrophobic surface with erasable and rewritable pattern, J. Am. Chem. Soc. *2006*, 128, 14458–14459.

18. Cheng, F. J.; Runshi, Y.; Jianguo, H. Cellulose substance with reversible photo-responsive wettability by surface modification, J. Mater. Chem. *2011*, 21, 17519–17525.

19. Tiffany, A.; Amit, R.; Yifeng, C.; Yuval, N.; Eldho, A.; Tal, B. S.; Shaul, L.; Oded, S. Nanocellulose a tiny fiber with huge applications, Curr. Opin. Biotechnol. *2016*, 39, 76–88.

20. Kazuya, N.; Hiroaki, K.; Munetoshi, S.; Tsuoshi, O.; Hideki, S.; Taketoshi, M.; Masahiko, A.; Akira, F. UV/thermally driven rewritable wettability patterns on TiO_2-PDMS composite films, ACS Appl. Mater. Interfaces. *2010*, 2, 2485–2488.

21. Masaya, Y.; Hiroyuki, Y. Transparent nanocomposites based on cellulose produced by bacteria offer potential innovation in the electronics device industry, Adv. Mater. *2008*, 20, 1849–1852.

22. Weixin, H.; Qihua, W. UV-driven reversible switching of a polystyrene/titania nanocomposite coating between superhydrophobicity and superhydrophilicity, Langmuir *2009*, 25(12), 6875–6879.

23. Cong, L.; Yan, S.; Jialu, L.; Xiansheng, W.; Su-Ling, L.; Wei, F.; Enhanced photochromism of heteropolyacid/ polyvinylpyrolidone composite film by TiO_2 doping, J. Appl. Polym. Sci. *2015*, 10, 41583.

24. Kesong, L.; Moyuan, C.; Akira, F.; Lei, J. Bio-inspired titanium dioxide materials with special wettability and their applications, Chem. Rev. *2014*,114, 19, 10044–10094.

25. Andrew, M.; Jishun, W.; Soo-Keun, L.; Morten, S. An intelligence ink for photocatalytic films, Chem. Commun. *2005*, 2721–2723.

26. Zi, F. Y.; Long, W.; Hua, G. Y.; Yong, H. S. Recent progress in biomedical applications of titanium dioxide, Phys. Chem. Chem. Phys. *2013*, 15, 4844–4858.

27. Jiaguo, Y.; Qinglin, L.; Zhan, S. Dye-sensitized solar cells based on double-layered TiO_2 composite films and enhanced photovoltaic performance, Electrochim. Acta *2011*, 56, 18, 6293-6298.

28. Noah, M. J.; Yuriy, Y. S.; Chris, S.; Daniel, H.; Masoud, S.; Kenneth, K. S. L.; Hai-Feng, J.; Photochromic dye-sensitized solar cells AIMS, Mater. Sci. *2015*, 2, 4, 503–509.

29. Brunauer, S.; Emmette, P.H.; Teller, E., Adsorption of gases in multimolecular layers, J. Am. Chem. Soc. *1938*, 60, 309.

30. Nishizava, H.; Katsube, M.; Preparation of $BaTiO_3$ thin films using glycolate precursor, J. Solid State Chem. *1997*, 131, 43.

31. Patsi, M. E.; Hautaniemi, J. A; Rahiala, H. M.; Peltola T. O.; Kangasniemi, I. M. O. Bonding strengths of titania sol-gel derived coatings on titanium, J. Sol Gel Sci. Technol. 1998, 11, 55–66.

32. Manoj, A. L.; Walid A. D. Achieving selectivity in TiO_2-based photocatalysis, *RSC Adv.* 2013, 3, 4130–4140.

33. Baiju, K. V.; Shukla, S., Sandhya, K. S.; James, J.; Warrier, K. G. K.; Photocatalytic activity of sol–gel-derived nanocrystalline titania, J. Phys. Chem. C 2007, 111, 21, 7612–7622.

34. Farook, A.; Jimmy, N. A.; Zakia, K.; Radhika, T.; Mohd, A. M. N. Utilization of tin and titanium incorporated rice husk silica nanocomposite as photocatalyst and adsorbent for the removal of methylene blue in aqueous medium, Appl. Surf. Sci. *2013*, 264 718–726.

35. Jothi, R. R.; Prabhakaran, A.; Radhika, T.; Anju, K. R.; Nimitha, K. C.; Al-Lohedan, H. A. Surface and electrochemical characterization of N-Fe-doped-TiO_2 nanoparticle prepared by hydrothermal and facile electro-deposition method for visible light driven pollutant removal, Int. J. Electrochem. Sci. *2016*, 11, 797–810.

36. Jothi, R. R.; Radhika, T.; Reshma, P. R.; Shaban, R. M.; Sayed, H. A. Al-lohedan, Meera, A.; Moydeen, D. M. Al-dhayan. Platinum nanoparticle decorated rutile titania synthesized by surfactant free hydrothermal method for visible light catalysis for dye degradation and hydrogen production study, Int. J. Hydrogen Energy *2019*, 44, 23959.

37. Irina, P.; Ekaterina, K.; Svetlana, C.; Sergei, T.; Andrey, R.; Yury, S. Titania synthesized through regulated mineralization of cellulose and its photocatalytic activity, RSC Adv. *2015*, 5, 8544–8552.
38. Gulsen, A.; Murvet, K.; Nursel, P. B. Effect of sonic treatment on the permeation performance of cellulose acetate membranes modified by n-SiO_2, Turkish J. Chem. *2015*, 39, 297–305.
39. Antoni, W. M.; Ewelina, K. N.; Jacek, P.; Roksana, K.; Juliusz, P. Cellulose-TiO_2 nanocomposite with enhanced UV–Vis light absorption, Cellulose *2013*, 20, 1293–1300.

CHAPTER 11

Effect of Thermal Treatment on Structure and Properties of Plasticized Starch–Polyvinyl Alcohol (PVA) Blend Films

SUBRAMANIAM RADHAKRISHNAN[*], SWAPNIL THORAT, ANAGHA KHARE, and MALHAR B. KULKARNI

Research Development and Innovation, Maharashtra Institute of Technology (MIT-WPU), S124, Paud Road Kothrud, Pune 411038, India

[*]*Corresponding author. E-mail: radhakrishnan.s@mitpune.edu.in*

ABSTRACT

Waste cassava starch was soaked in water for 24 h and the solution was heated at 70 °C that was then plasticized by using different plasticizer such as glycerine, sorbitol, and polyethylene glycol separately (0%–40% with respect to starch). This was blended with polyvinyl alcohol (PVA) matrix in aqueous solution. Plasticizer content (Plc) was varied from 10% to 40% with respect to starch. The solutions were cast into films in a plastic dish and dried at 55 °C for 24 h. These were subsequently given short-time heat treatment (150 °C for 2 min). The films were characterized by XRD, FTIR and tested for mechanical properties and seal strength. The two prominent XRD peaks at 19.6° and 23° merge in one broad peak for high Plc percentage. The crystallinity (Ci) decreased with the increase of Plc percentage in the order PEG < sorbitol< glycerine. Increase in the content of PVA led to a shift of onset degradation temperature to higher value while an increase of Plc percentage it lowered slightly. The FTIR bands in the region of 1130–1020 cm^{-1} indicated crystal modification of starch after heating. The tensile strength decreased with the increase in Plc percentage but increased after heat treatment (even for Plc percentage of 30%–40%). The elongation increased considerably with Plc percentage but dropped after heat treatment. The seal strength increased with the increase of plasticizer content. There was optimum PVA concentration (66.6%) at which best seal strength

284 *Natural Polymers: Perspectives and Applications for a Green Approach*

was achieved. These results correlated well with the structural changes noted with respect to composition and annealing. The above study helps to develop ecofriendly packaging films and improvements in the seal strength of such films in presence of modified starch.

11.1 INTRODUCTION

The use of petroleum-based plastics, namely, polyethylene, polypropylene, polystyrene, and others in packaging films has led to severe disposal problem causing their large accumulation in the environment. Many alternative materials that are biodegradable or compostable have been looked into so that long term solution can be provided for the plastic pollution. Among the biodegradable and or compostable polymers, polylactic acid, its derivatives/ copolymers, and blends have been extensively studied in the recent past [1–5]. There are many sources of biodegradable materials in nature that are having potentials for film production. Among the natural polymers, the availability of starch is second only to cellulose [6]. The most important industrial sources of starch are corn, potato, tapioca, rice, and cassava. The low price and the availability of starch is very favorable for making polymers and blends as cost-effective biodegradable materials. However, starch by itself is difficult to process/extrude as films due to its degradation prior to melting, no flow, and no melt strength. Hence, efforts have been made to make thermoplastic starch (TPS) by using plasticizers, modifiers, and graft copolymers [7–10]. Various types of starch have been investigated in the past for blending with thermoplastics and making packaging films. These include blends with polyethylene, polypropylene, polystyrene, and others that are not truly biocompostable. TPS starch has been blended with other biocompostable polymers as well to reduce the cost without loss of compostability [11–13]. Most of these reports deal with pure edible starch from corn, maize, potato, and cassava. The present studies are focused on the use of naturally available waste material (nonedible portion) so as not to disturb the food chain. Polyvinyl alcohol (PVA) is a water-soluble synthetic polymer that can be cast into films that are transparent with good mechanical properties. Also it is odorless and nontoxic due to which PVA films have been used for packing detergent, plant protection, food packaging, and so many other usage [14, 15]. Commercially PVA is produced by hydrolysis of vinyl acetate, yields different grades PVA depending upon hydrolysis number ranging from 70% to 99%. The degradation of PVA in the environment is slow depending upon the hydrolysis number. A few studies have been reported on starch blended with

Effect of Thermal Treatment on Structure   285

PVA [16–19] with and without plasticizers. Due to multihydroxy structure and presence of hydrogen bonding in starch, films casted from starch are brittle in nature and possess poor mechanical properties. To overcome such problems it is first plasticized and blended with other polymers. Upon blending these difficulties are eliminated. Although TPS is made with plasticizer initially and then blended with other polymers in the melt, it is not clear how the thermal cycling affects the structure and morphology. Many assume that the plasticization of starch is not hampered by the melting and solidification and blends are homogenous. Although some authors mention difficulties in film extrusion of plasticized starch and its blends [20, 21], detailed studies on the effect of heat treatment on the structure and properties are lacking.

In the present studies, cassava waste starch was plasticized with different plasticizers, blended with PVA, and films made by solution casting method. These were given short heat treatment and characterized thoroughly by different techniques. Mechanical properties, structural change, and effect of plasticizer have been investigated and these have been correlated with corresponding changes.

11.2 EXPERIMENTAL

11.2.1 MATERIALS

All the chemicals that were used for the preparation of the films were analytical grade received from a local chemical supplier. Cassava waste starch was provided by Speciality Starch Pvt. Ltd., Salem Tamil Nadu, India in the form of off-white powder. Amylose content in cassava starch ranged from 23.01% to 26.98%. PVA of M.W. 14,000 and degree of hydrolyzation 86–89 mole% was supplied by Thomas Baker. Plasticizer like glycerine, sorbitol, and polyethylene glycol (PEG) of make Sigma-Aldrich (Merck) from local supplier. Deionized water was used for making blends and casting films.

11.2.2 PREPARATION OF STARCH–POLYVINYL ALCOHOL BLEND FILMS WITH GLYCERINE AND POLYETHYLENE GLYCOL AS PLASTICIZERS

Starch–PVA blend films were prepared by solution casting. Waste cassava was first soaked in water (starch 10% solution) for a few hours to swell it. The mixture was heated to 70 °C with constant stirring. Plasticizers glycerine or

286   *Natural Polymers: Perspectives and Applications for a Green Approach*

PEG was added separately to this in different proportions ranging from 10% to 50% by weight with respect to starch. This was allowed to digest for a few hours. A solution of PVA (10% by weight) was prepared in deionized water separately. Such made solution was added to the plasticized starch suspension in large beaker and the solution is stirred for 1 h. Starch to PVA ratio was varied from 0% to 50% by weight and from that blended solution 25 mL solution was poured in a plastic Petri dish of diameter 120 mm because of this we will get the thickness of the films nearly constant. Such films were kept in an air circulating oven at 55 °C to form uniform films that were further dried for 24 h.

11.2.3 *PHYSICAL CHARACTERIZATION AND MEASUREMENTS*

FTIR spectra (600–4000 cm^{-1}) were recorded in Bruker Alpha spectrometer in both ATR and transmission modes. First thin films of the sample were prepared and then placed in sample holder to measure FTIR spectra in normal conditions (room condition). Differential scanning calorimeter (DSC) curves were recorded in Hitachi DSC7020 series with temperature range 30–180 °C and on cooling 180–30 °C with heating and cooling rate of 10 °C/min in nitrogen gas environment, by placing 3–4 mg sample in an aluminum pan and sealed it by aluminum lead and differentiate with reference pan that is kept empty. Thermogravimetric analysis (TGA) was studied using Hitachi Thermo gravimetric Analyser STA7000. A 7–8 mg sample was cut from films and placed in the platinum sample holder, temperature range 30–650 °C, in N$_2$ atmosphere keeping reference pan empty. X-ray diffraction (XRD) scans were recorded by Bruker D8 advance diffractometer. XRD specimens were prepared by cutting from the casted films in the dimension 2.5 cm. Scanning range set to 5°/min using CuKα radiation of wavelength 1.54058 Å from generator operates at 40 kV and 40 mA. Wide angle XRD was recorded from 2θ 5°–45° range. Crystallinity was determined in the usual way

$$X_c = I_c/(I_c + I_A)$$ (11.1)

where X_c = crystallinity, I_c = area under crystalline peak, and I_A = area under the amorphous peak.

11.2.4 *MECHANICAL PROPERTIES*

Strips were cut from the films (10 × 2 cm) for testing the mechanical properties such as tensile strength and elongation according to ASTM D638 method

*Effect of Thermal Treatment on Structure*287

on a universal testing machine (STS248) (Lloyd LR 50K). Tests were carried out on untreated and thermally treated samples. The thermal treatment was carried out by giving short exposure to 150 °C for 2 min and immediately cooling to room temperature that simulates the sealing cycle. The tests were performed at a crosshead speed of 50 mm/min using Universal Testing Machine (UTM). The test was carried out at PRAJ Metallurgical Laboratory.

11.2.5 PREPARATION OF SAMPLE FOR SEAL STRENGTH

Two separate rectangular specimens used to make test specimen having 25 mm width and 100 mm length. These are hot wire sealed using a conventional bag sealer with preset temperature (150 °C and time 3 s) and overlap of 10 mm from the one end. The seal strength was measured according to ASTM D638 method on the UTM (STS248) (Lloyd LR 50K) at a crosshead speed of 200 mm/min. Micrometre screw gauge is used to measure the thickness of the samples. The test was carried out at PRAJ Metallurgical Laboratory.

11.3 RESULTS AND DISCUSSION

11.3.1 STRUCTURE DEVELOPMENT AND CHARACTERIZATION

Figure 11.1 compares the XRD curves for modified starch blends with PVA plasticized with glycerine, sorbitol, and PEG differently for a same composition of starch and PVA. It is evident that the crystallinity present is in the order glycerine > sorbitol > PEG. The crystallinity estimated for these cases was 48%, 42%, and 27%, respectively. Crystallinity is determined by the given formula [Eq. (11.1)]. Starch is known to crystallize in two forms, namely, monoclinic (lattice parameters $a = 2.124$ nm, $b = 1.172$ nm, $c = 1.069$ nm, $\gamma = 123°$) and hexagonal (lattice parameters $a=b=1.85$ nm, $c=1.04$) with characteristic peaks at 19.3° and 23° in the XRD pattern [22]. The interesting observation is that the peak at 23° corresponding to V_A type starch crystals decreases much more than V_H type which has prominent XRD peak at 19.3° [22]. Also, the peak becomes broader in the case of PEG indicating lower crystallite size. The two extremes, namely, glycerine and PEG have been studied further in detail.

Figure 11.2 shows XRD for modified starch in PVA, before and after heat treatment. It is seen that the crystallinity in as made films (without heat treatment) decreases with the increase of plasticizer content. The peak

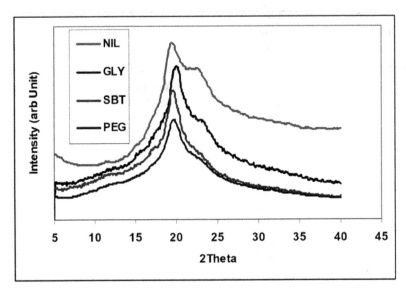

FIGURE 11.1 Comparison of different plasticizers on the crystallinity of starch in PVA blend Plasticizer content was 40% by weight with respect to starch. Starch-to-PVA ratio is 1:2 by weight.

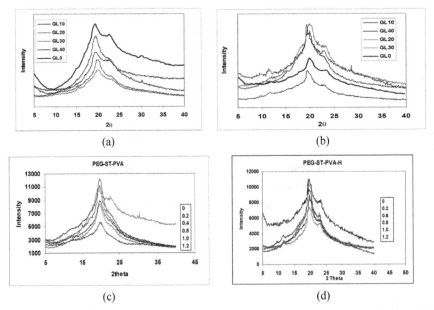

FIGURE 11.2 XRD of plasticized starch blended with PVA containing glycerine (a and b) and PEG (c and d) before and after heat treatment. Starch:PVA ratio was 1:2 in all cases and plasticizer concentration was varied from 0% to 50% by weight with respect to starch.

intensity decreases and it becomes broad suggesting that there is a considerable decrease in crystallinity and crystallite size. This is true for both the plasticizers. On the other hand, after heat treatment, the intensity of the peak at 19.6° (corresponding to V_H type crystals of starch) increases considerably and the peak becomes sharp. In the case of PEG as a plasticizer, there is a significant increase in the peak at 22° corresponding to V_A type crystals. This can be due to the increase in the chain mobility by the presence of plasticizer that allows the new crystals to form during heat treatment [23].

The effect of plasticizer concentration on the intensity of the major XRD peak of starch is seen in Figure 11.3. It is seen that there is a considerable decrease in intensity for plasticizer content > 30%. It should also be noted that the width of the XRD peak increases with the increase of plasticizer content. Since the crystallite size is inversely related to peak width [24], there is a large decrease of crystal size in plasticized starch as well as after blending with PVA.

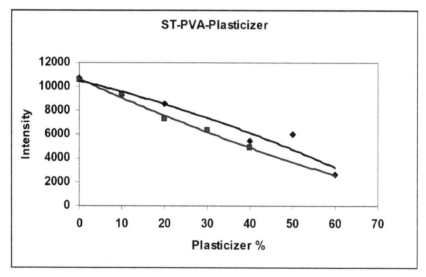

FIGURE 11.3 Change in XRD main peak intensity with the increase of plasticizer concentration. Starch to PVA was 1:2 by weight.

By Scherrer's equation, the size of crystallites was determined [Eq. (11.2)] [25]

$$d_{cryst} = (0.94\ \lambda)\ (\Delta\theta\ \cos\theta) \qquad (11.2)$$

where d_{cryst} = crystallite size(average) θ = Braggs angle, $\Delta\theta$ = line broadening in radians, and λ = X-ray wavelength (1.540 Å).

The crystal size of the V_H crystallites was estimated and found to decrease from the original 18 nm to 8 nm after plasticization. After heat treatment, the crystallite size increases again to 12 nm but only for one type of crystals. Thus there appears that there is selective crystals are retained after the addition of plasticizer.

Figure 11.4 shows FTIR spectra, mainly in the region of 1800–600 cm^{-1} before (a) and after heat treatment (b). The changes in the peaks in this region are associated with the crystalline modifications of starch [26]. It is seen that there are major changes in the peak intensities for the bands in the region of 1130–1000 cm^{-1} that are sensitive to the crystal modification of starch [27]. Also the sharp peak at 1640 and 1570 cm^{-1} becomes weak after heat treatment. This is associated with the crystalline form of starch. Thus these findings are in agreement with the observations made in the XRD studies.

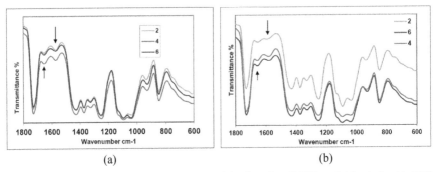

FIGURE 11.4 FTIR of starch plasticized with glycerine (30%) and blended with PVA (66%): (a) before heat treatment and (b) after heat treatment.

The TGA data is depicted in Figure 11.5 for starch–PVA blends. Figure 11.5A containing different concentrations of glycerine (10%, 20%, 30%, and 40% with respect to starch) at constant PVA content (66%) as well as Figure 11.5b that for samples containing different concentrations of PVA (50%, 66%, 75%, and 80% with respect to starch) at same glycerine content (40% with respect to starch). It is seen that the onset temperature of degradation decreases with the increase of plasticizer concentration. On the other hand PVA has more stabilizing effect in as much as the onset temperature shifts slightly to a higher temperature and there is more residue obtained at higher temperatures even at 550 °C (weight loss is decreased). This could be due to the virtual cross-linking of the PVA chains via plasticizer with starch giving more stability to the material as a whole.

Effect of Thermal Treatment on Structure 291

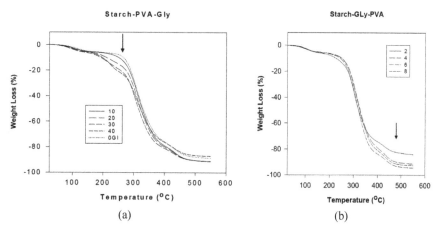

FIGURE 11.5 TGA curves for starch-PVA blends (a) containing different concentration of glycerine at same contents of PVA (66%) and (b) different concentration of PVA but same plasticizer content (40%).

11.3.2 MECHANICAL PROPERTIES

The mechanical properties such as tensile strength and elongation for these blends before and after heat treatment are shown in Figure 11.6a and b, respectively. The curves are for different concentration of glycerine indicated in Figure 11.7a and b. It is seen that the elongation increases with the increase of plasticizer content in the unheated samples.

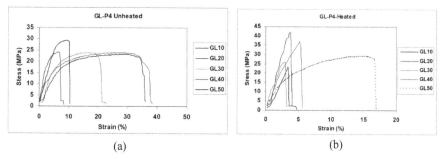

FIGURE 11.6 Stress–strain curves for PVA-starch plasticized with different concentration of glycerine (indicated by number %): (a) for unheated and (b) for heated films.

The tensile strength decreases with the increase of plasticizer content. The sharp break point that is seen in samples with low plasticizer concentration does not appear with high plasticizer concentrations (>30%). On the other hand, after heat treatment, the samples show a sharp break point, low elongation, and higher tensile strength for all samples up to 40% plasticizer. Only at very high plasticizer level, the curve remains more or less similar to the unheated sample. The increase in tensile strength after heat treatment can be associated with the increase in crystallinity and crystal size as noted above in the XRD. The corresponding decrease in elongation is also due to higher stiffness imparted by the recrystallization of starch.

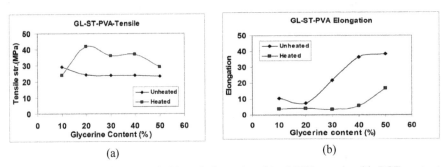

FIGURE 11.7 Tensile strength (a) and elongation (b) of PVA-starch with PGE content before and after heat treatment. PVA content 66%.

The stress–strain curves for the PVA-starch blend containing PEG as plasticizer are shown in Figure 11.8, whereas Figure 11.9 depicts the tensile strength and elongation for PEG containing films before and after heating. It is seen that the effect of heating on the tensile strength of PEG containing films is more pronounced at higher content of plasticizer. Also, the tensile strength is much higher in these samples than those containing glycerine as a plasticizer. This can be due to better plasticizing effect of PEG as well as its action in compatibilizing the blend components.

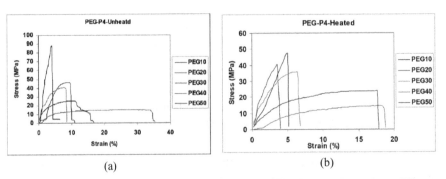

FIGURE 11.8 Stress–strain curves for PVA-starch plasticized with PEG at different concentration (indicated by numbers %): (a) for unheated and (b) for heated films.

Effect of Thermal Treatment on Structure 293

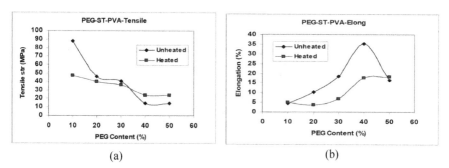

FIGURE 11.9 Tensile strength (a) and elongation (b) of PVA-starch blend with glycerine content before and after heat treatment.

The molecular interaction is depicted schematically in Figure 11.10. The longer molecules of PEG having linear array of –O–CH$_2$ groups and –OH groups only at the chain end give better advantage for compatibility with PVA and starch as compared to glycerine, which has three –OH groups juxtaposed to each other. The molecular interaction between starch–PEG–PVA is expected to be better due to many hydrogen bonding sites as represented in Figure 11.10.

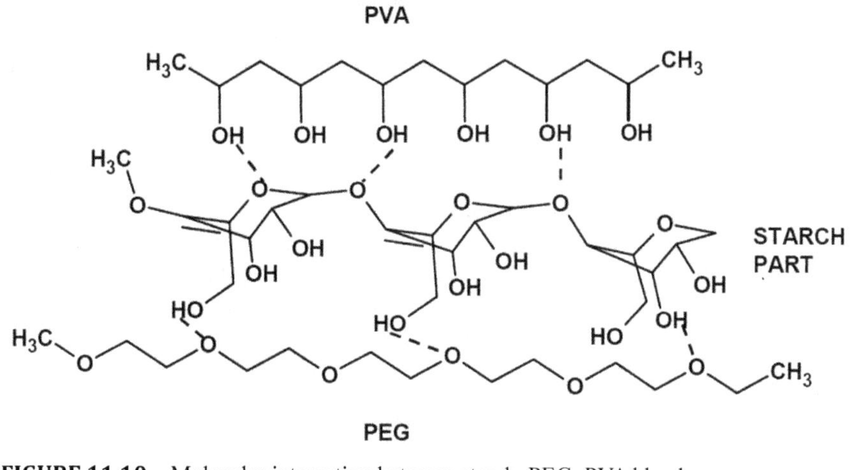

FIGURE 11.10 Molecular interaction between starch–PEG–PVA blends.

The mechanical properties of PVA-modified cassava starch blends have been reported by Guimaraes et al. [28] for as-cast films with different concentration of PVA and starch at 12% glycerine. They note a slight increase in the tensile strength in 20% PVA blend. They have not carried

294   *Natural Polymers: Perspectives and Applications for a Green Approach*

out a complete variation of plasticizer concentration. However, they have referred to an increase in compatibility in the polymers as the cause for an increase in the tensile strength in comparison to other compositions. Tian et al. [29] have reported the mechanical properties of starch/PVA blends with 30% plasticizer having a combination of urea and formaldehyde. They have varied the blend composition from 0% to 50% starch and the samples were made by compression moulding. They found a large decrease in tensile strength and elongation with the increase of starch concentration. They have not changed the plasticizer concentration. They associate this trend to the decrease of crystallinity with the change of starch content. Lani and Ngadi [30] have investigated the mechanical properties of starch/PVA blends with 30% glycerol and cast from solution. The plasticizer concentration was fixed while the starch to PVA concentration ratio was changed from 0% to 60%. They have reported that while the tensile strength decreased continuously with the increase of starch, the elongation goes through a maximum at the starch content of 30%. They explain this result on the basis of formation of hydrogen bonds between the chains and also poor dispersion of starch in PVA matrix. Siddaramiah et al. [31] have reported the mechanical properties of PVA-corn starch films cast from water without plasticizer. The starch concentration was varied from 0% to 10% for which the tensile strength and elongation were measured. The tensile strength did not show much change/any significant change (26.2–27.3 MPa) when starch concentration was increased. Since there was no plasticizer used by them, there was no compatibilizer for starch that would remain only as filler leasing to very little change in mechanical properties. The mechanical properties of PVA-corn starch blends plasticized with glycerol were reported by Sreekumar et al. [32]. The plasticizer content was varied from 15.75% to 78.75% of the total (31.5%–157.5% with respect to starch). It is the tensile strength of the blend containing 15.75% glycerol was found to be much higher than plasticized starch or PVA alone. For the higher concentration of glycerol they report a continuous decrease in tensile strength as well as elongation. These are much higher concentrations of plasticizer compared to present studies. However, it worth noting that in one case, they also observed higher tensile strength than others. It may be mentioned that the films produced by these authors were compression-molded by heating at 120 °C for 10 min that led to similar behavior as we have reported in this chapter.

The seal strength of the films is important for packaging applications. The films were sealed using hot-wire technique at constant temperature and time and then tested on UTM as per ASTM standard [33, 34]. The films get locally heated for short time and then cooled to room temperature. Figure 11.11 shows the typical curves for load versus displacement for samples joints containing different concentration of glycerine. Figure 11.12a and b shows the seal strength with PVA content and glycerine concentration respectively. The most interesting observation is that with the increase of glycerine concentration, the seal strength increases (Figure 11.12a) and also the displacement before break increases. Normally, with the increase of plasticizer, the tensile strength is expected to decrease and elongation to increase. However, in the present case, both seal strength as well as displacement at break increase with the increase of plasticizer. This may be explained as follows. As the plasticizer content is increased, the chain mobility increases and causes greater entanglement during heating for seal formation. At the same time, it is seen from the previous discussions in this chapter, there is recrystallization taking place after heat treatment in presence of plasticizer. This leads to higher tensile strength of the films themselves. These two factors lead to higher seal strength in these plasticized samples. It may be noted that there is an optimum concentration of PVA at which maximum seal strength is obtained for a given concentration of plasticizer as indicated in Figure 11.12b.

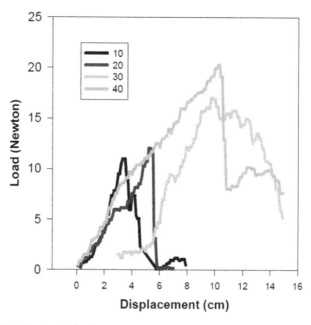

FIGURE 11.11 The load–displacement curves for different starch–PVA films sealed under the same condition by the hot wire unit.

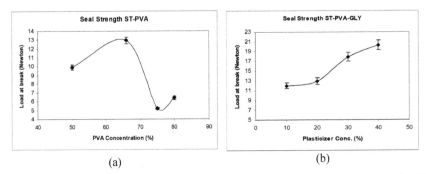

FIGURE 11.12 Blend composition effect on seal strength of the plasticized starch–PVA blend films: (a) for different glycerine content (b) for different PVA content.

It may be mentioned here that Lagos et al. [35] have studied the puncture strength (puncture force and puncture deformation) of plasticized cassava films and found that for glycerine as plasticizer the puncture force increased at a certain concentration of plasticizer. Puncture deformation increased with the addition of plasticizer. They have not measured other mechanical properties for these films. From the present results, the elongation increases with the addition of plasticizer that can be the reason for the increase of puncture deformation observed by them.

11.4 CONCLUSIONS

Cassava waste starch was first plasticized by soaking in water followed by the addition of plasticizer such as glycerine and PEG. This was blended with PVA in aqueous solution and then cast into films. The effect of heat treatment on the plasticized starch–PVA films on structure development was investigated by different characterization techniques. The crystallinity of starch was found to decrease with the addition of plasticizer and its chemical structure influenced the extent of lowering the X_c. The crystallinity was lowest for PEG plasticized starch and high in the case of glycerine containing starch. However, it was observed that recrystallization of starch takes place in presence of plasticizer when subjected to heat treatment. The V_H and V_A crystal forms were obtained by heating in presence of plasticizer. The crystallite size was very small (8 nm) as such in plasticized starch but it increased significantly after heat treatment (12–15 nm). These findings from XRD were also confirmed by other characterization tools. These structural modifications were reflected in the tensile strength and elongation in these films especially

Effect of Thermal Treatment on Structure

for glycerine as a plasticizer. All these structural modifications and changes in mechanical properties get reflected in the seal strength of the films, which involves short-time heating and cooling during seal formation, Increase of plasticizer in fact increased the seal strength up to certain concentration There was an optimum concentration of PVA at which best seal strength was obtained. These findings will help in the optimization of biocompostable film composition used for packaging.

ACKNOWLEDGMENT

The authors would like to acknowledge the Department of Biotechnology (DBT) of the Government of India for their financial support for this project under UK-INNOVATE Homi Bhabha scheme.

KEYWORDS

- **PVA**
- **sorbitol**
- **glycerine**
- **polyethylene glycol**
- **starch**
- **seals strength**

REFERENCES

1. Bastioli. C. *Handbook of Biodegradable polymers*, ChemTec Publishing: Toronto-Scarborough, Ontario, Canada, 2005.
2. Garlotta, D. A literature review of poly (lactic acid). *J. Polym. Environ.* 2005, 9 (2), 63–84.
3. Sedlarik, V.; Saha, V.; Sedlarikova, J.; Saha, P. Biodegradation of blown films based on poly (lactic acid) under natural conditions. *Macromol. Symp.* 2008, 272, 100–103.
4. Yang, K.K.; Wang, X.L.; Wang, Y.Z. Progress in nanocomposite of biodegradable polymer. *J. Ind. Eng. Chem.* 2007, 13, 485–500.
5. Jost, V. Packaging related properties of commercially available biopolymers—an overview of the status quo. *Express Polym. Lett.* 2018, 12, 429–435.
6. BeMiller, J.; Whistler, R. *Starch: Chemistry & Technology,* Academic Press Elsevier: Amsterdam, 2009.

7. Jane, J. Starch properties, modifications, and applications. *J. Macromol. Sci.* A, 1995, 32(4), 751–757.
8. Ma, X.F.; Yu, J.G.; Wan, J.J. Urea and formamide as a mixed plasticizer for the thermoplastic starch. *Carbohydr. Polym.* 2006, 64, 267–273.
9. Shanks, R. and Ing, K. Thermoplastic Starch Thermoplastic Elastomer, Adel el Sonbati, Ed., Intech, 2012; pp. 95–116 ISBN-978-953-51-0346-2.
10. Lui, H.; Xie, F.; Yu, L.; Chen, L.; Li, L. Thermal processing of starch-based polymers. *Prog. Polym. Sci.* 2009, 34, 1348–1368.
11. Yao, K.; Cai, J.; Liu, M.; Yu, Y.; Xiong, H.; Tang, S. l.; Ding S. Structure and properties of starch/PVA/nano-SiO hybrid films. *Carbohydr. Polym.* 2011, 86, 1784–1789.
12. Colivet, J.; Carvalho, R.A. Hydrophilicity and physicochemical properties of chemically modified cassava starch films. *Ind. Crops Prod.* 2017, 95, 599–607.
13. Nguyen, D.M.; Do, T.V.; Grillet, A.C.; Thuc, H.H.; Thuc, C.N.H. Biodegradability of polymer film based on low density polyethylene and cassava starch. *Int. Biodeter. Biodegrad.* 2016, 115, 257–265.
14. Lani, N.S.; and Ngadi, N. Preparation and characterisation of polyvinyl alcohol/starch blend film composites. *Appl. Mech. Mater.* 2014, 554, 86–90.
15. Wang, W.; Zhang, V.; Jia, R.; Dai, Y.; Dong, H.; Hou, H.; Guo, Q. High performance extrusion blown starch/polyvinylalcohol/clay nanocomposite films. *Food Hydrocoll.* 2017, 79,1-10.
16. Azahari, N. A.; Othman, N.; Ismail, H. Biodegradation studies of polyvinyl alcohol/corn starch blend films in solid and solution media. *J. Phys. Sci.* 2011, 22, 15–31.
17. Tang, X.; Alavi, S. Recent advances in starch, polyvinyl alcohol based polymer blends, nanocoposites and their biodegradability. *J. Carbohydr. Polym.* 2011, 85, 7–16.
18. Shafik, S.S.; Kawakib, J.M.; Kamil, M.I. Preparation of PVA/corn starch blend films and studying the influence of gamma irradiation on mechanical properties. *Int. J. Mater. Sci. Appl.* 2014, 3(2), 25–28.
19. Yu, F.; Prashantha, K.; Soulestin, J.; Lacrampe, M.F.; Krawczak, P. Plasticized-starch/poly(ethylene oxide) blends prepared by extrusion. *Carbohydr. Polym.* 2013, 91, 253–261.
20. Moad, G. Chemical modification of starch by reactive extrusion. *J. Progr. Polym. Sci.* 2011, 36, 218–237.
21. Brandelero, R.P.H.; Yamashita, F.; Grossmann, M.V.E. Films of starch and poly(butylene adipate co-terephthalate) added of soybean oil (SO) and Tween 80. *J. Carbohydr. Polym.* 2010, 62, 1102–1109.
22. Vliegenthart, Johannes, Van Soes5, Jeroen, S.H.D.; Hulleman, D de Wit. Influence of glycerol on the melting of potato starch. *Ind. Crops Prod.* 1996, 5. DOI: 10.1016/0926-6690995000047-x.
23. Iannace, S.; Nicolais, L. Isothermal crystallization and chain mobility of poly(L-lactide). *J. Appl. Polym. Sci.* 1997, 64, 911–919.
24. Leisen, J.; Beckham, H.W.; Sharaf, M.A. Evolution of crystallinity, chain mobility, and crystallite size during polymer crystallization. *Macromolecules* 2004, 37, 8028–80034.
25. Alexander. L.E. *X-Ray Diffraction Methods in Polymer Science.* Wiley: New York, 1969.
26. Henrique, C.M.; Teofilo. R.F.; Ferreira, M.M.C.; Cereda, M.P. Classification of cassava starch films by physicochemical properties and water vapor permeability quantification by FTIR. *Food Eng. Phys. Prop.* 2012, 72, 4.

27. Van Soest, J.G.; Tournois, H.; deWit, D.; Vliegenthart, J.F.G. Short-range structure in (partially) crystalline potato starch determined with attenuated total reflectance Fourier-transform IR spectroscopy. *Carbohydrate Res.* 1995, 201–214.
28. Guimaraes, M.; Botaro, V.R.; Novack, K.M.; Teixeira, F.G.; Tonoli, G.H.G. High moisture strength of cassava starch/polyvinyl alcohol-compatible blends for the packaging and agricultural sectors. *J. Polym. Res.* 2015, 22, 192–210.
29. Tian, H.; Yan, J.; Rajulu, A.V.; Xiang, A.; Luo, X. Fabrication and properties of polyvinyl alcohol/starch blend films: effect of composition and humidity. *Int. J. Biol. Macromol.* 2017, 96, 518–523.
30. Lani, N.S.; and Ngadi, N. Preparation and characterisation of polyvinyl alcohol/starch blend film composites. *Appl. Mech. Mater.* 2014, 554, 86–90.
31. Siddaramiah, Baldev, R.; Somashekar, R. Structure–property relation in polyvinyl alcohol/starch composites. *J. Appl. Polym. Sci.* 2004, 91, 630–635.
32. Sreekumar, P.A.; Al-Harthi, M.A.; De, S.K. Effect of glycerol on thermal and mechanical properties of polyvinyl alcohol/starch blends. *J. Appl. Polym. Sci.* 2012, 123, 135–142.
33. Hishinuma. K. *Heat sealing Technology and Engineering for Packaging: Principles and Applications.* DEStech Publications Inc.: Pennsylvania, USA, 2007.
34. ASTM F 88-00. *Standard Test Method for Seal Strength of Flexible Barrier Materials.* The American Society for Testing and Materials: New York, USA, 2015.
35. Lagos, J.B.; Vincentini, N.M.; Dos Santos, R.M.C.; Bittante, A.Q.B.; Sobral, P.J.A. Mechanical properties of cassava starch films as affected by different plasticizers and different relative humidity conditions. *Int. J. Food Stud.* 2015, 4, 116–125.

Index

A

Abiotic stresses, 78
Abrasions, 6, 155
Acetic acid, 99, 103, 104, 106, 107, 112, 131, 140, 258, 259
Acrylamide-acrylate copolymers, 79
Actin cytoskeleton, 162
Activated partial thromboplastin time (APTT), 154, 157, 171, 176–179
Active
  functional groups, 120
  nuclides, 126–129, 141
Additive manufacturing (AM), 91–93, 96, 97, 99, 104, 109–113
Adhesion, 28, 96, 154, 158, 161, 163, 171, 178–181, 186, 205, 220, 221
Adipic acid, 10
Adjacent glucose unit, 19
Adsorption field, 120
Aerobic conditions, 85
Aerospace, 42, 99, 109, 112, 194
Agarose, 23, 27, 78
Agrocompanies, 44
Agrowaste, 44
Alcoholic intoxication, 42
Alginate, 23, 24, 27, 78, 82, 104, 110, 112, 153, 154, 157, 164–168, 170–172, 175, 176, 179, 181, 182, 184, 186, 187
  chitosan polyelectrolyte complex, 165
Alginic acid, 23, 164
Alkali, 124, 193, 197, 200, 206, 207, 212–214, 217, 219
  earth metal ions, 124
Alkaline, 26, 99, 119, 122, 140, 205
  hydrolysis, 26
  insoluble fractions, 99
Alkalization, 196, 205, 206, 213
Ambient
  atmospheric conditions, 43
  temperature, 75
American society for testing and materials (ASTM), 1, 14, 43, 44, 49, 215, 286, 287, 294

Amine groups, 119, 124, 125, 137, 160, 163
Amino group, 22, 102, 119, 124, 126, 130, 132, 137, 139
Amorphous polymers, 10
Amphipod, 14
Amylopectin, 20, 71
Amylose, 71, 285
Angiogenesis, 29, 104, 110, 111
Anionic
  bacterial heteropolysaccharide, 72
  hydrogel, 79
  metal complex, 129
  species, 121, 124, 126, 134
Antibacterial
  characteristics, 110
  properties, 20, 153, 160, 164, 165
Antibiotics, 28, 165
Anticholesterolemic activity, 104
Antifibrinolytic agents, 182
Antigens, 34
Anti-inflammatory property, 29
Antimicrobial
  activity, 104
  agents, 20
Antimony
  ions, 122, 128
  nuclides, 127
Antiparallel beta-sheets, 27
Aquatic ecotoxicity, 6
Aqueous solutions, 22, 78, 126, 127, 134, 136–138
Aramid fibers, 95
Areca fruit husk (AFH), 41, 43
  fiber (AFHF), 41–43, 47–49, 58, 60–63, 65
    characterization, 43
    composite (AFHFC), 43–46, 49–53, 55–57, 59–63
    thermal analysis, 58
Aromatic structures, 2
Arthropod, 99, 100
Artificial
  scaffold printing, 97
  skin, 20, 28
  tissues, 109

302   *Index*

Ascophyllum nodosum, 164
Ascorbic acid, 135
Asparagine, 26
Atoms, 2, 130, 156, 209
Automobiles, 2, 8, 13, 203, 220, 231

B

Backbone chain, 71
Bacterial
  colonization, 5
  spores, 31
Beta-D glucose monomers, 19
Binding properties, 117, 119, 129, 135,
  137–140
Bio polyesters, 10
Bioabsorbancy, 81
Bioactive agents, 12
Bioadhesives, 27
Bioassays, 2
Bio-based
  composites multifunctional capabilities, 8
  medical implants, 9
  novel materials, 4
Bioceramics, 109
Biochemicals, 28, 29, 81
Biochemistry, 164
Biocompatibility, 12, 17, 20, 29, 34, 35, 98,
  99, 104, 106, 107, 109, 110, 112, 117,
  146, 254
Bio-composites, 41, 65
Biodegradability, 7, 21, 80–82, 85, 86, 95, 98,
  99, 104, 109, 112, 123, 204, 226, 246, 254
Biodegradable
  cyclic carbonates, 10
  materials, 76, 85, 252, 284
  plastics, 7, 14, 106
  polymeric materials, 18
  polymers, 7, 18
Biodegradation, 7, 14, 17, 30, 35, 78, 84,
  85, 194, 252
Biodiversity, 2, 5
Bioerosion, 7
Biofuel, 5, 10
Biohybrids, 10
Bioink, 20, 28
Biomass, 3, 9, 252–254
Biomaterials, 4, 17, 18, 20, 28, 30–32, 35,
  110, 123, 156, 157, 165, 181, 182

Biomedical
  applications, 17, 18, 21, 26–31, 35, 98,
    113, 254
  engineering, 17, 18, 35
Biomolecules, 74
Bioplastics, 2, 5, 7, 8
Biopolymer, 3, 9, 11, 19, 20, 26, 30, 82, 84,
  86, 91, 92, 98, 99, 113, 117, 118, 132,
  162, 164, 165, 252–255, 277, 278
  overview, 252
Biopolymeric
  materials, 82
  products, 2
Bioprinting, 104, 112
Biorecycling, 84
Biorefining, 9, 10
Biosorbent, 78
Biotechnology, 9, 10
Blocking groups, 6
Blood
  cell membranes, 162
  clotting assay, 154, 155, 157, 158, 163,
    167, 169–174, 187
    activated partial thromboplastin time, 176
    blood collection, 171
    clotting time, 171
    hemolysis assay, 173
    platelet isolation and adhesion, 178
    prothrombin time, 174
  coagulation, 26, 158, 161, 165, 176, 178
  surface interactions, 33
  vessel wall, 25
Body fluid, 30, 31
Bonding agent, 98
Bone wax, 153
Bottom-up approaches, 155, 156
Bovine collagen, 28
Broad-spectrum serum protease inhibitor, 183
Bursera bipinnata, 73

C

Cadmium cations, 126
Calcium carbonate, 101
Carbohydrates, 71, 72, 94, 195, 218, 253
Carbon
  chains, 2
  dioxide, 7, 9, 10, 20, 21, 32, 84, 85, 226
  fiber, 8, 95, 96

Index 303

footprint, 6, 10, 11
nanotubes, 141
Carbonyl compound, 138, 139
Carboxyl groups, 26
Carboxylate groups, 80
Carboxylic
  acid groups, 23
  compounds, 124
Carboxymethyl cellulose (CMC), 76, 82, 232
Cardiac tissue engineering, 28
Carrageenan, 72
Cartilage
  soft tissue, 30
  substitutes, 29
  tissue engineering, 28
Cascade reactions, 155, 181
Casein, 10
Catastrophic failure, 8
Cationic
  binding sites, 122
  sites, 122, 124
Cationicity, 102
Cauterization, 183
Cell
  culture, 97, 99
  differentiation, 109, 112
  division, 109
  growth, 18, 97, 107, 195
  populations, 30
Cellobiose, 19
Cellular therapy, 30
Cellulose, 9, 10, 12, 19, 20, 28, 29, 41, 43, 47, 71, 72, 81–83, 94, 98, 99, 103, 109, 110, 112, 118, 193–197, 199, 200, 202, 203, 205, 206, 208, 211, 214, 216, 217, 220, 221, 233, 242, 251–255, 258, 266, 278, 284
  acetate (CA), 98, 106, 175, 178, 251, 254, 255, 258, 259, 262–278
    metal oxide ceramics (CA-MOC), 251, 258, 259, 264–266
    modified metal oxide ceramics (CA-MMOC), 251, 258, 259, 262, 263, 265, 266
    preparation, 259
  fibrils, 19
  nanofiber (CNF), 12
  nanowhiskers, 9

Celox, 153–155, 157, 162, 171–174, 176, 178, 183, 186, 187
Center secondary layer, 195
Central pollution control board (CPCB), 13
Ceramic matrix composites, 229
Chain growth reactions, 19
Chelation, 124, 126, 129, 131, 138
Chemical
  additives, 1
  crosslinkers, 33
  energy, 10
  entities, 1
  fertilizers, 70, 74
Chemoattractant, 33
Chemotactic
  agents, 33
  factors, 34
Chitin, 10, 12, 22, 72, 91, 94, 98–104, 109, 110, 118, 121–124, 160
  extraction, 100
Chito-biose, 101
Chitonolysis, 121, 122
Chitosan (CS), 10, 12, 21, 22, 29, 35, 72, 76, 82, 91, 92, 96, 98–100, 102–113, 117–126, 129–142, 146, 153, 154, 157, 160–168, 170–173, 175, 176, 178, 179, 181, 182, 184–187, 252
  3D printing, 108
  alterations and enhancement, 162
  amine groups, 137, 163
  backbone, 123
  biopolymer, 92, 98, 113
  calcium phosphate, 104, 112
  cobalt ion imprinting, 131
  composite sorbents, 134
  conventional manufacturing processes, 105
  crosslinkers, 119
  derivation and properties, 99
  extraction, 100
  films, 106, 107, 161
  functional groups, 124, 129, 130
  matrix, 110, 119, 135–139
  metal ion binding properties, 122
  modifications, 118
  polymer, 108
  properties, 104
  receptors, 129
  sorbents, 121, 129, 137

304   *Index*

Chlorine atom, 119
Choline chloride (ChCl), 106
Chromatography, 12
Chromism, 255, 256
Chronic inflammation, 34
Chymotrypsin, 31
Citric acid (CA), 33, 106, 135
Clean-up circuit, 127
Coagulation factors, 182, 184
Coastal
  environments, 3
  regions, 5, 6
  sediment microcosms, 5
Cobalt
  ions, 130, 131
  naphthenate, 44
Collagen, 10, 24–29, 31, 32, 35, 71, 78, 97,
  98, 107, 153, 158
  fibers, 24
  matrices, 28
  microfibril, 25
  molecule, 25, 26
Collagenase, 31
Colorants, 1, 108
Committee for the purpose of control and
  supervision of experiments on animals
  (CPCSEA), 171
Complex
  matrices, 11
  metabolic processes, 19
Complexant, 124, 129, 131, 136
Complexing agents, 127, 129, 135
Composites classification, 229
  ceramic matrix composites, 232
  metal matrix composites, 231
  natural fiber-reinforced composites, 232
  organic matrix composites, 230
Compression molding, 91, 105–107
Controlled
  release (CR), 8, 18, 27, 69, 70, 73, 74, 76,
    86, 108, 254
    fertilizers (CRF), 74–78
Conventional
  cross-linking, 81
  fertilizers, 74
  manufacturing techniques, 91
Coolant
  circuits, 127
  surfaces, 129, 134
Copolyesters, 10

Copolymer, 79, 81, 100, 284
  matrix, 80
Copolymerization, 10, 138
Co-precipitation route, 12
Corrosion, 126–128, 226, 232
Cost-effective manner, 19
Cotyledons, 73
Covalent interactions, 33
Criteria for reporting and evaluating ecotox-
  icity data (CRED), 6
Critical parameter, 31
Crop
  productivity, 69, 78
  quality, 77
  yields, 70, 75
Crosslinker, 33, 119, 120, 130–132, 138, 139
Crosslinking, 27, 32, 33, 35, 119, 120, 126,
  130, 131, 138, 140, 161, 163, 207, 212
Crotalaria juncea, 193, 211
Crystalline, 19, 20, 22, 27, 41, 196, 197,
  206, 216, 217, 253, 254, 286, 290
Crystallinity, 20–22, 30, 104, 193, 206, 216,
  217, 220, 221, 283, 286–289, 292, 294, 296
  fibers, 20
  index (CI), 216, 217
Crystallization, 94
Cyamopsis tetragonolobus, 73
Cytocompatibility, 104
Cytokines, 34

D

Datura poisoning, 42
Deacetylation, 10, 21, 22, 100, 118,
  121–124, 161, 163
Decision-making, 3
Decontamination, 118, 126–129, 131, 135
Degradable
  plastics, 8, 13
  polymers, 13, 252
Degradation, 2, 9, 12, 15, 18, 30–32, 34, 35,
  41, 65, 76, 78, 84, 98, 106, 112, 123, 163,
  165, 166, 283, 284, 290
  kinetics, 32
  time, 31
Degree of,
  acetylation, 100, 102
  automation, 105
  deacetylation, 118
  ionization, 80

Index 305

N-acetylation (DA), 100, 104
polymerization, 19, 195
Demineralization, 101, 127
Democratization, 93
Denaturation, 32
Deoxyribonucleic acid (DNA), 71, 72, 94, 194
Depolymerization, 121
Deproteinization, 101
Dextran, 24, 98
D-glucosamine, 100, 160
D-glucuronic acid, 22
Diabetic foot ulcer (DFU), 169
Dialkylaminoalkyl chlorides, 122
Diatoms, 5
Diethylenetriaminepentaacetic acid (DTPA), 131
Differential scanning calorimeter (DSC), 104, 106, 259, 260, 278, 286
Dimethylacetamide, 101
Direct ink writing (DIW), 97, 99, 109, 110, 112
Disaccharide units, 22
Discarded fruit fiber, 41, 65
Disinfectants, 31
Disintegrant, 73
Downstream Raman spectroscopic analyses, 12
Drought, 70, 75, 78, 80, 81, 84, 85, 199
Drug
delivery, 4, 18, 27, 29, 30, 73, 184
encapsulation, 10
release, 10, 20, 72–74, 97

E

Ecological imbalances, 3
Edible nanocoating, 12
E-glass, 42, 49, 55, 63
fiber polymer, 42
Egyptian mummification practice, 154
Elastic modulus, 32, 98, 229, 254
Elasticity, 83, 204, 206, 232, 235
Elastomers, 94
Electron beam, 46
Electronic waste, 2
Electrospun
collagen fibers, 28
nanofibers, 169
scaffolds, 28

Electrostatic attraction, 124, 129
Encapsulation, 10, 23
Endocrine disruptors, 2, 5
Endogenous additives, 4
Endothelialization, 29
End-use applications, 102
Environmental
hazards, 6
quality standards, 6
scanning electron microscope (ESEM), 170, 179, 180
sustainability, 43
Enzymatic
degradation, 31
digestion, 25
treatment, 210
Enzyme
hyaluronidase, 31
plasminogen, 31
substrate ratio, 31
Epichlorohydrin, 119, 120, 135, 137, 138
Epithelial tissue, 25
Epoxide ring, 119
Epoxy matrix, 226, 227, 240, 246
Equilibrium, 83, 126, 129
Erythrocytes, 161, 162
Escherichia coli, 164
Espresso channels, 200
Ethylene oxide, 10, 31, 32
Ethylenediaminetetraacetic acid (EDTA), 131
Eucheuma, 72
European
Commission, 14
Union (EU), 1, 5, 14
Eutectic system, 106
Exoskeletons, 160
Extracellular matrix, 18, 22, 24, 30, 109
Extrinsic pathway, 158–160, 176

F

Fabrication, 12, 35, 45, 91, 92, 98, 105–108, 110, 211, 216, 226, 229, 255, 258
Ferrous ion, 127, 128, 131
Fertilizers, 69, 70, 74–77, 80, 82, 84, 203
Fiber
bonding, 41, 42
mass density, 43
matrix

bonding, 49, 56, 58
  debonding, 58
 orientation, 44, 49, 205, 212
 wastes, 42
Fibrillar technology, 25
Fibrin, 26, 27, 29, 31, 32, 35, 154, 158, 175, 176, 183
  clot, 27, 29, 176
  matrix, 26, 27, 29
  networks, 27
  stabilizing factor, 27
Fibrinogen, 26, 29, 158, 161, 175, 178
Fibrinolysis, 182
Fibroblast, 28, 97
Fibroin, 27, 29, 31
Fibrosis, 34
Field-flow fractionation, 12
Film-forming capacity, 103
Fire resistance, 8
First
  generation chitosan, 118
  series transition metal ions, 124
Flexible
  fabrics (FF), 62, 227
  film, 251, 278
Flexural
  modulus, 46, 53–55
  strength (FS), 51, 53–55, 46, 216, 219, 220
  tests, 44
Flocculants, 79
Fluctuation, 76
Fly ash, 225, 227, 233, 235, 240–242, 245, 246
Food
  chain, 3, 11, 284
  crops, 5
  packaging materials, 11
  sciences, 12
  security, 70, 78
  simulants, 11
  source, 10
Formaldehyde, 32, 294
Fourier transmission infrared (FTIR), 215, 218, 220, 283, 286, 290
Fractography, 58
Fragmentation, 4, 15
Fraudulent market behavior, 13
Freeze-drying, 32

Functional
  groups, 30–33, 81, 117–120, 123, 124, 129, 130, 163, 193
  monomer, 131
Fungal spores, 7
Fused deposition modeling (FDM), 94–96, 104, 113

G

Galactose, 72, 73, 197
Galactoxyloglucan, 72, 73
Gamma
  irradiation, 126, 131
  radiation, 76
  ray, 80
Gastrointestinal ailments, 42
Gaussian intensity, 111
Gelatin, 10, 26, 28, 78, 97, 98, 104, 109, 112, 153, 252
  matrices, 28
Gelation, 23
Gelling, 19, 23
Genetic engineering, 9
Geometric dimensions, 83
Germicides, 84
Gigartinastellata, 72
Glass
  fiber composites (GFC), 49, 51, 52, 55, 56, 58, 62–64
  transition temperature, 10, 32, 106
Global
  analysis, 100
  aquatic environments, 4
  climatic anomalies, 70
  industrialization, 70
  pollution, 41
  warming, 70
Glucosamine, 10, 21, 22, 99–101, 160
Glucose
  monomers, 19
  unit, 19, 195–197, 253, 254
Glucuronic acid, 73
Glutamine amino acids, 26
Glutaraldehyde, 32, 33, 119, 120, 130, 138, 139, 179
Glycerine, 283, 285, 287, 288, 290–293, 295–297
Glycine, 25

Index

Glycosaminoglycans (GAGs), 21
Glycosidic
  bonds, 19, 100
  linkages, 19, 195
Graft skin, 28
Gram-negative bacteria, 125
Granulation, 104
Graphene, 107
Green
  chemistry, 6, 11
  economy, 8
  polymers, 69
Greenhouse
  crops, 75
  gas, 6
Growth
  cycle, 19
  factors, 28, 29, 110, 111
Guar gum, 73
Gum acacia, 73

H

Hageman factor, 158
Health care
  products, 2
  sectors, 186
Heat
  dimensional stability, 12
  resistance, 8
  sensitive materials, 32
Hemicellulose, 43, 197, 199
Hemocompatibility, 157, 160, 161, 171,
    174, 186
Hemocytometer, 179
Hemoglobin, 71
Hemolysis, 171, 173, 174, 187
Hemophilia, 184
Hemorrhage, 153–155, 157, 158, 160,
    164–166, 169, 181, 183, 185, 187
Hemostasis, 153–158, 163, 164, 174, 183,
    186, 187
Hemostatic
  action, 104, 156, 164
  agent, 153–155, 160, 165, 169, 182, 183,
    185, 187
  application, 25, 164
  devices, 154
  pathway, 182, 183
  sponges, 28

Hemostats, 27, 28, 183
Hepatic tissue engineering, 28
Herbicides, 74, 82, 84
Heterogeneity, 169
High-intensity ultrasonication (HIU), 193,
    194, 212–215, 217–220
Hirsch's model, 42, 61–63, 65
Homopolymer, 195
Homopolymeric blocks, 164
Hyaluronic acid, 21–23, 27, 30
Hybrid clay composite adsorbent (HyDCA),
    124
Hydration radius, 138
Hydraulic press, 44
Hydrogel, 19, 23, 26, 28, 29, 69, 76, 78–86,
    97, 104, 109–111, 184
  characteristics, 82, 83
    drought stress reduction, 84
    enhanced fertilizer efficiency, 84
    physical-chemical characteristics, 83
    polymers and biodegradability, 84
  matrix, 109
  network, 80
  scaffolds, 109
  usage, 78
Hydrogen
  bonding, 19, 25, 26, 78, 79, 285, 293, 294
  peroxide, 12, 32, 121
Hydrolysis, 10, 26, 30, 163, 207, 284
Hydrophilic
  functional groups, 81
  gum, 73
Hydrophilicity, 20, 80, 178, 212
Hydrophobic
  interactions, 26
  interiors, 162
  matrix, 73
  nondegradable polyolefins, 10
  resin, 73
  sequences, 27
  surface, 199
Hydroxy butyl chitosan, 104, 112
Hydroxyl
  groups, 19–21, 33, 119, 131, 138, 196,
    214, 218, 253, 254
  methylcellulose, 82
Hypersensitivity, 35

I

Immobilization, 12
In situ soil release, 76
In vitro blood assay analysis, 170
In vivo conditions, 30
Incubation, 175, 177, 178
Infiltration, 81
Inflammatory cells, 30, 33, 34
Infrared, 12, 104
  spectra characterization, 104
Inhalation hazards, 7
Injection molding, 91, 105, 107
Inorganic
  mineral, 76
  oxides, 146
  phosphate, 158
Insecticides, 69
Interchain hydrogen bonding, 19
Interface, 58, 205, 211, 215, 221, 228
International
  normalized ratio (INR), 175, 186
  organization for standardization (ISO), 1, 14
Interpenetrating polymer networks (IPN), 82
Intracellular reactions, 161
Intraperitoneal implantation, 34
Intrinsic pathway, 158
Iodine, 32, 141
Ionic
  groups, 80, 83
  liquid, 140, 141
Irrigation frequency, 81
Isoelectric points, 26

K

Keratin, 71, 107
Keratinocytes, 28, 97
Kininogen, 159
Klimisch score, 6

L

Lacerations, 155
Laminar composite (LC), 229
Laser scanner parameters, 111
Leaching, 4, 75, 79, 82
Leukocytes, 34
Ligaments, 29
Lignin, 2, 3, 9, 10, 43, 47, 195, 197–199, 202, 218, 220, 233, 242

Lignocelluloses, 2, 3, 10
Lignocellulosic fiber anatomy, 195
  chemical components, 195
  fiber cell structure, 195
Linear
  chain polymer, 195
  dimension (LD), 237–239
  homopolymer, 19
  unsulfated glycosaminoglycan, 22
Linum usitatissimum, 203
Lipid, 19, 101
  peroxidation, 84
Locust bean gum, 73
Lymphocytes, 155
Lymphocytic cells, 34

M

Macro porous silk scaffolds, 29
Macrocystic pyrifera, 164
Macromolecular substances, 97
Macromolecule, 19, 76
Macroparticles, 156
Macrophages, 155
Macropolymer, 108
Macroscopic scale, 78
Magnetic
  cellulose nanofiber (mgCNF), 12
  chitosan microspheres, 137
  properties, 121
Maleic acid, 131
Marine
  biota, 1, 6
  brown algae, 23
  ecosystems, 14
  food webs, 14
  life, 3
  organisms, 5, 100
  red algae, 23
Mass loss (ML), 239
Material extrusion, 93, 94, 97, 112
Matrix, 29, 30, 42, 72, 97, 139, 205
  bonding, 47, 58
  materials (MMs), 228
  metalloproteinase, 31
  parameters, 47
  penetration, 58, 62
Mechanical
  resistance, 106
  strength, 12, 19, 27, 29, 33, 98, 165, 208

Index 309

stresses, 7
traction forces, 30
Medical application, 25, 91, 112, 113, 164
Membrane filtration, 32
Meniscal tissue engineering, 29
Metabolic pathways, 18
Metal
  ion
    binding, 117, 119, 139
    imprinting, 131, 132, 140
    matrix composites (MMCs), 229, 231
    oxide ceramics, 255, 257–259, 271, 275, 277, 278
Methacrylamide moieties, 97
Methyl ethyl ketone peroxide (MEKP), 44
Methylene blue (MB), 258
Micro fibrillated cellulose, 9
Microbeads, 4, 154, 166, 167, 173, 182, 187
Microbial
  activity, 81
  attack, 9, 254
  fermentation, 72
  infestation, 2
Microbiota, 7
Microenvironment, 28
Microfibrillar collagen, 153
Microfibrils, 58, 195, 200, 254
Microimplants, 17
Micron, 7, 170, 244
Micronutrients, 75
Microparticles, 153, 165, 166, 168, 170, 181, 182
Microplastic, 1, 3–6, 8, 11
  bottled water and salt, 5
  ecosystem impact, 5
  fibers, 4
  particles, 6
  pollution, 3, 4
Microspheres, 76, 104, 137, 184
Mimicking, 29
Modern analytical techniques, 4, 12
Modified metal oxide ceramics (MMOC), 251, 255, 258–263, 265, 266, 274, 277
Moisture
  retention capacity, 20
  stress, 82
Molar
  mass, 12
  ratio, 72

Molecular
  chain, 72
  cleavage, 85
  delivery, 30
  technology, 25
  weight, 10, 12, 22, 24, 30, 31, 72, 79, 85, 102–104, 121, 122, 159, 161, 163, 199
Mollusks, 98, 99
Monocytes, 34
Monomer, 1, 3, 4, 9, 19, 31, 71, 72, 121, 131, 252, 253
  fibrinogen, 26
  nucleotides, 72
  units, 24, 31
Monosaccharide, 19, 71
Montmorillonite clay, 76
Mucoadhesion, 104, 160
Mucoadhesive
  agents, 183
  applications, 73
Mucuna
  flagillepes, 73
  gum, 73
Multiangle light scattering (MALS), 12
Municipal solid waste, 13

N

N-acetyl-glucosamine groups, 21
Nanocoatings, 12
Nanocomplex, 156, 187
Nanocomposite, 9, 11, 251, 255, 258, 266, 272, 277, 278
  future, quantification, and innovations, 11
Nanofibers, 156, 169
Nanofibrous network structure, 20
Nanofillers, 9
Nanoindentation, 107
Nanomaterials, 11, 12, 185
Nanometer, 7, 168
Nanoparticles, 4, 8, 11, 104, 122, 138, 156, 165, 168, 255
NanoSpider, 169
Nanosystems, 156
Nanotitania, 134, 135
Natural
  coagulation cascade, 33
  fiber (NF), 8, 12, 42, 47, 49, 52, 55, 58, 95–97, 199, 204, 208, 210, 225, 226, 232, 227, 233, 246

composites, 55, 226
  polymer composite (NFPC), 232
  reinforced composite (NFRC), 95, 225, 226
functional tissues, 30
polymer, 17–24, 26–28, 30–35, 69–72, 78, 94–98, 100, 108–110, 112, 199, 252, 253, 284
  3D printing, 94
  agarose, 23
  alginate, 23
  alternative methods, 32
  applications, 27
  bamboo fiber, 203
  biocompatibility, 33
  biodegradation, 30
  cellulose, 19
  chitosan, 21
  classes, 19
  coir fiber, 201
  collagen, 24
  cotton fiber, 199
  dextran, 24
  ethylene oxide sterilization, 32
  fibrin, 26
  flax fiber, 203
  gamma irradiation sterilization, 32
  gelatin, 26
  heat treatment, 31
  hyaluronic acid, 22
  polysaccharide polymers, 19
  silk, 27
  sisal fiber, 201
  starch, 20
  sterilization, 31
N-carboxylation, 119
N-doped graphene sheets, 107
Nerve
  cell proliferation, 29
  innervation, 110
Neural tissue engineering, 29
Neutron, 127, 128, 134
  flux, 134
Neutrophils, 34, 155
Nitrilotriacetic acid (NTA), 124–126, 129, 130, 135
Nitrogen, phosphorous, and potassium (NPK), 75, 76, 83

Nonactive
  ions, 127
  metal ions, 128
Nonbiodegradability, 13
Noncovalent interactions, 26
Nonrenewable petroleum-based resources, 1
Nontoxic
  metabolic products, 28
  methods, 84
Nuclear
  applications, 127, 136
  field, 126
  industry, 117, 122, 126–128, 131, 136–138, 141, 142, 146
  plant
    effluent, 124
    operators, 126
  power plants, 127, 137
  reactor, 118, 124, 126, 127, 129, 131, 134, 136
    decontaminations, 124, 136
  waste solution, 138
Nucleic acid, 32, 72
Nucleophilic
  addition, 138
  hydroxyl groups, 119
Nuclides, 126–129, 141
Nutraceuticals, 12

O

Oleoresins, 73
Oligomers, 30, 85
Oligosaccharides, 121
Optical density (OD), 173, 174, 179
Optimization, 49, 65, 111, 206, 297
Organic
  acid, 99
  compounds, 71, 123
  polymer, 2, 252
  resins, 146
  solvents, 6, 103, 200, 253
Osmotic forces, 83
Osteo differentiation, 29
Osteoblasts, 29
Osteogenic activity, 110
Oxalic acid groups, 126
Oxidation, 30, 256, 258
Oxiranes, 10

Index 311

Oxo-plastics, 5
Oxygen radicals, 84

P

Pathogen detectors, 12
Pectic acid, 72
Pectin, 27, 72, 104, 112, 195, 203, 220
Peptide residues, 25
Peracetic acid, 32
Permanganate treatment, 208
Persistent organic pollutants (POPs), 4, 6, 14
Pesticides, 69, 70, 74, 203
Petrochemicals, 13, 252
pH, 12, 22, 26, 30, 76, 80, 82, 107, 108, 124, 126, 129, 130, 132–134, 136, 139–141, 160, 163, 166, 181
Phagocytic cells, 30
Phagocytosis, 34
Pharmaceutical, 164, 166, 182
    industry, 29
    potentials, 4
    use and additive strategies, 182
Pharmacopoeias, 1
Phenolic
    macromolecule, 2
    resin, 9
    skeletons, 2
Phosphate buffer saline (PBS), 106, 154, 179, 181, 182
Phospholipids, 175
Photochromic, 251, 255–260, 262, 266, 268–273, 275, 277, 278
    properties, 251, 260, 271, 272, 278
        evaluation, 266
Photopolymerization, 93, 94, 97
Photostability, 82
Photosynthetic activity, 20
Physicochemical properties, 11, 19, 26, 30, 104, 163
Plamae, 201
Planar films, 107
Plasma
    cells, 35
    fraction, 178
    sorption, 161
    treatment, 208, 209
Plastic
    compounds, 9
    debris, 4, 13, 14

footprint, 8
leaching, 4
marine litter, 3
Plasticizer, 1, 2, 284, 285, 288, 289
    content, 283, 287–289, 291, 294, 295
Platelet, 111, 154, 158, 162, 169, 175, 179, 181–183
    activation, 158, 165
    adhesion, 154, 158, 161, 171, 178–180, 186
    aggregation, 158
    plug formation, 158
    poor plasma (PPP), 175–177
Plethora, 182, 184
Pollutants, 4, 14, 118, 122, 124, 136
Poly(butylene succinate) (PBS), 106
Poly(D,L-lactic-coglycolic acid) (PLGA), 28
Poly(vinyl alcohol) (PVA), 76, 283–297
Polyacrylamide, 70, 79–81
Polyacrylates, 70
Polyamide, 19, 71, 230, 231
Polyanions, 156
Polyaspartic acid, 81
Polycations, 156
Polydispersity index, 162
Polyelectrolyte, 80, 103, 153, 156, 187
    complex (PEC), 103, 153, 154, 156, 157, 163, 165–167, 172, 175, 179, 187
Polyester, 10, 106, 231
    matrix, 60, 62
Polyethylene, 5, 12, 13, 104, 108, 283–285, 297
    glycol (PEG), 104, 112, 283, 285–289, 292, 293, 296, 297
    microplastics, 5
    plastic bags, 5
    terephthalate, 5
Polyhydroxybutyrate (PHB), 76
Polylactic acid (PLA), 10, 95, 96, 252, 253, 284
Polymer
    matrix, 8, 41, 42, 58, 63, 65, 95, 97, 119, 213, 220, 226, 229, 232
        composites, 42, 65, 229
    network, 82, 194
Polymeric
    films, 79
    materials, 1, 71
    resins, 127
    superabsorbent polymer, 80

312   *Index*

Polymerization, 2, 10, 19, 27, 29, 93, 131, 135, 197, 200, 230, 252
Polymorphonuclear (PMN), 33
Polynucleotides, 19
Polyolefin materials, 10
Polypeptide, 71, 94
   chains, 25, 26
Polypropylene, 10, 96, 107, 171, 173, 213, 214, 232, 284
Polysaccharide, 19, 21, 23, 24, 27, 31, 34, 70–73, 78, 82, 94, 95, 98, 101, 118, 153, 160, 164, 195, 253
Polystyrene, 108, 284
Polyvinyl
   alcohol, 80, 131, 283
   pyrrolidone (PVP), 76
Porosity, 75, 97, 98, 112, 163, 169
Potassium polyacrylate, 79, 84
Potential
   biomaterials, 17
   hazards, 17
   risks, 2
Powder density, 112
Procoagulant supplementers, 183
Proliferation, 28, 29, 98, 160, 163
Proline, 25
Propylene oxide, 10
Protein fibers, 10
Proteoglycan, 22
Proteolytic enzymes, 31, 25
Prothrombin, 154, 157–160, 175–178, 187
   time (PT), 154, 157, 171, 175–177, 187
Prothrombinase, 158–160, 175
Prussian blue, 141
Pseudomonas aeruginosa, 164

Q

Quality control, 1
Quantification, 4, 6
Quasi-hexagonal lattice, 25
Quaternization, 119

R

Radioactive, 117, 124, 127–129, 134, 137, 140, 146
   antimony, 134
   ion, 128, 129, 140
     exchange waste, 128

   isotopes, 127
   nuclide, 117, 137
   waste, 127–129, 137
Radiocarbon, 14
Raman spectroscopy, 259, 260, 278
Raw
   materials, 4, 6, 7, 14, 205
   Sunn Hemp (RSH), 214, 216–220
Reactive oxygen species, 30, 34
Reactor coolant surfaces, 134
Real-life applications, 120
Recycling, 5, 9, 10, 13
Red blood cells (RBCs), 154, 156, 161, 162, 168, 169, 172, 173, 181, 186
Regression equations, 62
Relative
   density, 112
   molecular mass, 19
Release mechanism, 73, 75
Renewable
   agricultural feedstocks, 10
   energy, 6, 14
   plastics, 8
   resources, 6, 11, 14, 19, 94
   sources, 12
Resolidification, 97
Respiratory burst, 33
Rhamnose, 73
Rheological behavior, 12
Rhodophyceae, 72
Ribonucleic acid (RNA), 71, 72, 94
Robotic dispensing, 97, 99, 112
Robustness, 2
Rockwell hardness, 44, 46
Round-robin testing, 12

S

Saccharomyces cerevisiae, 137
Sansevieria cylindrica, 49, 55
Scanning electron microscopy (SEM), 43, 46, 56, 58–60, 65, 104, 170, 206, 207, 209, 210, 215, 259, 278
Schiff
   base, 138, 139
   linkage, 33
   reaction, 119
Selective laser sintering (SLS), 94, 98, 111, 112

Index 313

Selectivity coefficient, 129, 137
Semiarid regions, 78, 79, 85
Semicrystalline, 21, 162, 216
Sericin, 27
Shear
  strength, 46
  thinning, 97
Sheet lamination, 93, 94
Shorea wiesneri, 73
Silanization, 206, 207
Silica gel, 131
Silk fibroin, 29
Size-exclusion chromatography, 12
Slow-release fertilizer, 76
Smart packaging, 12
Sodium
  borohydride, 123
  hydroxide, 99, 205, 214
Soft plastics, 10
Soil
  conditioners, 79, 80
  microorganisms, 81
  moisture, 79
  particles, 79
  sediments, 2
  structure, 79
Solar-powered chemical reactors, 9
Sol-gel formation, 26
Solid sorbent matrices, 139
Solidification, 94, 97, 285
Solution casting, 91, 105, 107, 251, 258, 285
Solvent
  evaporation, 108
  free
    olefin, 10
    processes, 11
Sorbent, 121, 125–129, 136, 140, 141, 146
  matrix, 130
Sorbitol, 283, 285, 287, 297
Sorption-desorption equilibrium, 126
Spanning gel network, 162
Spectroscopic characterizations, 130
Spray
  dried microbeads, 167
  drying, 165–168, 186
Stabilizers, 1, 175
Stakeholders, 8, 13
Standardization, 1, 5, 14

Stereolithography, 92, 97, 104
Sterilization, 17, 31, 32, 35, 163
Strontium, 137, 140, 141
Structural composites (SC), 228
Sulfhydryl groups, 33
Sulfuric acid, 138
Sunn hemp fiber, 193, 211, 212, 214, 215, 217, 218, 220
Superabsorbent, 76, 78, 80, 81, 83
  polymers (SAP), 69, 79–81
Super-paramagnetism, 168
Surface
  colonizers, 14
  topography, 167
Surfactants, 10
Sustainable
  development, 10, 252
  growing practices, 75
  growth, 69
  materials, 11, 109
Swelling
  behavior, 181, 182
  capacity, 80, 82, 83
  mechanism, 181
  ratio, 76, 154, 181
Synovial fluid, 22
Synthetic
  clothing, 4
  macromolecular materials, 70
  polymers, 70, 97, 100, 118, 252

T

Tamarind fruit (TF), 41–43
  fiber (TFF), 41–43, 45, 47–49, 60–63, 65
    characterization, 43
    composite (TFFC), 43–46, 49–52, 54–58, 60–64
    thermal analysis, 58
Tamarindus indica, 72
Tendons, 25, 29, 100
Tensile
  modulus, 46, 49–51, 96, 106, 213
  properties, 51, 61, 96, 213, 232
  strain, 58, 108
  strength (TS), 32, 41, 46, 47, 49–51, 62–64, 96, 97, 106–108, 165, 206, 213, 214, 219–221, 232, 235, 236, 254, 283, 286, 291–296

test gauge, 44
values, 62
Terephthalic acid, 10
Terminalia catappa, 73
Tetrabutylammonium bromide, 10
Tetraethylenepentamine (TEPA), 138
Therapeutic
  conjugations, 184
  medical devices, 17
  products, 18
Thermal
  analysis, 46, 104
  power plants, 126
  properties, 41
  resistance, 58, 93
  stability, 12, 33, 41, 46, 60
Thermogravimetric analysis (TGA), 12, 43, 46, 104, 259, 260, 278, 286, 290, 291
Thermoplastic, 8, 94, 106, 230, 232, 284
  composites, 96
  matrix, 95
  resin, 73
  starch (TPS), 284, 285
  wood, 9
Thermoset, 8, 106
  materials, 94
Thermosetting, 9, 205
Thiol groups, 130
Three-dimensional (3D), 20, 23, 28, 91–95, 97–99, 101–104, 107–113
  bioprinting, 20, 28, 92, 94, 97, 104, 112
  network, 23
  printing, 91–95, 97–99, 104, 108–110, 112, 113
  scaffold structures, 27
Thrombin, 26, 28, 154, 158, 175, 183, 186
Thromboplastin, 154, 157, 160, 175, 177–179
Time-dependent release kinetics, 72
Tissue
  engineering, 18, 27–30, 91, 97, 99, 106, 108, 109, 112, 113
  factor pathway, 160
  scaffolds, 98, 109
  thromboplastin, 160
Titania, 134, 135
Top-down approach, 155, 156
Toxic
  chemicals, 70, 86
  metabolites, 7
  pesticides, 70

Toxicity, 4, 6, 8, 15, 17, 33, 35, 79, 257, 258
Traditional
  chemical fertilizer, 74
  manufacturing methods, 91
  plastics, 7
Tranexamic acid (TXA), 183
Transition temperature, 23
Triethylene-tetramine, 137
Triple helical
  collagen molecule, 25
  structure, 25, 26
Trypsin, 31
Tumorigenesis, 33
Tumourigenic potential, 17
Tunable surface chemistry, 20
Twin-screw extrusion system, 96

U

Ultrasonication, 141, 163, 169, 193, 212, 213, 221
  treatment, 163, 221
Ultrasonification, 212
Ultraviolet (UV), 85, 208, 251, 256–261, 264, 265, 267–271, 274–278
  light, 32, 199
  rays, 85
Unsaturated polyester, 43, 46, 65
Uranyl ions, 138, 140
Urbanization, 78

V

Valorization, 3
Van der Waal forces, 19, 26, 79
Vascular
  endothelial growth factor, 111
  grafts, 29
  resistance, 158
  tissue engineering, 28, 29
Vascularization, 34, 111
Vasculature engineering, 29
Vasoconstriction, 155, 158
Versatile applications, 117, 277
Viscometry, 12
Viscosity, 46
Visible irradiation, 271–275
Volume expansive capacity, 24
Volumetric weight ratios, 110

Index

315

W

Waste
  management, 11
  recycling, 6
  treatment, 127
Water
  absorbency, 76
  efficient agriculture system, 78
  granules, 79
  holding capacity, 78–81, 86
  permeability, 76
  purification, 10, 254
  resources, 70, 78
  retaining agents, 81
  retention, 8, 75, 76, 78, 79, 82, 86, 197
  soluble
    chitosan derivatives, 119
    fertilizer, 75, 77
  stress, 81, 85
  vapor, 32, 106
    permeability (WVP), 106
Wound
  care applications, 29

dressing, 10, 20, 27, 28, 164, 165
healing applications, 20, 28

X

Xanthan gum (XG), 72
Xanthate moieties, 130
Xanthomonas campestris, 72
X-ray diffraction (XRD), 104, 118, 217,
  220, 262, 278, 283, 286–290, 292, 296

Y

Yield stress, 106
Young's modulus, 107

Z

Zea maize, 76
Zeolites, 138
Zero
  bioaccumulation property, 85
  error accommodation, 92
  order release kinetics, 73

CPSIA information can be obtained
at www.ICGtesting.com
Printed in the USA
JSHW022027090522
25519JS00001B/8